Agricultural Programs, Terms and Laws

AGRICULTURAL PROGRAMS, TERMS AND LAWS

FRANK J. HOUSE
EDITOR

Novinka Books
An imprint of Nova Science Publishers, Inc.
New York

For permission to use material from this book please contact us:
Telephone 631-231-7269; Fax 631-231-8175
Web Site: http://www.novapublishers.com

NOTICE TO THE READER

Library of Congress Cataloging-in-Publication Data:

Agricultural programs, terms, and laws / Frank J. House, (editor).
p. cm.
Includes index.
ISBN 1-59454-892-7
1. Agriculture--Terminology. I. House, Frank J.
S411.A3319 2005
630.1'4--dc22 2005037899

Published by Nova Science Publishers, Inc. ✢ New York

CONTENTS

Preface **vii**

Agriculture: A Glossary of Terms, Programs, and Laws, 2005 Edition **1**

Index **285**

PREFACE*

The complexities of federal farm and food programs have generated a unique vocabulary. Common understanding of these terms (new and old) is important to those involved in policymaking in this area. For this reason, the House Agriculture Committee requested that CRS prepare a glossary of agriculture and related terms (e.g., food programs, conservation, forestry, environmental protection, etc.). Besides defining terms and phrases with specialized meanings for agriculture, the glossary also identifies acronyms, abbreviations, agencies, programs, and laws related to agriculture that are of particular interest to the staff and Members of Congress. CRS is releasing it for general congressional use with the permission of the Committee.

The approximately 2,500 entries in this glossary were selected in large part on the basis of Committee instructions and the informed judgment of numerous CRS experts. Time and resource constraints influenced how much and what was included. Many of the glossary explanations have been drawn from other published sources, including previous CRS glossaries, those published by the U.S. Department of Agriculture and other federal agencies, and glossaries contained in the publications of various organizations, universities, and authors. In collecting these definitions, the compilers discovered that many terms have diverse specialized meanings in different professional settings. In this glossary, the definitions or explanations have been written to reflect their relevance to agriculture and recent changes in farm and food policies.

The glossary is in alphabetical order and contains an explanation for each term. Acronyms and abbreviations are not followed by an explanation. However, except for private organizations and associations, the abbreviations likely are defined under their full name.

The definitions and explanations are not legal in nature, but are explanatory. Hence, this document should not be used as a legal or administrative reference. For those purposes, the Statutes at Large, U.S. Code, and the Code of Federal Regulations are the more appropriate resources.

---

* Excerpted from CRS Report 97-905, dated June 16, 2005.

AGRICULTURE: A GLOSSARY OF TERMS, PROGRAMS, AND LAWS, 2005 EDITION

AAEA — American Agricultural Economics Association. [www.aaea.org].

AAFCO — American Association of Feed Control Officials. [www.aafco.org].

AAM — American Agriculture Movement. [www.aaminc.org].

AAMP — American Association of Meat Processors. [www.aamp.com].

AAR — Association of American Railroads. [www.aar.org].

AARCC — Alternative Agriculture Research and Commercialization Corporation.

Aarhus Protocol on Persistent Organic Pollutants — This 1998 protocol on persistent organic pollutants (POPs) is an addition to the 1979 Geneva Convention on Long-Range Transboundary Air Pollution (LRTAP). The Aarhus POPs Protocol seeks "to control, reduce or eliminate discharge, emissions and losses of persistent organic pollutants" in Europe, some former Soviet Union countries, and the United States. The protocol is an executive agreement that does not require Senate approval. However, legislation is needed to resolve inconsistencies between provisions of the protocol and existing U.S. laws (specifically the Toxic Substances Control Act and the Federal Insecticide, Fungicide, and Rodenticide Act).

ABA — American Bakers Association. [www.americanbakers.org]. American Bankers Association. [www.aba.com/default.htm]. American Bar Association. [www.abanet.org].

Abandoned wells — Abandoned drainage wells and abandoned water wells on vacant farmsteads are of particular concern for agriculture. Abandoned wells can present both safety risks and a direct conduit by which groundwater can be contaminated by surface runoff. A number of states have incentive and/or regulatory programs to cap or seal abandoned wells.

Able-bodied Adults Without Dependents (ABAWDs) — This term refers to low income working adults who do not have dependents. The 1996 welfare law (P.L. 104-193) set categorical requirements for food stamp participation. Among these were restrictions on legal alien participation and participation by low income persons without dependents. The latter (formerly eligible based solely on low income) were made ineligible for food stamps if they received food stamps for 3 months during the preceding 3 years without working or participating in a work program for at least 20 hours a week, or without participating in a workfare program.

ACA — Agricultural Credit Association.

ACE — Agriculture in Concert with the Environment.

Acid deposition / acid rain — Abnormally acidic (low pH) precipitation (or dry deposition) resulting from emissions of sulfur and nitrogen compounds that transform during chemical processes in the atmosphere. Acid deposition can affect the chemistry of soils and acidify lakes, adversely affecting forests and fish. It may adversely affect cropland. The Clean Air Act (42 U.S.C. 7401 *et seq.*) includes a program focused on controlling precursor emissions of acid deposition, primarily sulfur oxides from coal-fired electric utilities.

ACP Countries — African, Caribbean and Pacific countries are former European colonies associated with the European Union under the Cotonou Agreement negotiated in 2000. EU-ACP relations were formerly covered by the Lome Convention. ACP countries benefit from development assistance and trade preferences provided under the Cotonou Agreement.

ACPA — American Crop Protection Association changed its name to CropLife America. [www.croplifeamerica.org].

Acquired lands — Lands in federal ownership that were obtained by the federal government through purchase, condemnation, gift, or exchange. One category of public lands.

ACR — Acreage conservation reserve.

Acre — 1 acre = 43,560 sq. ft. = 208.71 ft.^{2} = 0.405 hectares. Note that 640 acres = 1 sq. mile (called a "section").

Acre-foot — The volume of water that would cover one acre of land (43,560 square feet) to a depth of one foot, equivalent to 325,851 gallons of water. An acre-foot is the basic measure of agricultural water use. On average, irrigators apply almost 2 feet of water on each acre through the crop growing season; the amount ranges from 4 feet in the Southwest to a half foot in some eastern states, and varies, depending on the crop grown. Water withdrawn for irrigation from ground and surface sources totals about 150 maf (million acre-feet) of water annually.

Acreage allotment — Under provisions of permanent commodity price support law, a farm's acreage allotment is its share, based on its previous production, of the national acreage needed to produce sufficient supplies of a particular crop. Under the 2002 farm bill (P.L. 101-171, Title I), acreage allotments are not applicable to the covered commodities, peanuts, or sugar. Subsequently, allotments and quotas and price support for tobacco were eliminated beginning in 2005 (P.L. 108-357, Title VI).

Acreage conservation reserve — The cropland acreage diverted from production under the acreage reduction program authorized prior to 1996.

Acreage diversion programs — Historically, commodity programs included provisions to reduce commodity supplies by diverting acreage to non-crop uses. Examples include paid diversion, unpaid diversion, set-aside, and acreage reduction programs. The 1996 farm bill (P.L. 104-127) eliminated authority for the USDA to implement annual acreage reduction programs. The Conservation Reserve Program pays farmers for the long-term conversion of fragile cropland land to conserving uses and is not considered to be an acreage diversion program.

Acreage limitation — With respect to commodity policy, acreage limitation might refer to planting constraints under an acreage reduction program, set-aside, or paid land diversion. These programs are no longer authorized. In relation to water policy, it is the maximum number of acres that may be irrigated with less than full-cost water from

Bureau of Reclamation projects. Generally, the acreage limitation for individuals or legal entities representing 25 people or fewer is 960 acres; however, amounts vary depending on a landowner's legal status. Also referred to as ownership limitation, ownership entitlement, or non-full-cost entitlement.

Acreage Reduction Program (ARP) — A no-longer- authorized annual cropland retirement program for wheat, feed grains, cotton, or rice in which farmers participating in the commodity programs (in order to be eligible for nonrecourse loans and deficiency payments) were mandated to idle a crop-specific, nationally-set portion of their base acreage during years of surplus. The idled acreage (called the acreage conservation reserve) was devoted to a conserving use. The goal was to reduce supplies, thereby raising market prices. Additionally, idled acres did not earn deficiency payments, thus reducing commodity program costs. ARP was criticized for diminishing the U.S. competitive position in export markets. The 1996 farm bill (P.L. 104-127) did not reauthorize ARPs. ARP differed from a set-aside program in that under a set-aside program reductions were based upon current year plantings, and did not require farmers to reduce their plantings of a specific crop.

ACS — Alternative conservation system.

Action levels — As opposed to tolerances (which are established for pesticide residues occurring as a direct result of proper usage), action levels are set for inadvertent residues resulting from previous legal use or accidental contamination. At the action level set by the EPA, Food and Drug Administration and USDA are required to take enforcement action against the contaminated food or agricultural commodity. The term is also used in other regulatory programs.

Active ingredient — In any pesticide product, the component that kills, or otherwise controls, target pests. Pesticides are regulated by the EPA primarily on the basis of active ingredients. See also Inert ingredient.

Active packaging — Technological advances make it possible for food packaging to prolong shelf life, monitor freshness, and display information on quality. Active packaging can interact with food to reduce oxygen levels, or add flavorings and preservatives. Also called intelligent packaging. The use of such packaging raises issues of regulation and labeling.

ACTPN — Advisory Committee for Trade Policy and Negotiations.

Actual Production History (APH) — A measure of an individual farmer's annual production of a commodity over a multi-year period. The APH serves as the basis for the farmer's "normal" crop yield in the crop insurance program. When the actual crop yield deviates by more than a certain percentage from the APH, an insured producer is eligible for an indemnity (loss) payment. The 2002 farm bill (P.L. 101-171, Sec. 1101-1102) allows producers of covered commodities participating in the Direct and Counter-cyclical Program (DCP) the option of updating acreage for direct payments, and acreage and yield for counter-cyclical payments, using 1998 through 2001 as the base period for actual acreage and yields.

Actual Production History (APH) — A record of an agricultural producer's crop yields over a multi-year period. Such records are used in the federal crop insurance program to determine "normal" production levels for a producer. The term Actual Production History insurance is used synonymously with Multi-Peril Crop Insurance.

Actuarially sound — The financial goal of any insurance program (including the federal crop insurance program) is to operate on an actuarially sound basis. That is, total premiums collected should more than offset total indemnities paid out.

Acute toxicity — The ability of a substance to cause harmful effects soon after a single exposure or dose. Also, any severe poisonous effect resulting from a short-term exposure to a toxic substance. See also Chronic toxicity.

Ad valorem duty — A tariff expressed as a fixed percentage of the value of the imported commodity or product. Generally, by contrast, a specific rate duty is applied as a charge on each unit or specified quantity of an imported item (i.e., $5 per ton).

AD — Anti-dumping duty.

ADA — American Dairy Association. [www.ilovecheese.com/default.htm].

ADA — American Dietetic Association. [www.eatright.org]; American Diabetic Association. [www.diabetes.org/home.jsp].

ADC — Animal Damage Control Program.

Additional peanuts — Prior to the complete redesign of the peanut program in the 2002 farm bill (P.L. 107-171, Sec 1301-1310), a two-tiered pricing scheme provided higher support for quota peanuts than for additional peanuts. The additional peanuts were those marketed above the quantity specified as peanut poundage quota for each farm. Additional peanuts had to be exported or crushed into oil and meal in contrast to quota peanuts that could be sold for domestic edible uses.

ADI — Acceptable daily intake; usually refers to dietary adequacy.

Adjusted gross income limitation — The 2002 farm bill (P.L. 101-171, Sec. 1604) established an eligibility limit on commodity program and conservation payments. For the first time, the law makes an individual recipient of commodity and conservation subsidies ineligible for payments if their 3-year average adjusted gross income exceeds $2.5 million. This limitation is waived if the 75% or more of the adjusted gross income is from farming, ranching, or forestry. Separately, there are annual payment limitations on how much a person can be receive under various farm programs.

Adjusted Gross Revenue (AGR) Insurance — A revenue insurance program implemented in 1999 as a pilot program by USDA and continuing on a limited basis. Where available, it allows farmers to receive a guarantee of a percentage of their revenue for multiple commodities, including some livestock revenue, rather than just the revenue from an individual commodity.

Adjusted Gross Revenue (AGR)-Lite — Similar to the AGR insurance program described above, except that AGR-Lite is available to smaller farmers (income below $512,821, and liability below $250,000). Where the basic AGR program limits eligible livestock coverage to 35% of expected allowable income, the AGR-Lite contains no limitations on the proportion of livestock income.

Adjusted world price (AWP) — As part of the upland cotton and the rice marketing assistance loan programs, USDA calculates and publishes, on a weekly basis, what is known as the adjusted world price (AWP). The AWP is the prevailing world price for upland cotton or rice, adjusted to account for U.S. quality and location. Producers who have taken out USDA marketing assistance loans may choose to repay them at either the lesser of the established loan rate, plus interest, or the announced AWP for that week. The AWP for cotton also is used for determining Step 2 payments.

Administered price — A price fixed by policy makers in order to determine, directly or indirectly, domestic market or producer prices. All of the administered price schemes set a minimum guaranteed support price or target price for a commodity, which is maintained by associated policy measures such as quantitative restrictions on production and imports; taxes and tariffs on imports; export subsidies; and public stockholding. Administered prices under the 2002 farm bill (P.L. 107-171) include loan rates and/or target prices, and price support levels for sugar, and dairy products.

Administrative Procedure Act of 1946 — P.L. 79-404, as amended, establishes, among other things, minimum procedural requirements or models for federal agency rulemaking and certain types of hearings. For instance, the APA establishes procedures for informal rulemaking, which may include notice-and-comment requirements, or formal rulemaking, which includes trial-type hearings. Exemptions from rulemaking requirements are included in the Act. The APA provides standards for judicial review of final agency action. The provisions of the APA apply to USDA rulemaking, unless exempted under the provisions of another statute. For example, hearings conducted by the USDA's National Appeals Division (NAD) are not governed by the APA. The final determination of the NAD is reviewable and enforceable by a U.S. District Court in accordance with the judicial review provisions of the APA.

Adulterated food — Generally, impure, unsafe, or unwholesome. However, the Federal Food, Drug, and Cosmetic Act (21 U.S.C. 321 *et seq.*), the Federal Meat Inspection Act (21 U.S.C. 601 *et seq.*), and the Poultry Products Inspection Act (21 U.S.C. 451 *et seq.*) contain separate language defining in very specific (and lengthy) terms how the term adulterated will be applied to the foods each of these laws regulates. Products that are adulterated under these laws' definitions cannot enter into commerce for human food use.

Advanced meat recovery (AMR) — A mechanical process that removes the last traces of usable meat from bones after the primal cuts have been carved off manually. In 1994, FSIS issued a rule allowing such meat to be labeled as meat for human consumption, providing that the bones from which it was removed were still intact after processing. In 1997, following tests indicating that central nervous system (CNS) tissue was showing up in mechanically removed meat, FSIS issued a directive to its inspectors instructing them to ensure that spinal cord tissue was removed from bones before the AMR process. Following the identification of a BSE-infected U.S. dairy cow in December 2003, FSIS issued new regulations expanding the definition of prohibited CNS tissue to include additional cattle parts. Furthermore, all AMR-processed product from cattle more than 30 months old now is prohibited from being used for food, and such product from younger cattle and from other livestock species also is prohibited if it contains CNS material. AMR meat typically is used as an ingredient in products requiring further processing, such as hot dogs.

Adventitious presence (AP) — The accidental or unintentional appearance of foreign material in a product. In the case of agriculture, usually this happens in the production, harvesting, storage, and marketing of grains, seeds, or food products, for example. Grain and seed companies argue that virtually all shipments contain some type and level of adventitious material, such as some weed material in a bin of soybeans or wheat. Generally, buyers recognize that some level of adventitious material is acceptable and foreign material limits are specified in purchase contracts. AP is now a key issue in the debate over regulation of biotechnology. As more and more crops and acres are devoted to genetically engineered (GE) varieties, it becomes increasingly difficult to segregate these from GE-free varieties, which some buyers and countries demand. The European Union is considering a proposal to require GE labeling of any food or feed product that contains more than 0.5% of material derived from genetically modified organisms.

Adverse Effect Wage Rate (AEWR) — The AEWR is the minimum wage that the U.S. Department of Labor (DOL) has determined "must be offered and paid to U.S. and alien workers by agricultural employers of nonimmigrant H-2A agricultural workers (Federal Register, February 10, 1999, p. 6690). Where agricultural employers offer employment to nonimmigrant foreign workers, payment of at least the AEWR is required. Published once a year, usually in early February, by DOL with the assistance of the U.S. Department of Agriculture, the AEWR sets a separate minimum wage rate (i.e., a rate that will not adversely effect the employment opportunities of U.S. workers) for each state (see 20 CFR 655).

Advisory Committee for Trade Policy and Negotiations (ACTPN) — A 45-member group appointed by the President to provide advice on matters of trade policy and related issues, including trade agreements. The 1974 Trade Act (P.L. 93-618) requires the ACTPN's establishment and broad representation of key economic sectors affected by trade. Below ACTPN are seven policy committees, including the Agricultural Policy Advisory Committee (APAC). The Agricultural Policy Advisory Committee is made up of farm sector and industry representatives.

AEWR — Adverse Effect Wage Rate

AFBF — American Farm Bureau Federation (Farm Bureau). [www.fb.com]. **AFDO** — Association of Food and Drug Officials. [www.afdo.org]. **AFFI** — American Frozen Food Institute. [www.affi.com].

AFIA — American Feed Industry Association. [www.afia.org].

Aflatoxin — Aflatoxin is a naturally occurring mycotoxin produced by two types of mold: *aspergillus flavus* and *aspergillus parasiticus*. Both species are common and widespread in nature, but A. flavus is more likely to infect grain, cotton seed, and peanuts grown under stressful conditions such as drought. Favorable conditions for mold growth include high moisture content and high temperature. At least 13 different types of aflatoxin are produced in nature with aflatoxin B1 considered as the most toxic. While the presence of Aspergillus flavus does not always indicate harmful levels of aflatoxin, it does mean that the potential for aflatoxin production is present. The Food and Drug Administration (FDA) has established action levels for aflatoxin present in food or feed to protect human and animal health. The FDA will consider action if aflatoxin levels exceed: 20 ppb for corn and other grains intended for immature animals (including immature poultry) and for dairy animals, or when its destination is not known; 20 ppb for animal feeds, other than corn or cottonseed meal; 100 ppb for corn and other grains intended for breeding beef cattle, breeding swine, or mature poultry; 200 ppb for corn and other grains intended for finishing swine of 100 pounds or greater; 300 ppb for corn and other grains intended for finishing (i.e., feedlot) beef cattle and for cottonseed meal intended for beef cattle, swine or poultry. All corn exported from the United States is required to be tested for aflatoxin. Aflatoxin testing is available nationwide upon request, for a fee. Grain Inspection Packers and Stockyards Administration has approved several different types of test kits for this service.

AFPA — Agricultural Fair Practices Act of 1967 (P.L. 90-288) (7 U.S.C. 2301 *et seq.*).

AFT — American Farmland Trust. [www.farmland.org].

Agenda 2000 — A set of measures involving changes to common EU policies, including the Common Agricultural Policy (CAP), for the 2000-2006 period agreed by EU Heads of State at the March 1999 European Summit in Berlin. Intended in part to prepare the EU's Common Agricultural Policy for the enlargement of the European Union in 2004 to include 10 Central and Eastern European Countries and for the next round of World Trade Organization (WTO) negotiations on agriculture which began in 2001, the reforms included reductions in market price support for beef and veal, milk and milk products, and grains and oilseeds. Agenda 2000 established rural development as a "second pillar" of the CAP (price and income support being the "first pillar") and set guidelines for spending on commodity support and rural development for the 2000-2006 period. Under the guidelines, spending on the first pillar would average around 40 billion Euros annually, while rural development spending would average around 4.3 billion Euros (1999 prices). Total spending on the CAP is to be held at these levels including after accession of the new member countries in 2004. The Berlin Summit also decided that a mid-term review of the reforms of the CAP would be carried out by 2003.

Aggregate measurement of support (AMS) — An indicator of the amount of domestic support for agriculture. As used in the Uruguay Round Agreement on Agriculture, the AMS refers to a measure of the gap between domestic and world prices multiplied by the quantity supported, plus any other commodity-specific transfers. Internal or domestic support reduction commitments in the Uruguay Round Agreement on Agriculture are expressed in terms of reductions in a total AMS covering all trade-distorting internal support measures for agriculture.

Agreement on Agriculture — The Uruguay Round Agreement on Agriculture reached in 1994 and implemented in U.S. law by the Uruguay Round Agreements Act of 1994 (P.L. 103-465) brings agricultural trade more fully under international trade rules and obligations. The Agreement provides for the conversion of quantitative barriers to trade to tariffs or tariff-rate quotas, and for reductions in export subsidies and trade-distorting domestic support policies.

Agreement on Technical Barriers to Trade — One of the Uruguay Round agreements, it is intended to ensure that countries' various technical standards and regulations, including those for food and agricultural products (other than sanitary and phytosanitary measures) do not unnecessarily restrict or distort trade and/or are imposed simply to protect domestic industries.

Agreement on the Application of Sanitary and Phytosanitary Measures — See Sanitary and phytosanitary (SPS) measures and agreements.

Agribusiness — Agriculturally related businesses that supply farm inputs (such as fertilizer or equipment) or are involved in the marketing of farm products (such as warehouses, processors, wholesalers, transporters, and retailers). Farms are not usually included when the term agribusiness is used.

Agricultural Act of 1949 — P.L. 81-439, along with the Agricultural Adjustment Act of 1938 (P.L.75-430), makes up the major part of the permanent law that mandates commodity price and farm income support. The original 1949 Act designated mandatory support for basic commodities and the following nonbasic commodities: wool and mohair, tung nuts, honey, Irish potatoes (later excluded in the Agricultural Act of 1954, P.L. 83-690), and milk, butterfat, and their products. Periodic farm bills (most recently the 2002 farm bill (P.L. 101-171)) make temporary changes in the levels and design of commodity programs.

Agricultural Act of 1954 — P.L. 83-690 established a flexible price support for basic commodities (excluding tobacco) at 82.5-90% of parity and authorized a Commodity Credit Corporation (CCC) reserve for foreign and domestic relief. Title VII was designated the National Wool Act of 1954 and provided for a new price support program for wool and mohair to encourage increased domestic production. Price support for wool and mohair continued through marketing year 1995, at which time it was phased down and terminated under the explicit mandate of P.L. 103-130 (November 1, 1993). Mandatory support for wool and mohair was restored by the 2002 farm bill (P.L. 101-171, Sec. 1201-1205).

Agricultural Act of 1956 — P.L. 84-540 created the Soil Bank Program (Title I was called the Soil Bank Act), addressed the disposal of Commodity Credit Corporation (CCC) inventories of surplus stocks, contained commodity support program provisions, and contained forestry provisions. The Soil Bank Act authorized short- and long-term removal of land from production with annual rental payments to participants (Acreage Reserve Program and Conservation Reserve Program, respectively). The Acreage Reserve Program, for wheat, corn, rice, cotton, peanuts, and several types of tobacco, allowed producers to retire land on an annual basis in crop years 1956 through 1959 in return for payments. The Conservation Reserve Program allowed producers to retire cropland under contracts of 3, 5, or 10 years in return for annual payments. The Soil Bank Act was repealed by Section 601 of the Food and Agriculture Act of 1965 (P.L. 89-321). The Conservation Reserve portion of the Soil Bank was a model for the subsequent Conservation Reserve Program (CRP), enacted in 1985.

Agricultural Act of 1970 — P.L. 91-524 initiated a significant change in commodity support policy. This 3-year farm bill replaced some of the more restrictive and mandatory features of previous law (acreage allotments, planting restrictions, and marketing quotas) with voluntary annual cropland set-asides and marketing certificate payments to achieve parity prices (the precursor to target prices and deficiency payments). For the first time, the law adopted an annual payment limitation per producer (set at $55,000 per crop). Among other things, the Act also amended and extended the authority of the Class I differential in federal milk marketing order areas.

Agricultural Adjustment Act (AAA) of 1933 — P.L. 73-10 was the New Deal initiative to assist the farm sector during the Great Depression. This was the first comprehensive effort to raise and stabilize farm prices and income. The law created and authorized the Agricultural Adjustment Administration to (1) enter into voluntary agreements to pay farmers to reduce production of designated "basic" commodities (cotton, wheat, corn, rice, tobacco, hogs, and milk), (2) to make advance payments to farmers who stored crops on the farm, (3) to create marketing agreements between farmers and middlemen, and (4) to levy processing taxes to pay for production adjustment and market development. The Commodity Credit Corporation (CCC) was incorporated under the laws of the state of Delaware on October 17, 1933, to carry out financial activities, including making nonrecourse loans on the basic crops. Support for other commodities was authorized upon recommendation by the Secretary with the President's approval. Commodity loan programs carried out by the CCC for 1933-37 included cotton, corn, rosin, turpentine, tobacco, peanuts, dates, figs, and prunes. The provisions for production control and processing taxes in the Act were later declared unconstitutional in the Hoosac Mills decision of 1936. Congress responded by adopting the Soil Conservation and Domestic Allotment Act of 1936 (P.L. 74-46), the Agricultural Marketing Act of 1937 (50 Stat. 246), and the Agricultural Adjustment Act of 1938 (P.L. 75-430), all of which remain as permanent law.

Agricultural Adjustment Act Amendment of 1935 — P.L. 74-320 made several important and lasting changes to the Agricultural Adjustment Act of 1933 (P.L. 73-10). Section 22 of the law gave the President authority to impose quotas when imports interfered with commodity programs designed to raise prices and farm income. Section 32 was designed to widen market outlets for surplus agricultural commodities by permanently appropriating funds (30% of annual gross customs receipts) to promote food consumption, reduce agricultural surpluses, and provide for the food needs of low income populations. Section 32 funds are used by the Secretary to purchase surplus commodities for donation outside normal channels of trade (e.g., to school lunch programs), and to support the costs of child nutrition programs. Section 22 has been superseded, but Section 32 continues to operate and is used primarily for child nutrition programs.

Agricultural Adjustment Act of 1938 — P.L. 75-430 was enacted as an alternative and replacement for the farm subsidy policies found unworkable in the AAA legislation of 1933. The 1938 Act was the first to make price support mandatory for corn, cotton, and wheat to help maintain a sufficient supply in low production periods along with marketing quotas to keep supply in line with market demand. It established permissive supports for butter, dates, figs, hops, turpentine, rosin, pecans, prunes, raisins, barley, rye, grain sorghum, wool, winter cover-crop seeds, mohair, peanuts, and tobacco for the 1938-40 period. Also, Title V of the Act established the Federal Crop Insurance Corporation. The 1938 Act is considered part of permanent law for commodity programs and farm income support (along with the Commodity Credit Corporation (CCC) Charter Act and the Agricultural Act of 1949). Provisions of this law are often superseded by more current legislation (such as the 2002 farm bill (P.L. 101-171)). However, if the current legislation expires and new legislation is not enacted, the law reverts back to the permanent provisions of the 1938 Act.

Agricultural Adjustment Act of 1980 — P.L. 96-213 amended the Food and Agriculture Act of 1977 (P.L. 95-113) primarily to raise the target prices for wheat and corn.

Agricultural Attache, Counselor, or Trade Officer — An agricultural expert, employed by the Foreign Agricultural Service, on the staff of a U.S. embassy, consulate, or agricultural trade office.

Agricultural commodity — No single official definition of this term exists. The term and others, such as agricultural product, basic commodity, nonbasic commodity, perishable commodity, and livestock, appear, as statutory language, throughout the U.S. Code and in various federal regulations. In each case (such terms are defined in at least 20 different places in the U.S. Code alone), the term is intended to have a specialized or unique meaning, most often to either exclude or include particular items/producers from eligibility or coverage under a program or activity. A term might be relatively general and/or expansive, as in the law authorizing foreign food aid under P.L. 480 (7 U.S.C. 1732): "Agricultural commodity, unless otherwise provided for in this chapter, includes any agricultural commodity or the products thereof produced in the United States, including wood and processed wood products, fish, and livestock, as well as value-added, fortified, or high-value agricultural products." Or, it may be deliberately exclusive, as in the Agricultural Fair Practices Act of 1967 (7 U.S.C. 2302): "Agricultural products shall not include cotton or tobacco or their products."

Agricultural Conservation Program (ACP) — Administered by the Farm Service Agency, this largest and oldest conservation cost-sharing program paid farmers up to $3,500 per year as an incentive to install approved conservation practices. It was terminated in the 1996 farm bill (P.L. 104-127) and replaced by a new Environmental Quality Incentives Program (EQIP).

Agricultural Credit Act of 1987 — P.L.100-233 was enacted in response to the severe financial crisis of the early- to mid-1980s, which affected both farmers and their lending institutions. The Act authorized a $4 billion financial assistance package for financially vulnerable institutions of the Farm Credit System (FCS), protected the full value of FCS borrower stock when retired, established a permanent insurance mechanism to ensure the repayment of funds borrowed by the FCS for lending purposes, required the FCS and USDA's Farmers Home Administration (now, the Farm Service Agency) to restructure severely delinquent farm loans that met certain criteria, mandated FCS consolidation, and established a secondary market for farm real estate loans.

Agricultural Credit Association (ACA) — An institution of the Farm Credit System that has direct lending authority to make short-, intermediate-, and long-term loans to agricultural producers, rural homeowners and some farm-related businesses.

Agricultural district — A planning term which defines an area within a local jurisdiction where farming is the preferred economic activity. Districts may be voluntarily created by landowners who receive benefits, usually in return for not developing the land for a certain number of years, or they may be designated in a local land use plan. An agricultural district is not a conservation district.

Agricultural diversification — A system of farming that encourages production of a variety of plants and animals and their products as opposed to monoculture or large-scale specialization. Advocates of diversification argue that it provides greater income stability. Specialized farms benefit from economies of size.

Agricultural Fair Practices Act of 1967 — This law (P.L. 90-288) was enacted to protect farmers from retaliation by handlers (buyers of their products) because the farmers are members of a cooperative. The act permits farmers to file complaints with USDA, which can then institute court proceedings, if they believe their rights under the law have been violated. Several bills have been introduced in recent years on behalf of producers (among them, some poultry growers who have contracts with large companies) to give them more bargaining power under the Act, which, some producers contend, lacks adequate enforcement authorities.

Agricultural literacy — A phrase being used by several universities (e.g., Texas Tech, the University of Arizona, and California State Polytechnic) to describe programs to promote the understanding and knowledge necessary to synthesize, analyze, and communicate basic information about agriculture to students, producers, consumers, and the public. These programs focus on assisting educators and others to effectively incorporate information about agriculture into subjects being taught or examined in public and private forums, and to better understand the impact of agriculture on society.

Agricultural Management Assistance Program — Authorized in the Agricultural Risk Protection Act of 2000 (P.L. 106-224, Sec. 133) and permanently authorized and amended in the 2002 farm bill (P.L. 107-171, Sec. 2501) to provide mandatory funding of $20 million annually from FY2003 through FY 2007 and $10 million in all other years (but annual appropriations laws have reduced the funding). Participants in 15 designated states, primarily in the northeast, that had been under-served by crop insurance are to receive financial assistance, not to exceed $50,000 per year, to help pay to install conservation practices and take other specified actions that will reduce their financial risk.

Agricultural Market Transition Act (AMTA) — Title I of the 1996 farm bill (P.L. 104-127). It allowed farmers who had participated in the wheat, feed grain, cotton, and rice programs in any one of the 5 years prior to 1996 to enter into 7-year production flexibility contracts for 1996-2002. Total national production flexibility contract payments (sometimes called AMTA payments, or contract payments) for each fiscal year were fixed in the law. The AMTA allowed farmers to plant 100% of their total contract acreage to any crop except fruits and vegetables, and receive a full payment. Land had to be maintained in agricultural uses. Unlimited haying and grazing and planting and harvesting alfalfa and other forage crops was permitted with no reduction in payments. AMTA commodity support provisions were replaced by the 2002 farm bill (P.L. 101-171, Title I), a 6-year farm bill.

Agricultural Marketing Agreement Act of 1937 — This law, signed June 3, 1937 (P.L. 75-137), provided authority for federal marketing orders, and also reaffirmed the marketing agreements provisions of the Agricultural Adjustment Act of 1933 (P.L. 73-10). Under the authority of this permanent law and subsequent amendments, marketing orders have been established for milk as well as numerous fruits, vegetables, and specialty crops.

Agricultural Marketing Service (AMS) — A USDA agency that establishes standards for grades of cotton, tobacco, meat, dairy products, eggs, fruits, and vegetables. AMS also operates inspection and grading services and market news services; provides supervisory administration for federal marketing orders; administers USDA's Section 32 budget account (including commodity purchasing using Section 32 funds); oversees the National Organic Program; provides support for various farm marketing activities; and conducts research and analysis of transportation problems affecting agriculture. [www.ams.usda.gov].

Agricultural pollution — Wastes, emissions, and discharges arising from farming activities. Includes runoff and leaching of pesticides and fertilizers; pesticide drift and volatilization; erosion and dust from cultivation; and improper disposal of animal manure and carcasses. Some agricultural pollution is point source, meaning that it is derived from a single discharge point, such as a pipe. Large feedlots are an example of point sources, and they require permits under the Clean Water Act (P.L. 92-500, 33 U.S.C. 1251-1387). However, much of the pollution from agriculture is from nonpoint sources, meaning that it derives from dispersed origins, e.g., blowing dust or nutrients leaching from fields. Most pollution control programs have focused on particular categories of point sources, although nonpoint and unregulated point sources account for an increasingly large proportion of remaining pollution. The EPA concludes that agricultural sources account for over one-half the pollution impairing surface water quality in the U.S. based on state surveys. The Clean Water Act mandates that states develop and implement management programs to control nonpoint sources of water pollution generally, and the Coastal Zone

Management Program requires participating states to develop similar programs for farms within state-designated coastal zones.

Agricultural product — The term agricultural products is often used as a collective expression and like agricultural commodity, is defined in different ways to meet different circumstances. For example, 7 U.S.C. 138 states "The term 'agricultural product' means any fresh fruit or vegetable or any commodity or product derived from livestock or fowl, that is marketed in the United States for human consumption." While 7 U.S.C. 451 says "the term 'agricultural products' means agricultural, horticultural, viticultural, and dairy products, livestock and the products thereof, the products of poultry and bee raising, the edible products of forestry, and any and all products raised or produced on farms and processed or manufactured products thereof, transported or intended to be transported in interstate and/or foreign commerce." And 7 U.S.C. 6502 says "The term 'agricultural product' means any agricultural commodity or product, whether raw or processed, including any commodity or product derived from livestock that is marketed in the United States for human or livestock consumption." Possibly the broadest definition may be contained in the Commerce and Trade title, Consumer Credit chapter of the Code (15 USC1602(t)), which states "The term 'agricultural products' includes agricultural, horticultural, viticultural, and dairy products, livestock, wildlife, poultry, bees, forest products, fish and shellfish, and any products thereof, including processed and manufactured products, and any and all products raised or produced on farms and any processed or manufactured products thereof."

Agricultural protection zoning — Local zoning codes that include provisions such as large lot size requirements and use limitations to separate farming and related activities from other land uses. Some jurisdictions further subdivide agricultural zones, so as to distinguish full time commercial farming from a mix of uses that might include rural residence farms and retirement farms on very large lots. See Agricultural zoning.

Agricultural purposes — The term agricultural purposes is sometimes used to categorize a collection of activities covered by a law or set of regulations. Yet, the agricultural laws assembled in Title 7 of the U.S. Code do not include a definition. A definition is contained in the Commerce and Trade title, Consumer Credit chapter, of the Code(15 USC1602(s)), which states: The term "agricultural purposes" includes the production, harvest, exhibition, marketing, transportation, processing, or manufacture of agricultural products by a natural person who cultivates, plants, propagates, or nurtures those agricultural products, including but not limited to the acquisition of farmland, real property with a farm residence, and personal property and services used primarily in farming.

Agricultural Quarantine Inspection (AQI) — A program of USDA's Animal and Plant Health Inspection Service (APHIS), part of which was transferred to the new Department of Homeland Security (DHS) under the Homeland Security Act of 2002 (P.L. 107-296). As of March 1, 2003, roughly 2,500 AQI border inspection personnel are part of the DHS Border and Transportation Security directorate. AQI inspects incoming passengers, luggage, and cargo at U.S. ports of entry in order to protect U.S. agriculture from foreign animal and plant pests and diseases that may enter by unintentional or intentional means. The plant and animal quarantine function of AQI remains in APHIS.

Agricultural Research, Extension, and Education Reform Act of 1998 — P.L. 105-185 was separate legislation that revised and reauthorized federally supported agricultural research, education, and extension programs from June 1998 through May 2002 (historically, these authorities have been part of an omnibus farm policy law enacted every 4 to 6 years). The 1998 Act built upon reforms that were made in the research title of the farm law in effect at the time, the 1996 farm bill (P.L. 104-127). Key provisions were new accountability measures for recipients of federal research funds, and a new competitive research grant program called the Initiative for Future Agriculture and Food Systems, for which mandatory funds were authorized (annually appropriated discretionary funds support most of USDA's research, education and extension programs). The 1998 law's provisions, as well as new revisions of research, education, and extension policies, are included in Title VII of the 2002 farm bill (P.L. 107-171).

Agricultural Research Service (ARS) — A USDA agency employing federal scientists to conduct agricultural research at more than 100 field locations in the United States, U.S. insular areas (e.g., Puerto Rico, the Virgin Islands), and several foreign countries. ARS scientists conduct basic, applied, and developmental research in the following fields: livestock; plants; soil, water and air quality; energy; food safety and quality; nutrition; food processing, storage, and distribution efficiency; non-food agricultural products; and international development. [www.ars.usda.gov].

Agricultural Resource Management Survey (ARMS) — ARMS is USDA's primary source of information on the financial condition, production practices, resource use, and economic well being of America's farm households. Sponsored jointly by the Economic Research Service (ERS) and the National Agricultural Statistics Service (NASS), ARMS began in 1996 as a synthesis of the former USDA cropping practice, chemical use, and farm costs and returns surveys, which dated back to 1975. ARMS data underpin USDA's annual estimates of net farm income and fulfills a congressional mandate that USDA provide annual cost-of-production estimates for commodities covered under farm support legislation. ARMS also provides data regarding chemical use on field crops required under environmental and food safety legislation.

Agricultural Stabilization and Conservation Service (ASCS) — This was the USDA agency once primarily responsible for administering the farm commodity price and income support programs, and conservation cost-sharing programs. Its functions were folded into a new Farm Service Agency (FSA) as a consequence of 1994 reorganization. A local field service center is maintained in nearly all farming localities. [www.fsa.usda.gov/pas/default.asp].

Agricultural Trade Development and Assistance Act of 1954 — P.L. 83-480 is commonly referred to as both P.L. 480 and Food for Peace. The law established what continues to be the primary U.S. overseas food assistance program. The program makes U.S. agricultural commodities available through long-term credit at low interest rates and provides food donations.

Agricultural Trade Office — The Agricultural Trade Act of 1978 (P.L. 95-501) directed the establishment of trade offices in major centers of commerce throughout the world. Agricultural trade offices are operated by the Foreign Agricultural Service (FAS) to develop, maintain, and expand international markets for U.S. agricultural commodities and serve as centers for export sales promotion and contact points for importers seeking to buy U.S. farm products.

Agricultural zoning — Designations made by local jurisdictions that are intended to protect farmland and farming activities from incompatible nonfarm uses. Agricultural zoning can specify many factors, such as the uses allowed, minimum farm size, the number of nonfarm dwellings allowed, or the size of a buffer separating farm and nonfarm properties.

Agriculture and Consumer Protection Act of 1973 — P.L. 93-86 was the 4-year farm bill that adopted target prices and deficiency payments as a tool that would support farm income but reduce forfeitures to the Commodity Credit Corporation (CCC) of surplus stocks. (Target prices were eliminated by the 1996 farm bill (P.L. 104-127), but restored by the 2002 farm bill (P.L. 101-171, Sec. 1104).) It reduced payment limitations to $20,000 (from $55,000 set in 1970) for all program crops. The Act might be considered the first omnibus farm bill because it went beyond simply authorizing farm commodity programs. It authorized disaster payments and disaster reserve inventories; created the Rural Environmental Conservation Program; amended the Food Stamp Act of 1964 (P.L. 88-525), authorized the use of commodities for feeding low income mothers and young children (the origin of the supplemental food program); and amended the Rural Development Act of 1972 (P.L. 92-419).

Agriculture and Food Act of 1981 — P.L. 97-98 was the 4-year omnibus farm bill that continued and modified commodity programs through 1985. It set specific target prices for 4 years, eliminated rice allotments and marketing quotas, lowered dairy supports, and made other changes affecting a wide range of USDA activities. The next year this farm bill was amended to freeze the dairy price support level and mandate loan rates and acreage reserve provisions for the 1983 crops (Omnibus Budget Reconciliation Act of 1982, P.L. 97-253). Again in 1984, amendments were adopted to freeze target prices, authorize paid land diversion for feed grains, upland cotton, and rice, and provide a wheat payment-in-kind program for 1984 (Agricultural Programs Adjustment Act of 1984, P.L. 98-258).

Agriculture in Concert with the Environment (ACE) — An EPA program, administered cooperatively with USDA's Sustainable Agriculture Research and Education (SARE) program, to fund research projects that reduce the risk of pollution from pesticides and soluble fertilizers.

Agriculture Innovation Centers — The 2002 farm bill (P.L. 107-171, Sec. 6402) directed the USDA to provide grants and to assist in the establishment of Agriculture Innovation Centers that provide information, training and direct assistance to agricultural producers in the production, processing, development and marketing of value-added agricultural commodities and products. In September 2003, the USDA announced $10 million in grants for the establishment of demonstration centers in Indiana, Iowa, Kansas, Michigan, Minnesota, Montana, New Jersey, New York, North Dakota, and Pennsylvania.

Agriculture Mediation Program — A program initially authorized by the Agricultural Credit Act of 1987 (P.L. 100-233, Title V, and recently amended by P.L. 106-472, Sec. 306; 7 U.S.C. 5101), to facilitate the use of mediation to settle disputes arising in conjunction with USDA actions. If agreement is not reached through mediation, all parties remain free to pursue other available administrative appeals or legal actions. Typical areas of dispute include farm loans, farm and conservation programs, wetland determinations, rural water loan programs, grazing on national forest lands, and pesticides. This program is administered by the Farm Service Agency (FSA).

Agriculture Risk Protection Act of 2000 — P.L. 106-224 made major revisions to the federal crop insurance program and provided emergency agricultural assistance. The crop insurance provisions: significantly increased the government subsidy of the program; improved coverage for farmers affected by multiple years of natural disasters; and authorized pilot insurance programs for livestock farmers and growers of other farm commodities that were not served by crop insurance, among many other provisions. The emergency provisions made available a total of $7.14 billion in emergency farm assistance, mostly in direct payments (called market loss payments) to growers of various commodities to compensate for low farm commodity prices.

Agroterrorism — The deliberate introduction of an animal or plant disease with the goal of generating fear, causing economic losses, and/or undermining stability. Agroterrorism is a subset of the more general issue of bioterrorism. An agroterrorism event would affect the production agriculture sector economically in terms of plant and animal health, and affect supply and demand. Humans could be at risk in terms of food safety or public health, especially if the chosen disease is transmissible to humans (zoonotic). Losses would accrue to individuals, businesses, and governments through costs to contain and eradicate the disease, and to dispose of contaminated products. The economic impact can spread to input suppliers, food processors, transportation, retailers, and food service providers.

AHI — Animal Health Institute. [www.ahi.org].

AI — Artificial insemination

AID — See United States Agency for International Development (USAID). [www.info.usaid.gov].

AIF — Animal Industry Foundation changed its name to Animal Agriculture Alliance. [www.aif.org].

AIIS — See Automated Import Inspection System.

Air pollution — Contamination of the atmosphere by substances that, directly or indirectly, adversely affect human health or welfare. Air pollution results from human activities, both deliberate releases (as from smokestacks) and fugitive emissions (as dust blown from streets or fields), and from natural sources (including sea spray, volcanic emissions, and pollen). The Clean Air Act (42 U.S.C. 7401 *et seq.*) authorizes the EPA to regulate air pollution (see National Ambient Air Quality Standards).

Alar — Trade name for daminozide, a plant regulator and therefore classed as a pesticide, that makes apples redder, firmer, and less likely to drop off trees before harvest. It was also used to a lesser extent on peanuts, tart cherries, concord grapes, and other fruits. Alar was suspended by the EPA in 1989 following a controversy over allegations of cancer risk to children from residues of Alar and its breakdown product UDMH on apples and in apple products.

Alcohol — The family name of a group of organic chemical compounds that includes methanol, ethanol, isopropyl alcohol, and others. Ethanol is produced from crops or residues with a high carbohydrate content. Alcoholic beverages contain ethanol, and ethanol is blended with gasoline to produce gasohol. Most industrial ethanol produced in the U.S. is from corn wet-milling and dry-milling.

ALF — Animal Liberation Front. [animalliberationfront.us].

Alien Species Prevention and Enforcement Act of 1992 — P.L. 102-393 makes it illegal to ship certain categories of plants and animals through the mail. The prohibited species are certain injurious animals, plant pests, plants and materials under federal quarantine, and certain plants and animals under the Lacey Act (16 U.S.C. 3371-3378), a law that pertains to illegal trade in fish, wildlife, and plants. These also may be referred to as invasive species.

Allotment — In conjunction with commodity support programs, acreage allotments and marketing quotas historically served to limit a farm's output or volume marketed. For federal lands grazing, an allotment is an area designated and managed for grazing of livestock. The Bureau of Land Management and the Forest Service stipulate the number of livestock and time period (season) of use for each allotment under their respective jurisdictions.

Allowable Sale Quantity (ASQ) — The maximum quantity of timber that may be sold from national forest lands under a Forest Service forest plan for a period of ten years.

Alternative Agricultural Research and Commercialization Corporation (AARCC) —As authorized by the 1990 farm bill (P.L. 101-624), AARCC was originally established as the Applied Agricultural Research Commercialization Center in the USDA to be a public venture capital agency that would invest in small businesses to help them develop and commercialize new nonfood products from agricultural and forestry commodities. The 1996 farm bill (P.L. 104-127) changed the Center from a government agency to a wholly-owned venture capital corporation of USDA. Congress repealed the authority for AARCC in the 2002 farm bill (P.L. 107-171, Sec. 6201).

Alternative agriculture — See Sustainable agriculture.

Alternative fuels — According to the Department of Energy, alternative fuels are substantially nonpetroleum sources of energy. As defined by the Energy Policy Act of 1993 (EP Act) DOE currently recognizes the following as alternative fuels: mixtures containing 85% or more by volume of alcohol fuel, including methanol and denatured ethanol; natural gas (compressed or liquefied); liquefied petroleum gas (propane); hydrogen; coal-derived liquid fuels; fuels derived from biological materials; electricity (including electricity from solar energy); and 100% biodiesel (B100). Renewable fuels also are a subset of alternative fuels. The U.S. Department of Energy maintains an information center for alternative fuels. [www.eere.energy.gov/afdc/index.html].

Alternative test methods — New laboratory procedures that reduce use of animals or reduce the suffering of animals in evaluating the potential toxicity of chemicals.

Amber box policies — Countries' agricultural policies that are considered under world trade rules to have the greatest potential effect on production or trade, and are therefore subject to review and reduction over time. Amber box policies include market price support, production-based direct payments, input subsidies, and similar programs. The expressions green box policies, amber box policies, and blue box policies were developed to categorize agriculture policies using a traffic light analogy in the Uruguay Round Agreement on Agriculture.

Amenable species — A term used within the context of USDA's meat and poultry inspection program to signify exotic species (livestock and fowl not covered by the statutes) that might be added to the laws and thus be eligible for mandatory federal inspection, which is taxpayer-funded. An exotic species is considered an amenable species if its anatomy and biology are substantially the same as the animals currently inspected. The Poultry Products Inspection Act (P.L. 85-172, as amended; 21 U.S.C. 451 *et seq.*) was expanded in 2001 to cover ostrich, rhea, and emu (ratites) because USDA determined that the hazards they present to food safety are essentially the same as those posed by chickens, turkeys, ducks, etc., and the existing contamination detection and prevention systems are sufficient to control them. Bison and buffalo have been considered for inclusion under the Federal Meat Inspection Act (21 U.S.C. 601 *et seq.*) because they are bovine species (like cattle). Deer and elk, on the other hand, are cervids, and pose hazards for food safety that are not yet fully known or controlled for under the existing meat inspection system. The term non-amenable sometimes is used to describe cervids and certain other exotics, like rabbits, for example.

American Heritage Rivers Protection Program — A Clinton Administration (1993-2001) initiative to deliver federal resources more efficiently and effectively in support of voluntary community efforts to enhance and protect designated rivers or river segments; the designations were selected based on proposals submitted by local sponsors. Portions of these designations are located in or affect agricultural lands.

AMI — American Meat Institute. [www.meatami.org].

Ammonia — A pungent alkaline gas, a compound of nitrogen and hydrogen (NH_3). It is formed naturally when bacteria decompose nitrogen-containing compounds, such as manures. Emissions of ammonia can be a problem in enclosed livestock facilities, and in the ambient air they may contribute to very fine particulate matter. Synthetic ammonia is used as a nitrogen fertilizer. Also called anhydrous ammonia, it is the basic feed stock for the production of all nitrogen fertilizers as well as being a direct application material. Synthetic ammonia is made through a reaction between natural gas and nitrogen.

AMR — Advanced meat recovery

AMS — Aggregate measure of support; Agricultural Marketing Service. **AMTA** — Agricultural Market Transition Act (P.L. 104-127, Title I). **ANCOM** — Andean Common Market.

Animal and Plant Health Inspection Service (APHIS) — A USDA agency originally established to conduct border inspections and regulatory and control programs to protect animal and plant health. The border inspection function became part of the Department of Homeland Security (DHS) in 2002 (P.L. 107-296). USDA retains the agency's other functions, including: animal and plant quarantine; pest and disease monitoring, control and eradication programs; trade facilitation through sanitary or phytosanitary (SPS) negotiations with foreign countries and certification of pest- or disease free status of U.S. exports; and horse protection and animal welfare programs, among others. [www.aphis.usda.gov].

Animal biologics — Vaccines, bacterins, antigens, diagnostic kits and other products of biologic origin. The Animal and Plant Health Inspection Service (APHIS) has responsibility for approving and regulating some animal biologics according to the Viruses, Serums, Toxins, Antitoxins, and Analogous Products Act (commonly called the Virus, Serum, Toxin Act (21 U.S.C. 151-159). Animal drugs are regulated by the FDA. See Animal biologics, Biologics, Veterinary biologics.

Animal byproducts – The parts of livestock other than meat cuts from muscle derived during processing (i.e., fat, organs, blood, edible and inedible offal, hide, bones, bones, horns hooves, etc.). Inedible animal byproducts typically are rendered for use in livestock feed, pet food, personal care products, and industrial ingredients.

Animal Damage Control (ADC) Program — Renamed in 1997 as the Wildlife Services (WS) program, it is an Animal and Plant Health Inspection Service (APHIS) effort to protect agriculture, natural resources, property or endangered species from unwanted and potentially harmful effects of wildlife species, including predators and invasive species.

The program also works to prevent wildlife/airplane collision hazards at civilian and military airports.

Animal drugs — Drugs intended for use in the diagnosis, cure, mitigation, treatment, or prevention of disease in animals. The Food and Drug Administration (FDA) has the broad mandate under the Federal Food Drug and Cosmetic Act (21 U.S.C. 321 *et seq.*) to assure the safety and effectiveness of animal drugs and their use in all animals, including farm animals. Before FDA formally approves an animal drug, the sponsor or manufacturer of the drug must document in scientific testing that the drug has been found "safe and effective." The testing data also must demonstrate that a methodology is available to detect and measure any residue left in edible animal products. Farmers and veterinarians using drugs on farm animals must adhere to guidelines about how much time must elapse before a treated animal can be slaughtered, and any other use constraints or warnings stated on the drug label. Animal biologics (e.g., vaccines and tests) are regulated by APHIS.

Animal feeding operation — Facilities where animals are kept and raised in confined situations. Feed is brought to the animals. Agriculture census data from 1997 show there were 238,000 such operations nationwide that year. When large enough, these facilities are designated as concentrated animal feeding operations (CAFOs) and they become subject to regulatory requirements to prevent point source pollution. USDA and the EPA issued a Unified National Strategy for Animal Feeding Operations on March 9, 1999. The goal was to minimize water pollution from confinement facilities and land application of manure through adoption of site-specific comprehensive nutrient management plans. EPA issued a final rule on December 16, 2002, that will require a total of about 15,500 operations to obtain permits by 2006. At the time the rule was issued, about 4,500 operations had permits.

Animal Health Protection Act (AHPA) — The AHPA (P.L. 107-171, Title X, Subtitle E; 7 U.S.C. 8301 *et seq.*) consolidates all of the animal quarantine and related laws on the books, some dating back to the late 1800s, and replaces them with one statutory framework. While most of the authorities contained in the consolidated AHPA were taken from existing laws, some new provisions were added to help fully protect U.S. animal agriculture due to gaps in legal authority.

Animal identification (ID) and traceback — Currently, the private marketing system, assisted by computerization of records, generally can trace products back to their original suppliers, although not necessarily all the way to the farm. Animal identification is one component of traceback (or traceability); it refers to marking individual animals so that they can be tracked from birth to slaughter. Livestock producers already frequently identify their animals using back-tags, ear tags, radio frequency identification (RFID) tags, tattoos, and other devices, often for production management and animal health purposes. Animal identification and traceback are not programs in themselves; rather they may be useful tools in animal health, food safety, quality assurance, country-of-origin labeling, or other regulatory or marketing programs. Interest in a national, possibly mandatory, animal ID program that can quickly respond to animal disease outbreaks has grown in the wake of the December 2003 discovery of a cow in the United States with bovine spongiform encephalopathy (BSE).

Animal protein — Protein used in livestock feed that is derived from meatpacking or rendering plants, surplus milk or milk products, and marine sources. The Food and Drug

Administration (FDA) has banned proteins derived from certain parts of mammals from inclusion in livestock feeds in order to protect U.S. consumers from exposure to the agent that causes Creutzfeldt-Jakob disease (related to mad cow disease, or BSE, in cattle).

Animal Protein Free Certification (APFC) Program — One of the AMS process verification programs. APFC is a voluntary, user-fee service available to poultry producers and processors to provide third-party verification that poultry and poultry products have never been fed animal protein, animal fats, or animal by-products.

Animal unit month (AUM) — An animal unit month (AUM) is the amount of forage needed to sustain one animal unit, or its equivalent, for one month. One animal unit is a 1,000 pound beef cow with a daily requirement of 26 pounds of dry matter forage. Therefore, one AUM is equal to 780 pounds of dry matter forage. The carrying capacity of a pasture is expressed in AUMs. Grazing fees for federal lands are charged by animal unit months or head-months.

Animal unit (AU) — A 1,000-pound beef cow is the standard measure of an animal unit. The dry matter forage requirement of one animal unit is 26 pounds per day. Animal unit equivalents (AUE) are calculated for various other animals. A 700-pound steer is 0.80 animal units. A 1,300 pound horse is 1.20 animal units. A 120-pound sheep is 0.20 animal units. Federal land management agencies, such as the Bureau of Land Management and the Forest Service, may use different standards for setting grazing fees. The Natural Resources Conservation Service uses animal units to estimate manure production and manure nutrient content when designing projects under the Environmental Quality Incentives Program (EQIP). EPA does not use the term animal units to define size classes for purposes of compliance with effluent limitations for Concentrated Animal Feeding Operations (CAFOs). Instead it sets thresholds by specifying the actual number of animals (40 CFR 122.23).

Animal waste management facility — Any structure, such as a waste treatment pond, used to store, process, or dispose of waste associated with the production of animals.

Animal Welfare Act of 1966 — P.L. 89-544 was enacted to curb the theft and mistreatment of dogs and cats for experimental and research purposes. The principal federal animal protection law, it has been amended several times to address specific concerns such as the shipping of pets on public transportation, dog fighting, and using other warm-blooded animals in biomedical experiments. Although administered by the Animal and Plant Health Inspection Service (APHIS), the law has always excluded farm animals from its coverage. Generally, USDA is authorized to "promulgate standards to govern the humane handling, care, treatment, and practices in experimental procedures to ensure that animal pain and distress are minimized...." The law excludes from the definition of animal "...horses not used for research purposes and other farm animals, such as, but not limited to livestock or poultry, used or intended for use as food or fiber, or livestock or poultry used or intended for use for improving animal nutrition, breeding, management, or production efficiency, or for improving the quality of food or fiber." Animal welfare has become more controversial in recent years as certain animal protection groups have argued for more extensive legal protections for animals. The debate over the meaning of animal welfare revolves around the most appropriate methods for taking care of animals, including farm animals. Legislation has been proposed (but not enacted) in recent years that would intervene in animal production operations by regulating confinement facilities; determining the diets of veal calves; specifying how poultry must be slaughtered; and prohibiting dealers from handling nonambulatory (downer) livestock unless they are humanely killed. (7 U.S.C. 2131 *et seq.*).

ANPR — Advance notice of proposed rulemaking.

Antemortem — Before slaughter. As used in the meat and poultry inspection program, the term refers to the examination that USDA meat inspectors are required to conduct of all live animals just before they are killed.

Antibiotics — Chemical substances produced by microorganisms or synthetically that inhibit the growth of, or destroy, bacteria. Antibiotics are used at therapeutic levels to fight disease in humans and animals. Since the 1950s they have been used at subtherapeutic levels in animal feeds, where they enhance growth and may help prevent disease in livestock and poultry. Rules guiding the use of veterinary drugs and medicated animal feeds, including tolerance levels for drug residues in meats for human consumption, are promulgated by the Center for Veterinary Medicine of the Food and Drug Administration (FDA). The Food Safety and Inspection Service (FSIS) enforces the FDA rules through a sampling and testing program that is part of its overall meat and poultry inspection program.

Antidumping duty — A duty or levy imposed under authority of Title VII of the Tariff Act of 1930 (P.L. 71-361). Title VII states that if the U.S. Department of Commerce determines that an imported product is being sold at less than its fair value, and if the International Trade Commission determines that a U.S. producer is thereby being injured, the Commerce Department shall apply antidumping duties equivalent to the dumping margin.

Antitrust — Term used to describe a policy or action that seeks to curtail monopolistic power within a market. Concentration and potential for monopolization have long been agricultural policy issues. The Sherman Act (15 U.S.C. 1 *et seq.*), the Clayton Antitrust Act (15 U.S.C. 12 *et seq.*), and the Federal Trade Commission Act (15 U.S.C. 41 *et seq.*) are the powers used to prevent monopolies.

APA — Administrative Procedure Act (P.L. 79-404).

APEC — Asian Pacific Economic Cooperation Forum. [www.apec.org].

APH — Actual production history

APHIS — Animal and Plant Health Inspection Service. [www.aphis.usda.gov].

Apple Market Loss Assistance Program — An FSA program that has made payments to apple producers to partially offset revenue losses from low prices caused by the loss of markets. The 2002 farm bill (P.L. 107-171, Sec. 10105) mandated the payment of $94 million by the Commodity Credit Corporation (CCC) for lost markets in crop year 2000. Earlier funding was mandated for the 2000 crop of apples by P.L. 107-76, Sec. 741 ($75 million), and for the 1998 and 1999 apple crops by P.L. 106-387, Sec. 811 ($100 million).

Applied tariff — The tariff rate actually levied on imports in contrast to the often higher bound tariff rate that is incorporated in schedules of commitments made in multilateral trade negotiations under WTO auspices. Because they are generally lower than bound rates, applied rates often are proposed by exporting countries as the basis from which tariff reductions should be made.

Appraised stumpage price (or appraised rate) — On national forests, the Forest Service estimate of the market price for timber to be cut and removed; it is the advertised minimum for competitive bidding by purchasers.

Appropriated entitlement — An entitlement program that is funded through annual appropriations rather than by a permanent appropriation. Examples within USDA are the Commodity Credit Corporation-funded programs; the federal crop insurance program; the food stamp program; and child nutrition programs. Because the authorizing statute for these programs requires the government to provide eligible recipients the benefits to which they are entitled, whatever the cost, Congress must appropriate the necessary funds. If the amount Congress provides in the annual appropriations act is not enough, it must make up the difference in a supplemental appropriation.

Appropriation — The amount of funding Congress provides for a federal program to spend in a given year. An appropriation provides legal authority for federal agencies to incur obligations and to make payments out of the Treasury for specified purposes. Thirteen regular appropriations bills are considered every year by Congress, and supplemental appropriations are considered from time to time. Appropriations for all of USDA (except for the Forest Service) is provided annually in the bill that funds USDA and Related Agencies. The Forest Service is funded through the Interior appropriations bill.

AQI — Agricultural Quarantine Inspection.

Aquaculture — The National Aquaculture Act of 1980 (P.L. 96-362; 16 U.S.C. 2801 *et seq.*) defines aquaculture as "the propagation and rearing of aquatic species in controlled or selected environments, including ocean ranching." The Act divides responsibility for most aquaculture research, regulatory, and related activities among the Departments of Agriculture, Commerce, and the Interior. Private aquaculture has grown rapidly and diversified in recent years; in the United States, aquaculture is dominated (80%) by catfish production.

Aquifer — An underground geological formation, or group of formations, containing usable amounts of groundwater that can supply wells or springs for domestic, industrial, and irrigation uses. Removing more groundwater from an aquifer than is naturally replenished is called overdrafting, and can result in a dropping water table, increased pumping costs, land subsidence (which reduces the future recharge capacity), saltwater intrusion, reduced streamflows in interconnected ground- and surface-water systems, and exhaustion of groundwater reserves. Overdrafting groundwater occurs primarily in the Plains States, especially in the Ogallala aquifer (which stretches from Texas to Nebraska) and in the West more generally.

Arable crops program — A consolidated support system operated under the EU Common Agricultural Policy for producers of major cereals, oilseeds, and protein crops. Production of these crops constituted 21% of farm income and 40% of agricultural lands in the EU in 2000. Main elements of the program include area compensatory payments, reductions in administered prices (also known as intervention prices), and annual land set-aside program requirements.

Area compensatory payments — Usually refers to EU subsidies that are tied to acreage and, sometimes more specifically, to the number of acres planted to a specific crop. Such payments might also be offered to producers in exchange for removing land from production, under certain supply control programs. The EU Common Agricultural Policy provides per-hectare area payments to producers for land in cereals, oilseeds, and protein crops as compensation for lower administered prices for these crops, so long as they meet land set-aside program requirements.

Area yield options contract — A contract entitling the holder to receive a payment when the area yield is below (above) the put (call) option strike yield. The strike yield is the yield at which the holder of an option contract can exercise the option.

Arid — A relatively dry climate in which annual precipitation is less than 10 inches, which generally is insufficient for crops to be grown without irrigation. Such areas usually are the focus of debate over federal water policies.

ARMS — Agricultural Resource Management Survey

ARP — Acreage reduction program.

ARS — Agricultural Research Service. [www.ars.usda.gov].

ASA — American Society of Agronomy. [www.agronomy.org].

ASA — American Soybean Association. [www.amsoy.org].

ASA — American Sugar Alliance. [www.sugaralliance.org].

ASAE — American Society of Agricultural Engineers. [www.asae.org].

ASCS — Agriculture Stabilization and Conservation Service, predecessor to the Farm Service Agency (FSA). [www.fsa.usda.gov/pas/default.asp].

ASEAN — Association of Southeast Asian Nations. [www.asean.or.id].

ASFSA — American School Food Service Association. [www.asfsa.org].

ASIA — American Sheep Industry Association. [www.sheepusa.org].

Asia-Pacific Economic Cooperation (APEC) forum — Established in 1989, APEC is a formal institution with a permanent secretariat located in Singapore. Its original 12 members include Australia, New Zealand, the United States, Canada, Japan, South Korea, Thailand, Malaysia, Indonesia, the Philippines, Singapore, and Brunei. In 1991, APEC admitted China, Taiwan (admitted as Chinese Taipei), and Hong Kong. Mexico and Papua New Guinea joined in 1993; Chile joined in 1994; Peru, Russia, and Vietnam joined in 1998. The 21-nation member APEC provides a forum for ministerial level discussion and cooperation on a range of economic issues including trade, investment, technology transfer, and transportation. According to APEC, a key feature that sets it apart from other international organizations is its commitment to business facilitation and the regular involvement of the private sector in a wide range of APEC activities. [www.apecsec.org.sg/apec.html].

Asian long-horned beetle — A serious pest of hardwood trees in its native China, that by 1998 had been found in 14 states in the United States, where it has no known natural enemies. The Animal and Plant Health Inspection Service (APHIS) is now working to detect and destroy the beetle, which is virtually impossible to eradicate with pesticides because it bores deep inside trees to lay its eggs; the only known suppression method is to remove and destroy infected trees. The agency reports that the beetle, which already has led to the destruction of many trees in parts of New York, could destroy millions of acres of hardwoods (including maples, horse chestnuts, poplars, willows, and elms) if it becomes established in the environment. APHIS believes that the beetle has been entering the United States in solid wood packing materials such as pallets and crates from China. For that reason, in late 1998, it banned all shipments from China containing such packing materials if they have not been treated to kill the pest.

ASP — American Society for Plasticulture. [www.plasticulture.org].

ASQ — Allowable sale quantity (timber).

Assessment — Under certain agricultural marketing orders or commodity promotion programs, assessments may be applied against farm sales receipts to help pay for generic commodity advertising or research. The term check-off is often used interchangeably with assessment. In a different context, federal deficit reduction marketing assessments were applied during much of the decade of the 1990s to certain commodity price support programs (dairy, peanuts, sugar, tobacco, and soybeans) to help reduce the federal budget deficit, which arguably was higher because of the programs. Additionally, tobacco quota buyout legislation included an "assessment" on tobacco product manufacturers and imports as the source of $10.14 billion in financing for the buyout program (P.L. 108-171).

Assimilative capacity — The ability of a body of water to cleanse itself; its capacity to receive waste waters or toxic materials without deleterious effects and without damage to aquatic life or humans who consume the water.

Association of Southeast Asian Nations (ASEAN) — A multilateral organization formed in 1967 by the governments of Indonesia, Malaysia, the Philippines, Singapore, and Thailand to promote economic, social, and cultural cooperation among nations in the Southeast Asian region. Brunei, Cambodia, Vietnam, Laos, and Myanmar joined later. The countries' total of more than 500 million people constitute the third largest overseas market for U.S. exports generally. [www.aseansec.org/home.htm].

ASTA — American Seed Trade Association. [www.amseed.com].

At-risk species — A term used by the Natural Resources Conservation Service (NRCS) in implementing the Environmental Quality Incentive Program and proposed to be used in implementing the Conservation Security Program (CSP) that includes any plant or animal species determined by a State Technical Committee to need direct intervention to halt a decline in the size of the population. At risk species is one of many factors used to decide who will receive funding under both programs. The at-risk species definition used by NRCS does not make reference to the terms threatened or endangered as used under the Endangered Species Act.

ATO — Agricultural Trade Office.

Atrazine — A selective herbicide, widely used on corn. Due to concerns about surface water and groundwater contamination and worker exposure, EPA initiated a special review of atrazine registration (as well as registrations of two similar pesticide ingredients, cyanazine and simazine) in November 1994. Atrazine is still subject to the conditions of the Special Review and is undergoing reregistration and tolerance reassessment under the Food Quality Protection Act (FQPA, P.L. 104-170). The review might be completed in 2006.

Attainment area — An area considered to have air quality as good as or better than the National Ambient Air Quality Standards as defined in the Clean Air Act (42 U.S.C. 7401 *et seq.*). An area may be an attainment area for one pollutant and a non-attainment area for others.

Attractant — A chemical or agent that lures insects or other pests by stimulating their sense of smell. Attractants are a nontoxic technique for luring insects into traps and are heavily used in orchard crops. Though distinct from toxic baits, attractants are regulated as pesticides.

AU — Animal unit.

Audubon Society — National Audubon Society. [www.audubon.org].

AUM — Animal unit month.

Australian Wheat Board (AWB) — A statutory marketing agency that handles Australia's domestic marketing of wheat and export marketings of wheat and flour. Under the Australian system, farmers take their wheat to elevators designated as official handling agents for the AWB. Following delivery, farmers receive an initial payment, then over a period of time (which can be over a year) they receive additional payments until the full price has been paid. AWB became a grower-owned and controlled company operating under Australian corporation laws on July 1, 1999, and is listed on the Australian Stock Exchange. [www.awb.com.au/awb/user/default.asp].

Automated Import Information System (AIIS) — The AIIS is a computerized tracking system introduced by FSIS in 2002 to monitor imports of meat and poultry. While all imported products are inspected in the country of origin and reinspected visually before being released by FSIS, the AIIS selects shipments for additional reinspection verification. The additional reinspection tasks could include testing for residues, microbiology or food chemistry. Whether import shipments are reinspected is statistically based on the annual volume of shipments from the exporting country.

Avian influenza — Avian flu is a form of the Influenza A virus that infects birds, and certain strains have been known to infect both animals and humans. Many different strains of avian flu exist throughout the world. Avian influenza has two forms in birds: a low pathogenicity (LPAI) form that causes mild illness, and a highly pathogenic (HPAI) form that is extremely contagious and causes severe illness. Avian flu is spread by contact with infected feces, nasal or eye excretions, or contaminated equipment, vehicles, or clothing. Although avian influenza viruses have rarely infected humans, officials are concerned that the virus could mutate and cause a pandemic through human-to-human transmission. International health organizations conclude that poultry products from affected countries do not pose any food safety risk for humans.

AVMA — American Veterinary Medical Association. [www.avma.org].

AWA — Animal Welfare Act (P.L. 89-544) (7 U.S.C.2131 *et seq.*).

AWB — Australian Wheat Board. [www.awb.com.au].

AWP — Adjusted world price.

AWT — Advanced wastewater treatment.

B&CMA — Biscuit and Cracker Manufacturers' Association. [www.thebcma.org].

B&I — Business and Industry Guaranteed Loan Program.

Backgrounding — An intermediate stage sometimes used in cattle production which begins after weaning and ends upon placement in a feedlot. Background feeding relies more heavily on forage (e.g., pasture, hay) in combination with grains to increase a calf's weight by several hundred pounds and to build up immunity to diseases before it enters a feedlot. Some cattle operations specialize in backgrounding.

BACT — Best available control technology.

Balance on merchandise trade — Sometimes referred to as balance of trade or trade balance, it is the difference in value between a country's merchandise imports and exports in a year. Agricultural imports and exports are components of merchandise trade. Since 1960, agricultural exports have exceeded imports every year.

Band application — The spreading of chemicals over, or next to, each row of plants in a field, as opposed to broadcast application.

Bank for Cooperatives (BC) — Lending institution within the Farm Credit System that provides credit to agricultural cooperatives and rural utility cooperatives nationwide.

Nationally chartered CoBank has the authority to finance U.S. agricultural exports and to provide international banking services to farmer-owned cooperatives. [www.cobank.com].

Bankhead-Jones Farm Tenant Act of 1937 — P.L. 75-210 authorized acquisition by the federal government of damaged lands to rehabilitate and use them for various purposes. Both the Forest Service and the Bureau of Land Management manage some Bankhead-Jones lands. Some Forest Service Bankhead-Jones lands are National Grasslands.

Bargaining association — A farmer cooperative intended primarily to influence farm prices or other terms of trade between the members and the buyers of the commodities they produce.

Barrows and gilts — A barrow is a young castrated male hog. A gilt is a young female hog. Both are raised for pork. Market news reports of prices paid for barrows and gilts are of keen interest to producers and packers alike, as those are the primary slaughter animals.

Barter — A form of countertrade in which goods having comparable values are exchanged under a single contract, within a specified period of time, and without any flow of money taking place. The U.S. government ran a barter program from 1950 to 1973, exchanging surplus agricultural commodities for strategic materials and for goods and services it otherwise would have purchased. In addition, barter agreements between the United States and Jamaica were signed in 1982 and 1983.

Base acreage — A farm's crop-specific acreage of wheat, corn, grain sorghum, barley, oats, upland cotton, soybeans, canola, flax, mustard, rapeseed, safflower, sunflowers, and rice eligible to enroll in the Direct and Counter-cyclical Program (DCP) under the 2002 farm bill (P.L. 101-171, Sec. 1101-1108). A farmer's crop acreage base is reduced by the portion of cropland placed in the Conservation Reserve Program (CRP), but increased by CRP base acreage leaving the CRP. Farmers have the choice of base acreage used to calculate Production Flexibility Contract payments for crop year 2002, or the average of acres planted for crop years 1998 through 2001.

Base acres (or acreage base) — For purposes of Direct and Counter-cyclical Program (DCP) payments under the 2002 farm bill (P.L. 107-171, Sec. 1101), a farm's average planted acreage of specific crops (covered commodities and peanuts) over the four years 1998-2001, plus land cropland prevented by disaster from being planted, or 2002 contract acreage under the 1996 farm bill (P.L. 104-127). Payments acres are equal to 85% of base acres.

Base period price — The average price for an item in a specified time period used as a base for an index, such as 1910-14, 1957-59, 1967, 1977, or 1982. Time series of data are often deflated to a base period price. Such deflated time series are referred to as constant dollar values (versus nominal dollar values).

Base property — For the Bureau of Land Management: land or water resources, owned or controlled by a holder of a grazing permit or lease, that are suitable to support livestock for a part of the year. For the Forest Service: lands and improvements owned and used by a permittee for a farm or ranch and designated by the permittee to qualify for a grazing permit. One must own or control base property to be eligible for permits or leases to graze private livestock on federal lands.

Baseline budget estimate (agriculture) — A 10-year projection of mandatory spending (commodity support, crop insurance, food assistance, and certain conservation and trade programs) based on current law and current policy. Critical for commodity and trade program spending are the long-run projections of supply/use, trade, and prices for major U.S. commodities, while participation estimates are critical for food assistance, crop insurance and conservation. The baseline projections are developed on specific assumptions regarding macroeconomic conditions, weather, and international developments. The baseline is most important as a reference scenario for evaluating changes in the underlying economic assumptions and in alternative policy proposals. The Congressional Budget Office (CBO) baseline is used in the preparation of the congressional budget resolution, in developing reconciliation instructions, and in estimating the cost of pending legislation as bills are debated. The USDA, as part of the President's annual budget process develops its own annual baseline, which is released with the budget request sent to Congress in early February. Both CBO and USDA re-estimate the baseline in mid-year, which is called the mid-session review. The CBO baseline takes on added importance when Congress imposes upon itself extraordinary spending constraints.

Baseline budget estimate — A projection of future revenues, expenditures, and other budget amounts under assumed economic conditions and participation rates, and assuming no change in current policy. The baseline is usually projected annually by the Congressional Budget Office for each of the subsequent five to ten years. It is used in the preparation of the congressional budget resolution and reconciliation instructions, and in estimating the amount of deficit reduction in reconciliation bills and the effects of legislation on the budget. USDA, in conjunction with the Office of Management and Budget also develops its own annual agriculture baseline budget. It projects what would be expected to happen under a continuation of current U.S. and international farm policy, trade agreements, and specific assumptions about external conditions. The baseline scenario provides a reference projection from which USDA analysts make comparisons of alternate scenarios by altering any of the assumed underlying policies or conditions.

Basic commodities — Six agricultural crops (corn, cotton, peanuts, rice, tobacco, and wheat) declared by permanent law (in the Agricultural Adjustment Act of 1938) as requiring federal price support. Nonbasic commodities are the others for which USDA is authorized to provide price support in permanent law.

Basic formula price (BFP) — Up until January 2000, the BFP was calculated monthly by USDA and used as the base price for all milk regulated by federal milk marketing orders. Subsequently, a different and more complex formula currently is used by USDA to establish minimum farm milk prices under federal milk marketing orders.

Basing point — A geographical site used to establish minimum fluid milk prices for federal milk marketing orders. Generally, minimum fluid farm milk prices increase according to the distance from the basing point. When federal milk marketing orders began in the 1930s, Eau Clare, Wisconsin was viewed as the principal surplus milk production region in the nation and hence served as the basing point for most milk priced under federal milk marketing orders. Generally, the further a region is from the Upper Midwest, the higher that region's minimum price for fluid farm milk. Some have argued that there currently are other surplus production regions in the country (e.g., in the northeast and southwest) which should serve as basing points. An attempt by USDA to establish a pricing structure using multiple basing points was thwarted by legislation in 1999 (P.L. 106-113, Sec. 1000(a)(8), which enacted H.R. 3428).

Basis risk — The possibility of unexpected variation in basis and a resulting loss of expected revenue when a futures contract is liquidated and the commodity sold on the cash market.

Basis — The difference between the current spot price (or cash price) of a commodity and the price of the nearest futures contract for the same or a related commodity. Basis is usually computed in relation to the futures contract next to expire and may reflect different time periods, product forms, qualities, or locations.

Baskets — Used in free trade agreements (FTAs) to categorize products by tariff reduction and/or quota elimination periods. For example, the U.S.-Chile FTA uses five baskets for removing trade barriers on most products traded between both countries. Tariffs on products placed in the A basket are eliminated immediately. Products in the B, C, D, and E baskets have phase-out periods of 4, 8, 10, and 12 years respectively. Sensitive agricultural products (many of which are currently protected by tariff-rate quotas) are usually placed in the last basket.

BAT — Best available technology.

bbf — Billion board feet.

BC — Banks for Cooperatives. [www.cobank.com].

BCS — Basic conservation systems.

Beef (cattle) price index (BPI) — An index of the weighted average annual price for beef cattle, excluding calves, for a 16 western state area as compared with a specific base period equal to 100. This index is used in calculating federal grazing fees.

Beef Export Verification (BEV) Program — USDA's Agricultural Marketing Service (AMS) initiated BEV in August 2003 as a voluntary, user-fee funded service. Under BEV, U.S. exporters desiring to sell beef to Japan (or any other country that may request similar documentation) can apply for BEV certification from AMS after satisfying a list of requirements enabling the agency to verify the origin of the beef. The program was in response to Japanese officials' demands that the United States verify that none its beef exports were of Canadian origin, in the wake of the May 2003 discovery in Canada of a cow with bovine spongiform encephalopathy (BSE). After the December 2003 discovery of a BSE cow in the United States, Japan was among the first of the many countries to suspend some or all imports of U.S. cattle,, beef and related products, so the future of BEV is clouded.

Beefalo — Beefalo are a cross between Bison (American Buffalo) and domestic cattle. This produces meat very low in fat and cholesterol. When emergency livestock assistance has been implemented, beefalo and buffalo have been eligible if maintained on the same basis as beef cattle. Also, beefalo and buffalo were eligible for Livestock Compensation Program payments.

Beginning farmer or rancher — An eligibility term used in some farm programs, usually to identify a subgroup who may benefit from additional assistance. Under the Conservation Security Program (CSP) and the Environmental Quality Incentives Program (EQIP), it includes any individual or entity who has been operating a farm or ranch for less than 10 years and materially participates in its operation. People meeting this classification may be eligible for additional incentives to encourage conservation stewardship, such as higher percentages of cost-sharing payments. Under Farm Service Agency (FSA) farm ownership and operating loan programs, the beginning farmer or rancher must have an operation be no larger than 30% of the average size farm in the county. For direct operating loans, a further requirement is that the applicant must have participated in the business operation of a farm for at least 3 years. People who meet these requirements get preferential access to federal financial assistance.

Below-cost timber sale — A timber sale from national forest lands in which the expected federal revenues are less than the estimated federal expenses to sell the timber.

Benefit/cost analysis — A quantitative and sometimes qualitative evaluation of the costs which would be incurred by some action (such as building a dam, or implementing an environmental regulation) versus the overall benefits to society of the proposed action. See Risk-benefit analysis.

Best management practices (BMP) — A conservation practice or combination of practices designed to maintain agricultural productivity while reducing point- and nonpoint- source water pollution. State water quality agencies (or their designees) determine BMPs to fit local conditions and to make the most efficient use of natural resources and purchased inputs.

BEV — Beef Export Verification Program

BFP — Basic formula price.

BFW — Bread for the World. [www.bread.org].

BGH — Bovine growth hormone (see Bovine somatotropin).

B&I — Business and industry.

BICO Report — The Foreign Agricultural Service's report of U.S. export and import data on Bulk, Intermediate, and Consumer-Oriented (BICO) agricultural commodities. In addition, the data base includes forest products and edible fish and seafood products. These trade data are further classified among dozens of separate product groups. Data are available in both calendar and fiscal year format and for numerous world regions and individual country markets. See also FATUS and U.S. Trade Internet System.

Bidding down — A process that has been used in several conservation programs to increase the cost-effectiveness to the government by allowing potential participants who would provide the same conservation benefits if they are accepted into a program to offer to participate at lower rental rates in the Conservation Reserve Program or less federal financial assistance in the Environmental Quality Incentives Program.

Bilateral aid — Development assistance provided directly by a donor country to a recipient country in contrast to multilateral aid that is provided through international agencies.

Bill Emerson Good Samaritan Act of 1996 — P.L. 104-210 (42 U.S.C. 1791) was named in honor of the late Missouri Representative who championed efforts to expand food donations to the poor and provide legal protections for those making food donations. This law makes permanent the Model Good Samaritan Food Donation Act (P.L. 101-610, Sec. 402) and incorporates it into the Child Nutrition Act of 1966 (P.L. 89-642, Sec. 22; 42 U.S.C. 1791). Good Samaritan laws are designed to encourage the donation of food and groceries to nonprofit charitable agencies by minimizing the risks of legal actions against donors and distributors of foods. The amended law excludes from civil or criminal liability a person or nonprofit food organization that, in good faith, donates or distributes donated foods for food relief. This does not supersede state or local health regulations and its protections do not apply to an injury or death due to gross neglect or intentional misconduct.

Bill Emerson Humanitarian Trust — A reserve of commodities and cash held in trust to supplement food aid made available under P.L. 480 programs. The Trust can hold up to 4 million metric tons of wheat, corn, sorghum, and rice; the authorizing statute also authorizes the Trust to hold cash in lieu of commodities. The Trust was first established as the Food Security Wheat Reserve in 1980 in P.L.96-494, Title III. Subsequently the authorization for this reserve was expanded to include corn, rice, and sorghum in addition to wheat by the 1996 farm bill (P.L. 104-127, Sec. 225). Renamed the Bill Emerson Humanitarian Trust in 1998 legislation (P.L. 105-385, Sec. 211) which also authorized it to hold cash in addition to commodities, it was extended through 2007 by the 2002 farm bill (P.L. 107-171, Sec. 3202). Commodities (or cash) can be released from the Trust to meet unanticipated needs for emergency food assistance or when domestic supplies are insufficient to meet P.L. 480 (7 U.S.C. 1736f-1) programming requirements.

Bioaccumulation — The increase in concentration of toxic chemicals, heavy metals, and certain pesticides in plants and animals as they take in contaminated air, water, or food, because the substances are very slowly broken down in the body or excreted. Toxicity can be expressed in several ways: lead that is ingested by calves can bioaccumulate in their bones, interfering with calcium absorption and bone development; stored chemicals may be released to the blood stream at a later time, for example, during gestation or weight loss; and, chemicals may concentrate to lethal levels at upper ends of the food chain. Bioconcentration is a synonym for bioaccumulation.

Biobased Products — Commercial and consumer products produced from biomass, including chemicals and plastics. The term generally applies to products that typically are produced from other sources (e.g. lubricants), as opposed to those traditionally produced from biomass (e.g. paper).

Biochemical oxygen demand (BOD) — A measure of the amount of oxygen consumed by natural, biological processes that break down organic matter, such as those that take place when manure or sawdust is put in water. High levels of oxygen-demanding wastes in waters deplete dissolved oxygen (DO) thereby endangering aquatic life. Sometimes referred to as biological oxygen demand. BOD is a standard measure of water quality. Chemical oxygen demand (COD) is a measure of the oxygen consumed when organic or inorganic matter is oxidized in water chemically, rather than biologically.

Bioconcentration — See Bioaccumulation.

Biodiesel — An alternative renewable fuel, produced from vegetable oils or animal fats through a refinery process called transesterification. Biodiesel contains no petroleum, but it can be blended at any level with petroleum diesel to create a biodiesel blend. Biodiesel is most commonly used as a blend of 20% biodiesel and 80% conventional diesel (called "B20"). Its use can result in substantial reduction of unburned hydrocarbons, carbon monoxide, and particulate matter. However, nitrogen oxide emissions tend to increase with biodiesel use. Provisions of the Energy Conservation Reauthorization Act (ECRA) of 1998 (P.L. 105-388) amended the Energy Policy Act (EPACT) of 1992 (P.L. 102-486) to allow that the use of biodiesel added to conventional diesel at blends of 20% and higher would produce credits to offset up to 50% each year of alternative fuel vehicle acquisition requirements. Farmers and processors anticipate that increased use of biodiesel will strengthen the market for soybean oil.

Biodiversity (or biological diversity) — In general, the variety and variation among plants, animals, and microorganisms, and among their ecosystems. It has 3 levels: ecosystem diversity, species diversity, and genetic (within species) diversity. Genetic diversity provides resources for genetic resistance to pests and diseases. In agriculture, biodiversity is a production system characterized by the presence of multiple plant and/or animal species, as contrasted with the genetic specialization of monoculture. Advocates of maintaining biodiversity hold that civilization should preserve the greatest possible number of existing species so that a highly diverse genetic pool, which might be tapped for useful and beneficial characteristics, will be available into the future; and argue further that damage to the planet's biodiversity risks harm with effects on humans that can not be predicted with current knowledge.

Bioenergy Program — An initiative of the Commodity Credit Corporation (CCC) in 2000 that was codified into law by the 2002 farm bill (P.L. 107-171, Sec. 9010). The program makes payments to ethanol and biodiesel producers who expand their production capacity. In the year of the expansion, the program payments help offset the cost of the additional commodity feedstocks (usually corn for ethanol and soybeans for biodiesel) needed for the expansion. Spending for the program is capped at $150 million annually. The Congressional Budget Office estimates that $204 million total will be spent between FY2002 and FY2006.

Bioenergy — Electricity, motor fuels (e.g., ethanol, biodiesel), or other energy products produced from biomass.

Bioengineering — See Genetic engineering.

Biofuels — Fuels made from biomass, which in the United States largely include corn-based ethanol (blended into gasoline and called gasohol) and soybean-based biodiesel. Biofuels are a subset of renewable fuels, which are a subset of alternative fuels.

Biological control — The practice of using beneficial natural organisms to attack and control harmful plant and animal pests and weeds is called biological control, or biocontrol. This can include introducing predators, parasites, and disease organisms, or releasing sterilized individuals. Biocontrol methods may be an alternative or complement to chemical pest control methods. Biocontrol is part of the Animal and Plant Health Inspection Service (APHIS) program to control several economically important pests of food and fiber crops; it also is researched and used by other USDA agencies that promote integrated pest management.

Biological monitoring — Using living organisms to test the quality of either effluent to be discharged into receiving waters, or waters downstream from a discharge.

Biological oxygen demand (BOD) — See Biochemical oxygen demand.

Biologics — Immunization vaccines, bacterins, antigens, and antitoxins and other preparations made from living organisms and their products, intended for use in diagnosing, immunizing, or treating humans or animals, or in related research. The Animal and Plant Health Inspection Service has responsibility for approving some animal biologics. See Veterinary biologics.

Biomagnification (or biological magnification) — The increase in the concentration of bioaccumulated toxic chemicals in organisms higher on the food chain due to preferential storage of the toxic chemical in edible body parts. For example, chlorinated pesticides concentrate in the fat and skin of fish in contaminated lakes and streams and are biomagnified when those fish are eaten by larger fish, and perhaps eventually by mammals or birds of prey.

Biomass — The Biomass Research and Development Act of 2000 (P.L. 106-224, Title III) defines biomass as "any organic matter that is available on a renewable or recurring basis, including agricultural crops and trees, wood and wood wastes and residues, plants (including aquatic plants), grasses, residues, fibers, and animal wastes, municipal wastes, and other waste materials." The objective of this Biomass Act is to encourage the development of biomass as a feedstock in the production of fuels, chemicals, and other products.

Biopesticide — A pesticide that is biological in origin (i.e., viruses, bacteria, pheromones, natural plant compounds) in contrast to synthetic chemicals. Transgenic Bt cotton and corn are biopesticides because *Bacillus thuringiensis* (Bt) is a naturally occurring soil bacterium that has been genetically engineered into the plants.

Biopharmaceutical (biopharm) crops — Crops that are bioengineered to produce pharmaceuticals, industrial chemicals and other medical and industrial products (sometimes called plant-made pharmaceuticals). Controversy surrounds the effect that these crops might have on native plants if cross pollination takes place. The flow of genetic modifications from bioengineered crops to their conventional native counterparts is a matter of concern to scientists and regulators. FDA, EPA, and USDA have a coordinated effort to assure that field tests are designed to prevent biotechnology-derived genes and gene products from appearing in commercial seed, commodities, and processed food and feed.

Biorefinery — A facility that converts biomass into fuels and chemicals.

Biosafety Protocol — Adopted in January 2000 by parties to the U.N. Convention on Biological Diversity, this protocol (sometimes referred to as the Cartagena Protocol) "seeks to protect biological diversity from the potential risks posed by living modified organisms resulting from modern biotechnology." It establishes a procedure for advanced notification so that countries are provided with the "information necessary to make informed decisions before agreeing to the import of such organisms into their territory." It is tied to the so-called "precautionary approach" and also establishes a biosafety clearinghouse. The United States is not among the more than 100 countries that signed the protocol.

Biosecurity — Biosecurity refers to agricultural practices intended to reduce or prevent the introduction of infectious diseases on a farm or other production facility, and includes practices such as limiting access by personnel and vehicles, reviewing and screening introduced items such a seed, feed and new animals, and controlling vermin. More recently, biosecurity programs have incorporated elements to protect against terrorism, vandalism, and other intentional acts that could compromise disease control, whether or not they were the primary aim of the illicit acts.

Biotechnology and Agricultural Trade Program — The 2002 farm bill (P.L. 107-171 Sec. 3204) authorizes appropriations of up to $6 million annually for technical assistance and public and private sector project grants to remove or mitigate significant foreign regulatory nontariff barriers to U.S. exports involving: agricultural commodities produced through biotechnology. Funds can also be used to address trade-related food safety, disease, and other sanitary and phytosanitary trade concerns.

Biotechnology — Broadly defined as transferring genes from one living entity (plant, animal, or microbe) into another (or into a synthetic compound) using advanced recombinant DNA technology (also called genetic engineering (GE)). Biotechnology has been used to develop Bt crops, herbicide-tolerant crops, and quality enhanced crops, and currently is being investigated as a way to develop crops (or animals) containing pharmaceuticals for use in animals or humans (called farmaceuticals). Prior to the emergence of biotechnology, scientists made improvements in plants or animals through selective breeding.

Bioterrorism — Bioterrorism, with regard to agriculture, refers to threats to the nation's food supply and agricultural resources (including agricultural research and testing laboratories) from deliberate acts of terrorism, such as introducing pests intended to kill U.S. food crops; spreading a virulent disease among animal production facilities; or poisoning water and food supplies. Congress passed the Public Health Security and the Bioterrorism Preparedness and Response Act of 2002 (P.L. 107-188) in order to bolster protections against bioterrorism aimed at the nation's food and water supplies, and prevent unauthorized access to certain animal and plant disease organisms in laboratories.

Blair House Agreement — The November 1992 agreement between the United States and the European Union on export subsidy and domestic subsidy reduction commitments in the Uruguay Round of multilateral trade negotiations. The agreement also dealt with some bilateral agricultural trade issues.

Blend price — Primarily used in the federal milk marketing order program. It is the weighted average price of milk, per hundred pounds, paid to each farmer based on how Grade A (fluid grade) milk is allocated by processors to different usage classes (e.g., fluid, manufacturing).

Blending — In grain marketing, the combining of two different qualities of grain in order to change the total value of both lots. For example, it is common to blend grains of differing moisture or different foreign material content to achieve the requirements of a contract order.

BLM — Bureau of Land Management. [www.blm.gov].

Blocked stocks — See Overall allotment quantity.

BLS — Bureau of Labor Statistics. [www.bls.gov].

Blue box policies — Direct payments, under the definition of "production-limiting" measures as defined in Article 6 of the Uruguay Round Agreement on Agriculture, that are not subject to the commitment to reduce domestic support. To qualify for the exemption, payments must be based on fixed areas or yields, on a fixed number of livestock, or on 85% or less of the base level of production. See Green box policies, and Amber box policies.

BMP — Best management practice(s).

Board foot — A measure for lumber, equal to a 1-inch thick board that is 12 inches wide and 1 foot long in nominal dimensions (a 2x6, for example, is less than 2 inches thick and 6 inches wide, but a 1-foot long 2x6 is still counted as 1 board foot); typically reported in thousands of board feet (mbf). Also used to estimate the volume of lumber that can be produced from logs and standing trees.

Board on Agriculture and Natural Resources — See National Academy of Sciences. [www7.nationalacademies.org/banr].

BOD — Biochemical oxygen demand; biological oxygen demand.

Boll weevil — An insect pest of cotton whose eradication is the object of an Animal and Plant Health Inspection Service program cooperatively funded and managed by cotton producers.

Bonus commodities — From the agricultural perspective, these are commodities donated to domestic feeding programs that USDA acquires for unexpected surplus removal reasons or because the Commodity Credit Corporation (CCC) holdings acquired under its price support programs are not needed for other purposes, or are in danger of waste or spoilage. For example, if meat prices fall, USDA may buy beef and donate it to the National School Lunch Program, or if the CCC is holding an excess of cornmeal that is in danger of spoiling, it might donate this to the Emergency Food Assistance Program (EFAP/TEFAP). From the food program perspective, these are commodities that are donated in addition to the commodities that must be provided under mandatory requirements in food program statutes.

BOR — Bureau of Reclamation. [www.usbr.gov].

Border price — See Reference price.

Botanical pesticides — Pesticides whose active ingredients are plant-produced chemicals such as nicotine, rotenone, or strychnine. Also called plant-derived pesticides. Being natural pesticides, as distinct from synthetic ones, they are typically acceptable to organic farmers.

Bottom — Usually synonymous with vessel or ship. A ship of American registry may be referred to as a U.S. bottom, whereas if registry is other than U.S., the ship, in U.S. usage, may be called a foreign bottom.

Bound tariff rate — The most-favored-nation tariff rate resulting from negotiations under the General Agreement on Tariffs and Trade (GATT) and incorporated as an integral component of a country's schedule of concessions or commitments to other World Trade Organization members. If a country raises a tariff to a higher level than its bound rate, those adversely affected can seek remedy through the dispute settlement process and may obtain the right to retaliate against an equivalent value of the offending country's exports or the right to receive compensation, usually in the form of reduced tariffs on other products they export to the offending country. See Tariff, and Applied tariff.

Bovine (cattle) Tuberculosis (TB) — TB is a contagious disease that can affect all warm blooded vertebrates, including humans and cattle, and can cross species. Bovine TB, which has affected animal and human health since antiquity, was once the most prevalent – and costly – infectious disease of cattle and swine in the United States. Starting in 1917, the Cooperative State-Federal Tuberculosis Eradication Program, which is administered by USDA's Animal and Plant Health Inspection Service (APHIS), State animal health agencies, and U.S. livestock producers, has nearly eradicated bovine TB from the Nation's livestock population. Its presence in humans has been reduced as a result of the eradication program, advances in sanitation and hygiene, the discovery of effective drugs, and pasteurization of milk.

Bovine growth hormone (BGH) — See Bovine somatotropin (bST).

Bovine somatotropin (bST) — Also called bovine growth hormone, bST is a naturally occurring protein that has been genetically engineered as a synthetic compound (now manufactured in large quantities and commercially available to farmers) that causes cows to increase the efficiency of milk production per unit of feed consumed. Its use has caused public controversy, and some states require retail dairy product labels to identify the use of synthetic bST.

Bovine spongiform encephalopathy (BSE) — Commonly known as mad cow disease, BSE is a slowly progressive, incurable disease affecting the central nervous system of cattle. It was first diagnosed in 1986 in Great Britain, where most of the world's estimated 187,000 known cases have occurred. The first indigenous North American cases were discovered in a Canadian cow in May 2003 and in a U.S. cow in December 2003. Prior to these findings, USDA already was banning the imports of live cattle from Great Britain (since 1989) and later the rest of Europe, and the Food and Drug Administration (FDA) was partially banning the use of ruminant protein in animal feed (since 1997). Consumption by cattle of BSE-contaminated ruminant proteins in animal feed has been cited as the most likely means of transmission, and the North American cows are believed to have contracted the disease prior to the feed ban (although that has never been verified). Despite the link scientists have made between BSE in cattle and several dozen European cases of a human variant of BSE, Creutzfeldt-Jakob Disease, BSE is not viewed by most experts as a major public health threat here. However, the recent outbreaks have had major implications for the U.S. and Canadian beef industries, because most foreign markets have closed their borders to their beef and cattle exports (as the United States itself routinely has done). By early 2004, U.S. officials had announced additional regulatory actions to test for BSE in the cattle herd, to contain any possible spread, and to keep it out of the food supply, all in hopes of maintaining consumer confidence in beef and in regaining foreign markets.

Boxed beef — Beef that a packer cuts into relatively small pieces, seals in vacuum packs, and ships in cardboard boxes, often ready for retail sale. Prior to the 1970s, most beef left the packer as partial carcasses.

BPI — Beef (Cattle) Price Index.

BPT — Best practicable technology, best practicable treatment.

Bracero — Term often applied generically to alien agricultural workers and to programs allowing them into the United States for agricultural work. Originally this term was given to a series of formal agreements between the U.S. and Mexico in 1942 that allowed Mexican nationals to enter the U.S. for short term agriculture and other types of work. These guestworkers were afforded some labor standard protections and a portion of their earnings was set aside to be paid to them upon their return to Mexico. Controversial in both countries, the *bracero* program, as it popularly was termed, remained in place in varied forms until the mid-1960s.

Breastfeeding promotion — Relates to activities required to be carried out by state and local agencies using federal funds provided for nutrition education and administrative services under the WIC program. States are required to use a portion of funds they receive to promote breastfeeding by postpartum mothers participating in the program.

Broadcast application — The spreading of pesticides or fertilizers over an entire area. See Band application.

Broiler — A young chicken, usually 5 to 8 weeks old and 3 to 5 pounds, raised primarily for its meat. Broilers, unlike laying hens, are raised on the floor of chicken houses and not in cages. Most broiler production in done by growers under contract with processing plants, following prescribed management plans. Broiler production and marketing is one of the most vertically coordinated livestock systems.

Brownfields — Under the Brownfields Revitalization and Environmental Restoration Act of 2001 (Title II of P.L. 107-118), the term brownfield site means "real property, the expansion, redevelopment, or reuse of which may be complicated by the presence or potential presence of a hazardous substance, pollutant, or contaminant." This EPA program provides grants for brownfield assessment and cleanup, and for job training. Other agencies also have brownfield programs; worth noting is the Department of

Housing and Urban Development's Brownfield Economic Development Initiative (BEDI).

Brucellosis — A highly contagious disease of cattle, goats, sheep, and swine that weakens livestock and causes spontaneous abortion in breeding herds. The Animal and Plant Health Inspection Service (APHIS), as part of its brucellosis eradication program, destroys affected herds and makes payments to farmers in partial compensation for their losses. APHIS also has been working with the U.S. Park Service and western states to eliminate the threat to ranch cattle of becoming infected with brucellosis from free-ranging bison. Brucellosis also can be transmitted to humans (undulant fever).

BSE — Bovine spongiform encephalopathy.

bST — Bovine somatotropin.

Bt crops — Crops that have been genetically engineered to carry the gene from the soil bacterium *Bacillus thuringiensis*. This bacterium produces a protein that is toxic when ingested by certain moth and butterfly caterpillars that are destructive to crops. Crops containing the Bt gene are able to produce this toxin, thereby providing protection throughout the entire plant. For example, Bt corn is genetically engineered to provide protection exclusively against the European corn borer and other caterpillar pests and Bt cotton is genetically engineered to control exclusively tobacco budworms, bollworms, and pink bollworms, and other caterpillar pests.

Bt — *Bacillus thuringiensis* is a naturally occurring soil bacterium, commonly known as Bt, that produces a natural protein that kills insects after the protein is ingested. Bt is a biological pesticide (biopesticide) that can be used as a spray or dust. Alternatively, the Bt gene can be genetically engineered into a crop variety and produce the toxic protein, which is specific to a small subset of insects, within the plant. Cotton has been genetically altered to control the tobacco budworm, bollworm and pink bollworm. Potatoes have been altered to control the Colorado potato beetle. A new hybrid of Bt corn, altered to be resistant to the European corn borer, has been available since 1997. Bt degrades rapidly to non-toxic compounds. It is not known to present any human or animal hazards. Pest resistance management (PRM) plans are required by the EPA as part of the registration.

bu. — bushel.

Budget authority — Authority provided by law to incur financial obligations that will result in immediate or future outlays of federal government funds. Budget authority may be provided in an appropriation act or authorization act. The laws requiring mandatory spending for commodity support, several conservation programs, crop insurance, and food stamps create budget authority. Most other programs operated by the USDA receive budget authority from appropriation acts.

Budget resolution — The annual framework that Congress uses to set targets for total spending and revenues. These targets, which are initially reported by the House and Senate Budget Committees, guide the appropriations committees' deliberations. It also may include binding instructions on the level of savings each committee must produce (see Reconciliation). A budget resolution does not become law and is not binding on the Executive Branch.

Buffer Initiative — An initiative led by the Natural Resources Conservation Service, with numerous partners, and started in 1997 to enroll 2 million miles (up to 7 million acres) in conservation buffers by 2002, using the Conservation Reserve Program's continuous enrollment option, and drawing on the resources of several other agricultural conservation programs, including EQIP, Wildlife Habitat Incentive Program, Wetlands Reserve Program, and Emergency Watershed Protection Program.

Buffer strips — Slender areas of permanent vegetation, often planted along the edge or the contour of a field, usually to slow the flow of water or the velocity of wind, in order to trap sediment and other materials (nutrients, pesticides, and certain pathogens) before they leave the farm and become pollutants. Types of buffers include filter strips, field borders, grassed waterways, field windbreaks, shelterbelts, contour grass strips, and riparian buffers. Also called conservation buffers.

Bulgur — Wheat that has been parboiled, dried, and partially debranned for later use in cracked or whole grain form.

Bulk carrier — Refers to two types of cargo ships: the dry-bulk carrier; and the liquid-bulk carrier, better known as a tanker. Bulk cargo is a shipment such as oil, grain, or ore, that is not packaged, baled, bottled, or otherwise packed and is loaded without counting or marking.

Bulk commodities — Generally, high volume, low value unprocessed agricultural commodities that are treated as though they are homogeneous (fungible) prior to processing. Grains, oilseeds, and cotton are considered bulk commodities. Contrasting categories are high value or value-added commodities, semiprocessed and processed commodities, and consumer ready commodities. Historically, bulk commodities constituted the major portion of U.S. agricultural exports. Beginning in 1991, high value commodities have surpassed bulk commodities as the leading agricultural export category. See Intermediate agricultural products, and Consumer-oriented agricultural products.

Bureau of Land Management (BLM) — A bureau within the Department of the Interior that has jurisdiction over about 264 million acres of land. A majority of this land is in the western states, and approximately one-third of the land is in Alaska. BLM also is responsible for 700 million acres of federal subsurface mineral resources, supervises mineral operations on 56 million acres of Indian trust lands, and carries out a fire management program on 370 million acres of land. [www.blm.gov].

Bureau of Reclamation (BOR) — A bureau within the Department of the Interior, whose mission is to manage, develop, and protect water and related resources. The agency replaced the Reclamation Service, which was established to implement the Reclamation Act of 1902 to "reclaim (arid lands) throughout the West through irrigation." The Bureau built, operates, and maintains more than 300 storage dams on rivers throughout the western United States. [www.usbr.gov].

Burley tobacco — The main type of air-cured tobacco; a cigarette tobacco that together with flue-cured tobacco accounts for more than 90% of total U.S. production. Burley production is centered in Kentucky. Burley tobacco production, until 2005, was limited by national marketing quotas and was eligible for nonrecourse price support loans.

Bushel — A dry volume measure of varying weight for grain, fruit, etc., equal to four pecks or eight gallons (2150.42 cubic inches). A bushel of wheat, soybeans, and white potatoes each weighs 60 pounds. A bushel of corn, rye, grain sorghum, and flaxseed each weighs 56 pounds. A bushel of barley, buckwheat, and apples each weighs 48 pounds.

Business incubator — A facility that supports the development and operation of a number of small start-up businesses. Tenants of the facility share a number of support services including computers, support staff, telecommunications equipment, and janitorial services. Occupants also may receive technical assistance, business planning, legal, financial, and marketing advice.

Butter-Powder Tilt — The farm bill requires USDA to support the farm price of milk at $9.90/cwt. by standing ready to purchase surplus butter, cheese and nonfat dry milk when wholesale prices for these commodities fall below administratively set levels. The 2002 farm bill (P.L. 107-171, Sec. 1501) allows USDA to adjust the government purchase price of butter and nonfat dry milk twice annually in order to better manage its inventories of surplus milk products and minimize government costs. However, whenever the purchase price of one commodity is reduced by USDA, it must increase the purchase price of the other commodity so that the overall support price of milk remains at $9.90 per cwt. This price adjustment is referred to as the butter-powder tilt.

Buy-up coverage — The portion of crop insurance coverage for which a participating farmer pays a premium. Current law offers catastrophic (CAT) crop insurance coverage without any premium payments required of the farmer. Any coverage that is purchased above the CAT level is referred to as buy-up coverage, and is partially subsidized by the federal government.

Buying-in price — The percentage of the intervention price at which EU national intervention agencies actually purchase commodities into intervention.

Buyout — In the context of commodity and farm support policy, the term buyout relates to compensation for the loss or decline in value of assets due to a change in policy or program design. For example, the 2002 farm bill, (P.L. 107-171, Sec. 1309) established a buyout of peanut quotas in conjunction with the complete redesign of the peanut support program. A buyout of tobacco quotas was enacted in 2004 (P.L. 108-357, Title VI). A whole-herd dairy buyout program was authorized by the Food Security Act of 1985 (P.L. 99-198) and was called the Dairy Termination Program.

Byrd Amendment — Formally the Continued Dumping and Subsidy Offset Act of 2000, this amendment to the FY2001 agricultural appropriations law (P.L. 106-387, Sec. 1003) requires that anti-dumping (AD) and countervailing duties (CVD), whether non-agricultural or agricultural, be redistributed to the domestic industries that were found to be injured by the imports. Such duties previously went into the general U.S. Treasury. The World Trade Organization (WTO) in 2004 determined that the amendment violates its rules. The 108th Congress did not enact legislation to comply with the WTO decision.

CAA — Clean Air Act (42 U.S.C. 7401 *et seq.*).

Cabotage — Trade or transport in coastal waters between ports within the same country. U.S. cabotage legislation, notably the so-called Jones Act (Sec. 27 of the Merchant Marine Act of 1920) is designed to support the maritime industry.

CACFP — Child and Adult Care Food Program.

CAFO — Concentrated animal feeding operation.

Cairns Group — A coalition of 17 agricultural exporting countries, formed in 1986 at Cairns, Australia, for the purpose of influencing the outcome of agricultural negotiations in the Uruguay Round of multilateral trade negotiations. During those negotiations and subsequently in the Doha Development Agenda negotiations, the Cairns Group has pressed WTO member countries to eliminate agricultural export subsidies, and to substantially reduce agricultural tariffs and trade-distorting farm subsidies. Members are Argentina, Australia, Bolivia, Brazil, Canada, Chile, Colombia, Costa Rica, Guatemala,

Indonesia, Malaysia, New Zealand, Paraguay, the Philippines, South Africa, Thailand, and Uruguay.

Call option — A contract that entitles the buyer the right, but not the obligation, to purchase an underlying futures contract at a stipulated basis or strike price at any time up to the expiration of the option. The buyer pays a premium to the seller for this contract. A call option is bought with the expectation of a rise in prices. See Put option.

Campylobacteriosis — A diarrheal disease caused by the type of bacteria known as *Campylobacter jejuni* (*C. jejuni*) associated with poultry, raw milk, and water. There are an estimated 2.5 million cases annually in the United States with 13,000 hospitalizations and more than 100 deaths. Campylobacteriosis has been linked to Guillain-Barre syndrome (a disease which paralyzes limbs and breathing muscles). USDA has estimated that this disease costs the United States between $1.2 to $1.4 billion annually in medical costs, productivity losses, and residential care.

Canadian Dairy Commission (CDC) — A Crown corporation established under the Canadian Dairy Commission Act (1966-1967) and accountable to Parliament through the Minister of Agriculture. The CDC has dual responsibilities: (1) the dairy support program operations financed by the government through parliamentary appropriation; and (2) marketing operations financed by milk producers under the provisions of the National Milk Marketing Plan. The CDC also chairs the Canadian Milk Supply Management Committee, which coordinates the management of industrial milk and cream supplies in Canada. [www.cdc.ca].

Canadian Wheat Board (CWB) — A quasi-governmental self-financed agency, established in 1935, that markets Canadian wheat, oats, and barley on behalf of producers. Commercial grain is put into annual marketing pools by grade, with the pool period lasting 12 months and ending July 31. The CWB markets the grain to domestic and foreign buyers, with unsold grain transferred to the pool established for the next year. The overall procedure ensures a uniform per-bushel return, excluding storage costs, to all producers for each grade, regardless of the time they deliver their grain to elevators. The flow of grain from farm to terminal is closely regulated. The CWB also works to develop new markets for Canadian wheat and has authority to enter into long-term supply contracts with foreign countries. [www.cwb.ca/en/index.jsp].

Cancellation — Refers to an action taken under Section 6(b) of the Federal Insecticide, Fungicide and Rodenticide Act (FIFRA;7 U.S.C. 136d) to cancel a pesticide registration for one or more specific uses when the EPA finds the use results in unreasonable adverse effects to the environment or public health when a product is used according to widespread and commonly recognized practice, or if its labeling or other material required to be submitted does not comply with FIFRA provisions.

Caneberries — The 2002 farm bill (P.L. 107-171, Sec. 10601) authorizes marketing orders for caneberries, which the Act defines to include "raspberries, blackberries, and loganberries."

Canola — Canola is one of the "other oilseed" crops eligible for support from marketing assistance loans and loan deficiency payments, direct payments and counter-cyclical payments under the 2002 farm bill (P.L. 107-171, Title I). North Dakota accounts for about 98% of national production (based on 2003 crop data). Canola is the name used in Canada and the United States for a rapeseed with low levels of two anti-nutritional compounds, erucic acid in the oil and glucosinolates in the meal, that were present in all rapeseed at one time. Canola oil was granted GRAS status in 1985. Canola oil is widely used as a cooking oil, salad oil, and for making margarine. Of all edible vegetable oils widely available today, it has the lowest saturated fat content, making it appealing to some health-conscious consumers. In contrast to this edible rapeseed, industrial rapeseed has a high content of erucic acid in the oil.

CAP — Common Agricultural Policy.

Capper-Volstead Act — P.L. 67-146 (February 18, 1922), the Co-operative Marketing Associations Act (7 U.S.C. 291, 292). The law was passed in response to challenges made against cooperatives using the Sherman Act (15 U.S.C. 1 *et seq.*), the Clayton Antitrust Act (15 U.S.C. 12 *et seq.*), and the Federal Trade Commission Act (15 U.S.C. 41 *et seq.*). It gave "associations" of persons producing agricultural products certain exemptions from antitrust laws. The law carries the names of its sponsors, Senator Arthur Capper of Kansas and Representative Andrew Volstead of Minnesota. It is sometimes called the Magna Carta of Cooperation.

Captive supply — Products that manufacturers or processors own or contract to purchase for future delivery to have a predictable source of raw materials for their plants. In agriculture, the term often is used, for example, to refer to the cattle that beef packers own or contract to purchase 2 weeks or more before slaughter. Examples of such contracts include an exclusive agreement with an individual feedlot in which the price is based on market prices at time of slaughter; or a contract in which the price is specified in advance or is based on some other formula. At issue is the effect that captive supplies have on prices paid to cattle producers in cash markets.

Carbon sequestration — Retention of carbon through physical or biological processes that prevent or delay its emission to the atmosphere as carbon dioxide by holding it in a carbon sink. This may help mitigate climate change by reducing the amount of carbon dioxide in the atmosphere. Silvicultural practices that encourage rapid, long term tree growth are an example. Crop residue retention practices designed to prevent erosion and improve the productivity of soil, such as conservation tillage, also retain larger amounts of carbon compared to many traditional cultivation practices.

Carbon sink — A process or activity that absorbs, or takes up, released carbon from another part of the carbon cycle. The four types of sinks, within which carbon behaves in a systematic manner, are the atmosphere, the terrestrial biosphere (including agricultural, forest, and freshwater systems); oceans, and sediments (including fossil fuels).

Carcass weight — The weight of an animal after slaughter and removal of most internal organs, head, and skin. On average the carcass weight of beef is about 60% of the live animal weight, for hogs it is about 73%.

Carcass-by-carcass inspection — See continuous inspection.

Carcinogen — Any substance that produces or promotes cancer. This is a key consideration in evaluating the safety of pesticides and other chemicals.

CARD — Center for Agricultural and Rural Development, Iowa State University. [www.card.iastate.edu].

Cargo preference — The Cargo Preference Act (P.L. 83-664) requires that whenever the federal government pays for equipment, material, or commodities shipped to other countries, a minimum percentage of the gross tonnage shipped by sea must go by U.S. flag vessels. Cargo preference requirements have been an issue in U.S. international food aid and export subsidy programs. Because U.S. vessels generally cost more to use, agricultural and food aid groups contend that resources otherwise available for commodities must be diverted to transportation costs.

Cargo Preference Act — P.L. 83-644 (August 26, 1954), as amended, contains permanent legislation concerning the transportation of waterborne cargoes in U.S.-flag vessels. The Act requires that 75% of the volume of U.S. agricultural commodities financed under P.L. 480 and other concessional financing arrangements be shipped on privately owned U.S.-registered vessels. Maritime interests generally support cargo preference, but proponents of P.L. 480 argue that it increases the costs of shipping U.S. commodities to poor countries and potentially reduces the volume of food aid that is provided.

Caribbean Basin Economic Recovery Act of 1983 (CBERA) — P.L. 98-67 (August 5, 1983), Title II, authorized unilateral preferential trade and tax benefits for eligible Caribbean countries, including duty-free treatment of eligible products. Often referred to as the Caribbean Basin Initiative (CBI). Amended several times, the last substantive revisions were made in the Caribbean Basin Economic Recovery Expansion Act of 1990 (P.L. 101-382, Title II, August 20, 1990). This made trade benefits permanent (repealing the September 30, 1995 termination date). The law gives preferential trade and tax benefits for eligible Caribbean countries, including duty-free entry of eligible products. To be eligible, an article must be a product of a beneficiary country and imported directly from it, and at least 35% of its import value must have originated in one or more CBERA beneficiaries. Slightly different import value rules apply to articles entering from Puerto Rico and the Virgin Islands. The duty-free import of sugar and beef products is subject to a special eligibility requirement intended to ensure that increased production of sugar and beef will not adversely affect overall food production. Preferential tariff treatment does not extend to imports of: textiles and apparel subject to textile agreements, specified footwear, canned tuna, petroleum and its products, and watches and watch parts containing any material originating in countries denied normal trade relations (most-favored-nation) trade status. Special criteria applied to the duty-free import of ethanol through FY2000. Import-sensitive products, not accorded duty-free tariff treatment, are eligible to enter at lower than normal trade relations tariff rates. These products include handbags, luggage, flat goods (such as wallets, change purses, and key and eyeglass cases), work gloves, and certain leather wearing apparel.

Carrier — An inert material added to an active ingredient in a pesticide to enhance its delivery or effectiveness.

Carrying capacity — The maximum stocking rate for livestock possible without damaging vegetation or related resources. Carrying capacity may vary from year to year on the same area, due to fluctuating forage production. Used by the government in decisions about how many livestock will be allowed on an allotment on public lands. This term also is used in ecology, wildlife management, recreation facility planning, and other subjects to describe the maximum use level that can be sustained without resulting in an unacceptable deterioration in quality.

Carryover — The supply of a farm commodity not yet used at the end of a marketing year and carried over into the next marketing year. An excessively large carryover is typically described as a surplus condition that causes prices to fall. When the carryover falls below normal, there may be concerns of a shortage contributing to price escalation.

Cartagena Biosafety Protocol — See Biosafety Protocol.

Cartagena Protocol — See Biosafety Protocol.

Cartel — An alliance or arrangement among industrial or commercial enterprises or nations aimed at limiting competition or exercising monopoly power in a market.

Casein — The major portion of milk protein, manufactured from skim milk and used in processed foods (such as dessert toppings and coffee whiteners) and in industrial products such as glue, paint and plastics. Casein may be blended with nonfat dry milk to produce milk protein concentrate.

Cash commodity — The physical or actual commodity as distinguished from the futures contract. Sometimes called spot commodity, or actuals.

Cash forward sale — See Forward contracting.

Cash grain farm — A farm where corn, grain sorghum, small grains, soybeans, or field peas and beans account for at least 50% of the value of farm products sold.

Cash in lieu of commodities — Refers to cash provided to food program operators (e.g., elderly nutrition programs, child care food programs, and some school food programs) in lieu of mandated commodity assistance. Meal program operators receive funding in lieu of commodities to buy whatever foods they need to operate their meal service programs.

Cash market — The market for the cash commodity (as contrasted to a futures contract), taking the form of; (1) an organized, self-regulated central market (e.g., a commodity exchange); (2) a decentralized over-the-counter market; or (3) a local organization, such as a grain elevator or meat processor, which provides a market for a small region.

Cash price — The price in the marketplace for actual cash or spot commodities to be delivered via customary market channels.

Cash settlement — A method of settling certain futures contracts or option contracts whereby the seller (or short position) pays the buyer (or long position) the cash value of the commodity traded according to a procedure specified in the contract.

CAST — Council for Agricultural Science and Technology. [www.cast-science.org].

CAT — Catastrophic crop insurance.

Catastrophic crop insurance (CAT) — A component of the federal crop insurance program, originally authorized by the Federal Crop Insurance Reform Act of 1994 (P.L. 103-354). CAT coverage compensates farmers for crop yield losses exceeding 50% of their average historical yield at a payment rate of 55% of the projected season average market price. CAT coverage requires that a farmer realize a yield loss of more than 50% and only makes payments on losses exceeding the 50% threshold. Producers pay no premium for CAT coverage, but except for cases of financial hardship must pay an administrative fee of $100 per crop. A producer has the ability to purchase additional insurance coverage (or buy-up coverage) beyond CAT coverage, but must pay a premium, partially subsidized by the government.

Cattle cycle — The approximately 10-year period in which the number of U.S. beef cattle is alternatively expanded and reduced over several consecutive years in response to perceived changes in profitability by producers. Generally, low prices occur when cattle numbers (or beef supplies) are high, precipitating several years of herd liquidation. As cattle numbers decline, prices gradually begin to rise, causing producers to begin adding cattle to their herds. The cycle is relatively long due to the long period of time it takes between the time a cow-calf operator decides to expand a cow herd to breed more beef cattle and the time those animals reach slaughter weight.

CBI — Caribbean Basin Initiative.

CBT — Chicago Board of Trade. [www.cbot.com].

CCA — See U.S.-Canada Consultative Committee on Agriculture.

CCC — Commodity Credit Corporation.

CCHP — Comprehensive Conservation Enhancement Program, which replaces ECARP in the 2002 farm bill (P.L. 107-171).

CCI — Cotton Council International. [www.cottonusa.org/cci.htm].

CCP — Counter-cyclical payments.

CD — Conservation district.

CDC — Canadian Dairy Commission. [www.cdc.ca].

CDC — Centers for Disease Control and Prevention [www.cdc.gov]. Canadian Dairy Commission. [www.cdc.ca/cdc/index.asp].

CED — County Executive Director.

CEFTA — Central European Free Trade Agreement.

Census of Agriculture — A comprehensive set of quantitative information on the agricultural sector of the U.S. economy, broken down to the state and county levels (i.e., number of farms, land in farms, crop acreage and production, livestock numbers and production, production expenses, farm facilities and equipment, farm tenure, value of farm products sold, farm size, type of farm, among other data). Special reports are issued on such subjects as irrigation, land ownership, economics, and an atlas of agriculture. The Census, conducted every 5 years, was the responsibility of the Commerce Department's Bureau of the Census. However, the FY1997 USDA appropriations act (P.L. 104-180) transferred funding for the Census of Agriculture to USDA's National Agricultural Statistics Service (NASS). Data from the 2002 Census was released throughout 2004. [www.nass.usda.gov/census].

Center for Food Safety and Applied Nutrition (CFSAN) — The agency within the Food and Drug Administration responsible for developing and overseeing enforcement of food safety and quality regulations and coordinating FDA and states' surveillance and compliance programs, among other activities. FDA's roughly 800 field inspectors (located administratively within FDA's Office of Regulatory Affairs) enforce CFSAN's food safety regulations at 53,000 processing facilities. Among other activities, this center is engaged in surveillance of imported fruits and vegetables, investigating the risk of *Listeria*, approving additives to safeguard the nation's food supply and adopting HACCP rules for manufacturers of fruit juices, seafood and shell eggs. [www.cfsan.fda.gov].

Center for Veterinary Medicine — An agency within the Food and Drug Administration (FDA) that is responsible for assuring that all animal drugs, feeds (including pet foods), and veterinary devices are safe for animals, are properly labeled, and produce no human health hazards when used in food-producing animals. [www.fda.gov/cvm].

Center pivot irrigation — A self-propelled irrigation system in which a single pipeline supported on towers rotates around a central point. These systems are typically about one-quarter mile long and serve 128- to 132-acre circular fields.

Centers for Disease Control (CDC) and Prevention — An agency within the U.S. Department of Health and Human Services that monitors and investigates foodborne disease outbreaks and compiles baseline data against which to measure the success of changes in food safety programs. The CDC operates the FoodNet and PulseNet survey systems for tracking and identifying the causal organisms of foodborne disease outbreaks. [www.cdc.gov].

CEQ — Council on Environmental Quality. [www.whitehouse.gov/CEQ].

Cereals — Generally, grains suitable for human or animal food. The EU Common Agricultural Policy, for example, recognizes the following cereals as eligible for support under its arable crops program: durum wheat, rye, barley, oats, maize, grain sorghum, buckwheat, millet, and canary seed.

Certificates (commodity) — Legal instruments, entitling a qualified bearer to a specific dollar value of USDA surplus commodities. Payment-in-kind (PIK) "certs" either can specify the types of commodities or be generic. Certificates were heavily used during the 1980s as a means of meeting financial obligations and simultaneously disposing of Commodity Credit Corporation (CCC)-owned commodities. The USDA, in 1999, approved the sale to farmers of commodity certificates at posted county prices that could be used to pay off marketing assistance loans. This administrative action was followed by explicit legislative authority in P.L. 106-78 (Sec. 812). The use of commodity certificates to repay commodity loans is not subject to payment limitations that otherwise constrain the amount of marketing loan gain or loan deficiency payments persons may receive each year.

Certified crop consultant (advisor) — Individuals who are certified by a professional group as having the expertise needed to advise producers on how to improve production of a crop, whether a crop is threatened by pests or disease, or how to protect a crop from pests and disease. Crop consultants are paid by producers for their services. They are likely to be third party providers for USDA programs requiring such technical expertise when Natural Resources Conservation Service (NRCS) implements the 2002 farm bill (P.L. 107-171, Sec. 1242).

Cervid — A member of the *cervidae* family of animals that includes deer, elk, moose, caribou, reindeer, and related species (those with solid deciduous antlers). As more producers begin to raise these animals in livestock enterprises, they have sought to include them among the other livestock species as eligible for federal aid, such as disaster payments, federally-funded meat inspection, and animal health programs. Deer, elk, and reindeer, which currently are considered exotic species under the Federal Meat Inspection Act (21 U.S.C. 601 *et seq.*). Exotic species are eligible for a voluntary, fee-for-service inspection program under the Agricultural Marketing Act

CFCs — Chlorofluorocarbons.

CFO — Conservation farm option; Chief Financial Officer.

CFR — Code of Federal Regulations. [www.access.gpo.gov/nara/cfr].

CFSAN — Center for Food Safety and Applied Nutrition. [vm.cfsan.fda.gov/list.html].

CFTC — Commodity Futures Trading Commission. [www.cftc.gov].

Channelization — Modifying watercourses by straightening, widening, or deepening them. Agricultural benefits include improving drainage and removing water from nearby farm land faster, and improving navigation so that farm products can be shipped by boat more efficiently. While improving drainage, this process can interfere with waste assimilation capacity, disturb fish and wildlife habitats, and aggravate flooding in other areas.

Charter Act — See Commodity Credit Corporation.

Check-off program — Usually, a reference to the generic research and commodity promotion programs for farm products that are financed by assessments applied to sales of those products by producers, importers, or others in the industry. See Generic advertising and promotion.

Chemigation — The application of a pesticide and/or fertilizer through any irrigation system. This delivery technique raises some concern that it may increase pollution.

Chemosterilant — A chemical that controls pests by preventing reproduction, thereby causing the population to collapse. This contrasts with chemicals that directly kill pests.

Chicory — Chicory (Cichorium intybus, and called succory by some) is a hardy perennial that has a variety of uses but is best known for its association with coffee. When coffee has become unavailable or very costly, as during World War II, roasted chicory root has been used as a substitute or a blend, though there is no caffeine in chicory. Chicory also is said to offer health benefits and is used as a dietary supplement by some people. The young leaves can be used in salads, and the root can also be boiled and eaten like a vegetable (it is related to endive and radicchio). It is also grown for cattle food in Europe. Farmers in Nebraska and Wyoming began commercial production of chicory as and alternative crop in 2003.

Child and Adult Care Food Program (CACFP) — This child nutrition program provides cash and commodity assistance to support meal service programs in child care centers, headstart facilities, outside of school programs, and family and group home day care homes for children, the elderly, and disabled. It is permanently authorized under Section 17 of the National School Lunch Act (P.L. 79-396), administered by the Food and Nutrition Service (FNS), and funded annually by agricultural appropriations. [www.fns.usda.gov/cnd/Care/CACFP/cacfphome.htm].

Child nutrition programs — A grouping of programs funded by the federal government to support meal and milk service programs for children in schools, residential and day care facilities, family and group day care homes, and summer day camps, and for low-income pregnant and postpartum women, infants, and children under age 5 in local WIC clinics. Programs include school lunch, school breakfast, summer food service, special milk, commodity distribution, after-school care and Department of Defense overseas dependents school programs, and the special supplemental nutrition program for women, infants and children (WIC). These programs are authorized under the Richard B. Russell National School Lunch Act (P.L. 79-396, as amended) and the Child Nutrition Act of 1966; (P.L. 89-642, as amended, 42 U.S.C. 1771 *et seq.*) are financed by annual agricultural appropriations laws; and are administered by the Food and Nutrition Service (FNS) of USDA. Changes to the authorizing statutes generally are made by the Agriculture Nutrition and Forestry Committee in the Senate. In the House, the Education and the Workforce Committee deals with most changes to child nutrition program authorizing statutes, although the Agriculture Committee usually is involved when proposed changes concern agricultural interests such as commodity distribution, food restrictions, and the Farmers Market Nutrition Program.

Child Nutrition Act of 1966 — P.L. 89-642 was an anti-hunger initiative begun by the Johnson Administration (1963-69) as part of its War on Poverty. The Child Nutrition Act (42 U.S.C. 1771 *et seq.*) permanently authorizes the special milk program and the school breakfast program. Programs with temporary (usually 3-5 year) authorizations include the special supplemental nutrition program for women, infants, and children (WIC), which provides federal grant funds to states for monthly food packages and nutrition education for low-income mothers and young children; federal spending for state administrative expenses (SAE) associated with the operation of child nutrition meal service programs; and the nutrition education and training (NET) program. This law has been amended numerous times, and its expiring provisions generally are reauthorized in conjunction with those in the Richard B. Russell National School Lunch Act (P.L. 79-396, as amended).

Chlorinated hydrocarbons — Also known as organochlorines, these synthetic organic compounds contain chlorine. They tend to be persistent in the environment and to biomagnify in the food chain. Chlorinated hydrocarbons that are pesticides include DDT, aldrin, dieldrin, heptachlor, chlordane, lindane, endrin, mirex, hexachloride, and toxaphene. Most chlorinated hydrocarbon pesticide uses have been canceled because of their persistence, propensity to bioaccumulate, and toxicity to nontarget species.

Chlorofluorocarbons (CFCs) — Non-toxic, non-flammable chemical containing atoms of carbon, chlorine and fluorine used in many aerosol sprays, blowing agents for foams, packing materials, solvents and refrigerants. Although safe to use in most applications and inert in the lower atmosphere, CFCs undergo significant reaction in the upper atmosphere that has been linked by some scientists to depletion of the ozone layer. The ozone layer protects the earth's surface from harmful ultraviolet radiation that can cause biological damage to plants and animals.

Chlorophenoxy herbicides — A class of pesticides that includes 2,4-D. They mimic plant hormones. Uses of some have been canceled because of concerns about adverse health effects.

Cholinesterase inhibitors — A class of chemicals that includes numerous insecticides, such as parathion or carbaryl. They inhibit an enzyme found in animals that regulates nerve impulses. Cholinesterase inhibition is associated with a variety of acute symptoms such as nausea, vomiting, blurred vision, stomach cramps, and rapid heart rate.

Chronic toxicity — The capacity of a substance to cause long-term or delayed adverse health effects. For example, a cancer resulting from exposure to a carcinogen may not appear for years or decades.

Chronic wasting disease (CWD) — A fatal neurological disease of farmed and wild deer and elk that belongs to the family of diseases known as transmissible spongiform encephalopathies (TSEs). Scientific experiments suggest that the disease cannot be transmitted to cattle that come in contact with infected wild animals, and likely is not transmissible to humans who consume meat from wild deer or elk that have been killed by hunting, although the Centers for Disease Control and Prevention (CDC) recommend that such meat be avoided. USDA's Animal and Plant Health Inspection Service (APHIS) has been working cooperatively with state agriculture and wildlife agencies to eliminate farmed elk and deer herds that have tested positive. A joint working group of USDA and Department of the Interior officials exists to develop and operate a coordinated plan of federal and state activities.

CIF (or cif) — Cost, insurance, and freight.

CIGAR — Consultative Group on International Agricultural Research. [www.cgiar.org].

CIPs — Commodity import programs.

Circuit breaker — Commonly-used term for Section 1601 of the 2002 farm bill (P.L. 107-171). Under this section, if the Secretary of Agriculture determines that U.S. commodity program expenditures will exceed allowable levels under the Uruguay Round Agreement on Agriculture domestic subsidy limits, the Secretary shall, to the maximum extent feasible, make adjustments in such expenditures to ensure they remain within limits. (For the United States, the current annual level of trade-distorting subsidies permitted is $19.1

billion.) Prior to such actions, a report to Congress is required on such a determination and the extent of such adjustments.

CIS — Commonwealth of Independent States.

CJD — Creutzfeldt-Jakob disease (see Bovine spongiform encephalopathy).

CLA — CropLife America. [www.croplifeamerica.org].

Class I differential — Under federal milk marketing orders, the minimum price a processor must pay for milk used for fluid consumption (Class I milk) is a monthly base price plus the Class I differential. The Class I differential varies by about $3.00/cwt. between the Upper Midwest and Southeast Florida. The Class I differential accounts for the costs of transporting milk, the added costs of marketing milk going into fluid milk products, and the higher cost of producing Grade A milk required for fluid products.

Class I equivalency — The amount of less productive land in a water district receiving Bureau of Reclamation water (Classes 2, 3, and 4) that would be necessary to be equivalent in productive potential to Class I land. This equivalency rating is made to adjust the number of acres that may be irrigated (see Acreage limitation) so that less productive lands are equivalent in productive potential to 960 acres of Class I land.

Class I land — For agriculture, the Natural Resources Conservation Service (NRCS) uses a land capability (classification), in which Class I lands are the best soils with few limitations restricting their use. Under reclamation law, Class I land is defined as irrigable land within a particular agricultural economic setting that is productive enough to yield the highest level of suitability for continuous, successful irrigation farming, and has the highest relative productive potential as measured in net income per acre.

Class I Railroad — Any railroad with annual gross revenues of at least $255.9 million (in 2001), according to the Surface Transportation Board. These are the largest long-distance U.S. railroad systems and include Union Pacific, Norfolk Southern, CSX, Burlington Northern-Santa Fe, and Kansas City Southern, which own most of the track in the United States. Since passage of the Staggers Rail Act of 1980, aimed at deregulating the once highly regulated industry to make it more efficient and cost-competitive, the number of Class I railroads has declined through consolidation and merger, from more than 30, to seven currently, including the Canadian National and Canadian Pacific. This consolidation has concerned many agricultural shippers (particularly those who lack access to nearby markets or to water transportation) fearful of higher freight rates due to lack of competition. See Regional railroad, and Short line railroad.

Classified pricing — The pricing system of federal milk marketing orders, under which milk processors pay into a pool for fluid grade (Grade A) milk. The price that processors have to pay into the pool is based on how the milk ultimately is used. Milk used for fluid (Class I) consumption generally receives the highest price and lower minimum prices are paid for the three classes of milk used for manufactured dairy products: Class II (yogurt, cottage cheese, ice cream, and other soft manufactured products), Class III (cheese), and Class IV (butter and nonfat dry milk).

Clayton Act — A 1914 law (15 U.S.C. 12 *et seq.*) that supplemented the Sherman Act of 1890 (15 U.S.C. 1 *et seq.*) by clarifying market activities (including those in agriculture) considered to be monopolistic or trade-restraining. The Capper-Volstead Act (7 U.S.C. 291, 292) later exempted agricultural cooperatives from certain Clayton and Sherman Act provisions.

Clean Air Act — The primary federal law governing efforts to control air pollution (42 U.S.C. 7401 *et seq.*). Federal legislation addressing air pollution was first adopted in 1955 (Air Pollution Control Act, P.L. 84-159) to provide research and technical assistance. Subsequent amendments, most notably the Clean Air Act Amendments of 1970 (P.L. 91-604), 1977 (P.L. 95-95), and 1990 (P.L. 101-549), strengthened the federal role. The Clean Air Act seeks to protect human health and the environment from emissions that pollute the air. The Environmental Protection Agency is required to establish minimum National Ambient Air Quality Standards (NAAQS), while states are assigned primary responsibility for developing compliance. Areas not meeting the standards (nonattainment areas) are required to implement specific control measures. There is no direct federal regulation of agriculture under the Clean Air Act. Two of the NAAQS (for particulates and ozone) could affect agriculture: particulates, because certain agricultural practices, such as prescribed burning and tilling, create airborne particles that might be targeted for control in State Implementation Plans; and ozone, because concentrations of ozone above the standard can adversely affect crop yields. Ozone is formed in the atmosphere when nitrogen oxides and volatile organic compounds (from manufacturing, transportation, and utilities) react in the presence of sunlight (agriculture rarely if ever represents significant sources of ozone precursors).

Clean Water Act — This is the principal law governing pollution of the nation's rivers, lakes, estuaries, and coastal waters (33 U.S.C. 1251-1387). Originally enacted in 1948 as the Federal Water Pollution Control Act (P.L. 80-845), it was totally revised by amendments in 1972 that gave the Act its current name and shape (P.L. 92-500). The objective of the Act is the restoration and maintenance of the chemical, physical, and biological integrity of the nation's waters. The Act is implemented by the EPA in partnership with state and local governments. Programs in the Act have been primarily directed at managing point source pollution (wastes discharged from industrial facilities, sewage treatment plants, and municipal storm sewer systems). Agricultural activities are generally exempt from the Act's regulatory requirements, but some may be affected. Large confined animal feeding operations are treated like industrial sources and are subject to permit requirements. Programs in the Act to manage nonpoint source pollution (rainfall runoff from farms, rangelands, forests, etc.) may affect agriculture, especially where nonpoint sources are determined to contribute to water quality impairment. However, irrigation return flows are specifically exempt from regulation. A program in the Act that regulates discharges of dredged and fill material into wetlands (Section 404) requires permits for activities on agricultural wetlands.

Climate change — Changes in the climate that are alleged to be the result of the changing concentration of greenhouse gases in the Earth's atmosphere, most predominantly carbon dioxide. Carbon dioxide in the atmosphere has increased by about one third over the past 150 years, due in part to increased emissions of greenhouse gases from growing energy use. A warming of Earth's average global temperature of nearly 1 degree Fahrenheit has been measured over the past 100 years, and climate models have predicted additional warming of between 3 to 10 degrees Fahrenheit by the end of this century. As the over-all climate warms, regional and national impacts will vary widely, as weather responds to a wide array of feedback mechanisms, and to the rate of change, which is currently very rapid by historical standards. Among the possible projected impacts that would be relevant to agriculture are increased droughts in some areas, heavier and more intense rainfall in others, stronger storms, more volatile weather patterns, continuing rise in sea levels, and greater melting of ice caps and glaciers. Among measures discussed to mitigate future climate change are carbon sequestration through greater planting of trees and soils management, and appropriate changes in forest and land use management.

CLOC — Commodity letters of credit.

CME — Chicago Mechantile Exchange. [www.cme.com].

CMS — Conservation management system.

CNMP — Comprehensive Nutrient Management Plan.

CNP — Child nutrition programs.

CNPP — Center for Nutrition Policy and Promotion. [www.usda.gov/cnpp].

CO — Conservation operations.

COAP — Cottonseed Oil Assistance Program.

Coarse grains — Generally refers to cereal grains other than wheat and rice (i.e., those used primarily for animal feed or brewing, including corn, barley, grain sorghum, oats, rye). In the United States, coarse grains usually are referred to as feed grains.

Coastal Zone Management Program — P.L. 92-583 (October 27, 1972) created the Coastal Zone Management Program in 1972 to provide grants to eligible states and territories as an incentive to prepare and implement plans guiding the use of coastal lands and resources. Thirty-four of the 35 eligible states and territories are implementing federally-approved plans. Amendments in 1990 require participants to develop agricultural nonpoint pollution programs. These programs must specify and implement management measures to restore and protect coastal waters. Management measures are specified for erosion, sediments, nutrients, pesticides, grazing, and animal waste. Participants must implement these management measures after they have been approved by whatever means necessary, including regulation. EPA and NOAA have conditionally approved all these programs; only a few states have received final approval.

CoBank — National Bank for Cooperatives. [www.cobank.com].

COC — County Office Committee (sometimes referred to as the County Farm Committee).

COD — Chemical oxygen demand.

Code of Federal Regulations (CFR) — The codification of the general and permanent rules published in the Federal Register by the Executive departments and agencies of the federal government. The Code is divided into 50 titles that represent broad areas subject to regulation. Most regulations directly related to agriculture are in title 7. Each title is divided into chapters that usually bear the name of the issuing agency, followed by subdivisions into parts covering specific regulatory areas. For example, 7 CFR 1410 are the regulations that apply to the Conservation Reserve Program. [www.access.gpo.gov/nara/cfr].

Codex Alimentarius Commission — A joint commission of the Food and Agriculture Organization (FAO) and the World Health Organization, comprised of some 146 member countries, created in 1962 to ensure consumer food safety, establish fair practices in food trade, and promote the development of international food standards. The Commission, often referred to simply as Codex, drafts nonbinding standards for food additives, veterinary drugs, pesticide residues, and other substances that affect consumer food safety. It publishes these standards in a listing called the Codex Alimentarius. Codex, which is explicitly cited in the WTO Agreement on Sanitary and Phytosanitary Measures, can play an important advisory role when issues of food health and safety arise in WTO negotiations or dispute settlement. [www.codexalimentarius.net].

Coliform index — A rating of the purity of water based on a count of fecal coliform bacteria. The presence of fecal coliform bacteria, which are harmless bacteria that live in the intestines of humans and other vertebrate animals, indicates contamination by human or animal feces, and hence the potential presence of disease pathogens.

Colonia — A substandard housing area defined in the Housing Act of 1949 (42 U.S.C. 1479) as any identifiable community that: (1) is in the states of Arizona, California, New Mexico, or Texas; (2) is in an area that is within 150 miles of the border between the United States and Mexico (except for standard metropolitan statistical areas that have a population exceeding 1 million); (3) is designated by the state or county as a colonia; and, (4) is determined to be a colonia based on criteria such as lack of potable water supply, lack of adequate sewage systems, and lack of decent, safe, and sanitary housing.

Colorado River Basin Salinity Control Program — This program was authorized in the Colorado River Basin Salinity Control Act and was repealed and replaced by the Environmental Quality Incentives Program (EQIP) in the 1996 farm bill (P.L. 104-127). Administered by the Natural Resources Conservation Service, it was used to implement salinity control measures, primarily to manage irrigation water using financial and technical assistance to landowners. This program supported U.S. efforts to meet international treaty obligations for downstream water quality in Mexico. The Department of the Interior's Bureau of Reclamation administers its own Colorado River Salinity control Program.

Colorado River Basin Salinity Control Act — P.L. 93-320 (June 24, 1974), and the laws authorizing three other conservation cost-sharing programs, were repealed in the 1996 farm bill (P.L. 104-127) and replaced by a new cost-sharing program, the Environmental Quality Incentives Program (EQIP). Until it was replaced, the Colorado River Basin Salinity Control Program provided cost-sharing assistance to producers to install on-farm irrigation system improvements to prevent irrigation water heavily charged with salts and minerals from reentering the Colorado River. Participating farmers received up to 70% of total project costs and technical assistance. Participation was concentrated at sites where problems existed. This program was available to producers in the seven states of the Colorado River watershed. The law was administered by the Farm Service Agency until FY1996, when management was transferred to the Natural Resources Conservation Service.

Combine — A self-propelled combination grain harvester. In one operation it combines cutting, threshing, separation, cleaning, and straw dispersal.

Commingling — The mixing of crops or other commodities from multiple fields or farms in the distribution and marketing system. Commingling is viewed as an efficient, cost-competitive method for shipping and exporting large quantities of U.S. grains. However, with commingling, marketers cannot segregate individual crops in order to preserve any unique characteristics that may enhance their market value or meet other goals, such as segregating crops produced through biotechnology from those produced conventionally.

Commission on 21st Century Production Agriculture — An 11-member panel authorized by Title I-G of the 1996 farm bill (P.L. 104-127) was directed to conduct a comprehensive review of the farm economy, including the impact of the 1996 law; and a follow-up review that included recommendations for changes in federal agricultural policy. The commission submitted its final report to Congress in January 2001. [www.usda.gov/oce/21st-century/index.htm].

Commission on the Application of Payment Limitations for Agriculture — The 2002 farm bill (P.L. 107-171, Sec. 1605; 7 U.S.C. 7993), required the creation of a commission to study various economic consequences from a further tightening of the limits on per person farm subsidy payments. The Commission was directed to deliver its report within one year, making it due in May 2003.

Committee on Surplus Disposal (CSD) — Known formally as the Consultative Subcommittee on Surplus Disposal (CSSD). A subcommittee of the Committee on Commodity Problems of the UN Food and Agricultural Organization (FAO), the CSSD monitors international shipments of agricultural commodities provided as food aid in order to minimize any adverse effects of these shipments on commercial trade and agricultural production.

Commodity absorption arrangements — A type of nontariff barrier used by some Central and South American countries that requires domestic end users and processors to buy the domestic output of key agricultural commodities before they can obtain a license to import such commodities.

Commodity Assistance Program — A term used by appropriators to refer to a variety of domestic programs receiving food in the form of USDA supplied commodities. The term was formalized for the first time in FY1996 appropriations law (P.L. 104-37, October 21, 1995) to refer to the consolidation for funding purposes of three commodity donation programs that are authorized under two separate statutes: the Emergency Food Assistance Program (EFAP), Soup Kitchen-Food Bank Program, and the Commodity Supplemental Food Program (CSFP).

Commodity certificates — Payments issued by the Commodity Credit Corporation (CCC) in lieu of cash payments to participants in farm subsidy or agricultural export programs. Holders of certificates are permitted to exchange them for commodities owned by the CCC. Alternatively, farmers may buy certificates and use them to settle marketing assistance loans as a way of avoiding per person payment limits on marketing loan gains and loan deficiency payments (LDPs). See Certificates, and Payment limitation.

Commodity Credit Corporation (CCC) — A wholly owned government corporation created in 1933 to stabilize, support, and protect farm income and prices (federally chartered by the CCC Charter Act of 1948 (P.L. 80-806)). The CCC, which has no staff, is essentially a financing institution for USDA's farm price and income support commodity programs, and agricultural export subsidies. It is authorized to buy, sell, lend, make payments and engage in other activities for the purpose of increasing production, stabilizing prices, assuring adequate supplies, and facilitating the efficient marketing of agricultural commodities. The 1996 farm bill (P.L. 104-127) expanded the CCC mandate to include funding for several conservation programs (including the Conservation Reserve Program) and made conservation one of the purposes of the CCC. The programs funded through CCC are administered by employees of the Farm Service Agency. The CCC has the authority to borrow up to $30 billion from the U.S. Treasury to carry out its obligations. Net losses from its operations subsequently are restored through the congressional appropriations process.

Commodity distribution — Direct donation of food products by the federal government to states for distribution to needy persons, schools, and food service institutions. This may include bonus commodities or entitlement commodities, or both.

Commodity Distribution Program — This program under Section 14 of the Richard B. Russell National School Lunch Act (NSLA) (P.L. 79-396, as amended) requires the Secretary of Agriculture to use agricultural surplus removal funds (Section 32) and Commodity Credit Corporation (CCC) funds to buy commodities for child and elderly nutrition programs. The Secretary is directed to use Section 32 funds not needed for other purposes and CCC funds (if stocks are not available) to buy commodities for donation to maintain the annually programmed level of commodity assistance for Child and Elderly Nutrition programs mandated by the National School Lunch Act, Child Nutrition Act (P.L. 89-642, as amended; 42 U.S.C. 1771 *et seq.*), and Older Americans Act (P.L. 89-73; 42 U.S.C. 3001 *et seq.*).

Commodity Distribution Reform Act and WIC Amendments of 1987 — P.L. 100-237 (January 8, 1988) established a free-standing law requiring the USDA to improve the distribution and quality of commodities donated to child nutrition programs. It also established a foodbank demonstration project making use of Section 32 agricultural surplus commodities, amended the National School Lunch Act (P.L. 79-396) to permit certain pilot projects receiving cash in lieu of commodities or commodity letters of credit to continue receiving them, and amended the Child Nutrition Act of 1966 (P.L. 89-642)to make a variety of changes to the WIC program to expand coordination with other programs, conduct studies, and convert certain food funding to use for administrative costs.

Commodity exchange — An organization operating under a set of bylaws aimed at promoting trade in one or more commodities by providing services and rules for the conduct of trade. Commodity exchanges may deal in cash and/or futures contracts. Commodity futures contracts are regulated federally by the Commodity Futures Trading Commission (CFTC). The National Futures Association is the industry's self-regulatory institution. Major active agricultural commodity exchanges include the New York Board of Trade (NYBOT), the Chicago Merchantile Exchange (CME), the Kansas City Board of Trade (KCBT), and the Minneapolis Grain Exchange (ME).

Commodity Exchange Authority — A former regulatory agency of USDA established to administer the Commodity Exchange Act of 1936; the predecessor to the Commodity Futures Trading Commission (CFTC).

Commodity Futures Trading Commission (CFTC) — The independent federal regulatory agency established by the Commodity Futures Trading Commission Act of 1974 (P.L. 93-463) to administer the Commodity Exchange Act of 1936, as amended. It regulates trading on the futures exchanges in the United States. The CFTC also regulates the activities of numerous commodity exchange members, public brokerage houses, commodity trading advisors, and commodity pool operators. [www.cftc.gov].

Commodity Futures Trading Commission (CFTC) Act of 1974 — P.L. 93-463 created the Commodity Futures Trading Commission, to replace the U.S. Department of Agriculture's Commodity Exchange Authority, as the independent federal agency responsible for regulating the futures trading industry. The Act made extensive changes in the basic authority of Commodity Exchange Act of 1936, which itself had made extensive changes in the original Grain Futures Act of 1923. (7 U.S.C. 1 *et seq.*). [www.cftc.gov/cftc/cftchome.htm].

Commodity Import Programs (CIPs) — The U.S. Agency for International Development uses a small portion of U.S. foreign aid funds to make grants and loans to countries judged important to U.S. foreign policy objectives. These CIPs, by making dollars available, help these countries finance purchases of U.S. commodities (which have included agricultural commodities) or other inputs needed to meet their development objectives and also provide balance-of-payments support to countries with very limited foreign exchange.

Commodity letters of credit (CLOC) — Food instruments issued in lieu of commodities to certain designated schools participating in the National School Lunch Program. These letters of credit specify the types of foods that schools must buy, which are the same types of foods being donated to other schools by USDA under the commodity distribution program.

Commodity loan rate — Price per unit (pound, bushel, bale, or hundredweight) at which the Commodity Credit Corporation (CCC) provides commodity loans to farmers to enable them to hold commodities for later sale, to realize marketing loan gains, or to receive loan deficiency payments (LDPs). Marketing assistance loan rates for the "loan commodities" and peanuts for crop years 2002 through 2007 are specified in the 2002 farm bill (P.L. 107-171, Sec. 1202, 1307). Nonrecourse loans also are available from the Commodity Credit Corporation for refined beet and raw cane sugar.

Commodity programs — This term is usually meant to include the commodity price and income support programs administered by the Farm Service Agency and financed by the Commodity Credit Corporation (CCC). The commodities now receiving support are: (1) those receiving Direct and Counter-cyclical Program (DCP) payments, specifically wheat, corn, grain sorghum, barley, oats, upland cotton, rice, soybeans and other oilseeds, and peanuts; (2) those eligible for nonrecourse marketing assistance loans, which includes the previous mentioned commodities plus wool, mohair, honey, dry peas, lentils, and small chickpeas; and (3) those having other unique support, including sugar, and milk. A broader phrase that includes these commodity programs and other assistance is farm programs. (See also Target price, and Loan commodity.).

Commodity promotion programs — Programs that advertise and promote an agricultural commodity or product without reference to the specific farmer, brand name, or manufacturer. Producers can and do organize voluntary commodity promotion programs, but most are operated under the authority of either federal or state laws, frequently with the objective of requiring that all members of the industry participate. At the federal level, the programs are authorized by law, implemented by industry groups (after USDA review, rulemaking and approval), and financed by assessments (also called check-offs) of industry members such as producers, importers, and/or handlers. In the past, Congress enacted separate laws permitting producers of specifically-designated commodities to create such programs. The 1996 farm bill (P.L. 104-127) also gives USDA general authority to create programs for any commodity at the request of a group of producers. In 2002, 16 federal promotion programs were operational: beef, blueberries, cotton, dairy products, eggs, fluid milk, hass avocados, honey, lamb, mushrooms, peanuts, popcorn, pork, potatoes, soybeans, and watermelons. In addition to the federally authorized programs, there are hundreds of state-legislated promotion programs covering a wide variety of farm commodities. Nine out of ten U.S. farmers contribute to one or more of these efforts, which, collectively, raise and spend hundreds of millions of dollars annually.

Commodity Supplemental Food Program (CSFP) — The CSFP provides funding for monthly food packages consisting of USDA commodities (juice, egg mix, and canned fruits and vegetables), and for administrative costs of local agencies serving low-income pregnant and postpartum women, infants, children up to age 6, and persons 60 years of age or older. The precursor of the WIC program, the CSFP operates in some 80 project areas located in 20 states. Over one-half of the beneficiaries are elderly. CSFP is authorized through FY2007 under the Agriculture and Consumer Protection Act of 1973 (P.L. 93-86), as amended by the 2002 farm bill (P.L.107-171).

Common Agricultural Policy (CAP) — The set of legislation and practices jointly adopted by the nations of the European Union (EU) in order to provide a common, unified policy framework for agriculture. Its stated purposes are to increase farm productivity, stabilize markets, ensure a fair standard of living for farmers, guarantee regular supplies, and ensure reasonable prices for consumers. The CAP rests upon four basic principles: common import restrictions, common financing, common pricing, and common treatment of surpluses.

Common carrier — A freight or passenger carrier that offers transportation services to the general public in exchange for compensation.

Common external tariff (CXT) — A tariff rate applied by a group of countries organized as a common market or customs union to imports from non-member countries. For example, the European Community allows free trade among member countries, but applies common external tariffs against many farm products imported from non-member or third countries. Countries that are members of a free trade agreement (FTA) do not usually maintain a common external tariff.

Community Development Corporation (CDC) — Tax-exempt, non-profit organizations whose primary mission is the economic and social revitalization of distressed urban and rural areas. A CDC is a community based organization carrying out its activities within a geographically defined area. CDCs may support or undertake such activities as housing development and rehabilitation, job training and counseling, and business development activities.

Community Facilities Program (CFP) — Administered by the Rural Housing Service (RHS) of USDA, the CFP provides grants, loans, and loan guarantees to local governments, federally recognized native tribes, and nonprofit organizations. Funds are used to construct, expand, or rehabilitate such community facilities as hospitals, clinics, nursing homes, ambulatory care centers, police and fire stations, rescue and fire vehicles, communication centers, telecommunications, distance learning and telemedicine, child and adult care centers, jails, courthouses, airports, and schools. The program is now one of the funding accounts administered under the Rural Community Advancement Program (RCAP).

Community Fire Protection Program — Enacted in the 2002 farm bill (P.L. 107-171, Sec. 8003), as an amendment to the Cooperative Forestry Assistance Act of 1978 (P.L. 95-313), to provide assistance to communities for fire protection, especially in the wildland-urban interface.

Community food projects — A program administered by the Cooperative State Research, Education and Extension Service providing one-time matching grants to private non-profit entities to establish and carry out multi-purpose projects designed to increase food security on a local, community-based level. Project objectives are to meet the needs of low-income people by increasing their access to fresher, more nutritious food supplies; to increase the self-reliance of communities in providing for their own food needs; and to promote comprehensive responses to local food, farm, and nutrition issues. Congress has provided from $1 million to $2.5 million annually for the program in recent years, which USDA has used to make grants ranging from $10,000 to $250,000 each.

Community supported agriculture — A form of risk management whereby a farmer supports his (usually small) operation by selling shares in the farm's annual production. Share-holders pay a certain amount of money at the beginning of the growing season and are entitled to a portion of each week's harvest until the end of the season. Share-holders are not refunded their money if some or all of the crops fail. Also known as subscription farming.

Comparative advantage — Refers to the economic theory that in international trade it is more advantageous for a country to devote its resources not to all lines of production in which it may have superiority (least cost production), but to those in which its relative superiority is greatest. Two countries may find trade mutually profitable even if one of the countries could produce all goods at lower cost than the other.

Compensatory payments — See Area compensatory payments.

Competitive advantage — A situation in which one country, region, or producer can produce a particular commodity more cheaply than another country, region or producer.

Competitive bidding (for WIC) — This refers to the method for controlling WIC program costs, particularly for infant formula contained in WIC food packages. It requires state WIC agencies to solicit bids to infant formula companies for the sale of their product. This process is recommended but not required for other products sold through the program.

Competitive foods (in meal service) — Term used in reference to a provision in the National School Lunch Act (P.L. 79-396, as amended; 42 U.S.C. 1751 *et seq.*) that prohibits the sale of foods of minimal nutritional value in competition with federally funded school meal programs unless they contain some nutritional value. For example, the competitive food rule allows the competitive sale of chocolate and nut candy bars because the contents (i.e., milk and peanuts) contain some nutrients. Carbonated non-juice sodas, chewing gum, and spun candies are not permitted for sale because they contain no nutrients. The law defines competitive foods to mean foods sold in the cafeteria or meal service area during meal service periods.

Competitive imports — A term used by the Economic Research Service to describe imports that are similar to and therefore competitive (in contrast to non-competitive) with those produced in the United States. Examples are beef, wheat, cotton, and sugar.

Composting — The controlled biological decomposition of organic material, such as sewage sludge, animal manures, or crop residues, in the presence of air to form a humus-like material. Controlled methods of composting include mechanical mixing and aerating, ventilating the materials by dropping them through a vertical series of aerated chambers, or placing the compost in piles in the open air and mixing or turning them periodically.

Comprehensive Conservation Enhancement Program (CCHP) — The 2002 farm bill (P.L. 107-171, Sec. 2006) created CCHP, the new name for the Environmental Conservation Acreage Reserve Program (ECARP). Like ECARP, the CCHP is not a program, but an umbrella authority that encompasses the Conservation Reserve Program, Wetland Reserve Program, and Environmental Quality Incentives Program. Good faith provisions (P.L. 107-171, Sec. 1613), first established by the Agriculture Appropriations Act for FY2001 (P.L. 106-387, Sec. 755), apply to conservation programs as well.

Comprehensive Environmental Response, Compensation, and Liability Act (CERCLA) of 1980 (P.L. 96-510) — See Superfund.

Comprehensive nutrient management plan (CNMP) — Comprehensive nutrient management plans (CNMPs) have become an integral part of the regulatory permitting and environmental stewardship for large, and in some cases medium and small, animal feeding operations (AFOs). A CNMP is developed to assist an AFO owner/operator in meeting applicable local, tribal, state, and federal water quality goals or regulations. CNMPs are developed in accordance with NRCS conservation planning policy and EPA guidance. The CNMP fits within the total resource management objectives of the entire farm/animal feeding operation. It identifies management and conservation actions necessary to meet clearly defined nutrient management goals aimed at reducing excess nutrients in soil and water. Conservation practices may include manure and wastewater handling and storage, land treatment practices, and feed practices. [www.nrcs.usda.gov/programs/afo/cnmp_guide_index.html]

Con Act — Consolidated Farm and Rural Development Act of 1972 (P.L. 92-419; 7 U.S.C. 1921 *et seq.*).

Concentrated animal feeding operation (CAFO) — Generally, a facility where large numbers of farm animals are confined, fed, and raised, such as dairy and beef cattle feedlots, hog production facilities, and closed poultry houses. The EPA has developed a specific regulatory definition of CAFO for the purposes of enforcing the Clean Water Act (P.L. 92-500, 33 U.S.C. 1251-1387). The Act requires individual places that are potential sources of water pollution to obtain point source discharge permits that specify the allowable levels of effluent from each of these places. The regulations define animal feeding operations as those confining livestock or poultry for 45 days or more in a 12-month period in a facility that has no vegetative ground cover. Such places are further considered concentrated, and therefore required to have an EPA permit, if they exceed a size threshold or meet other criteria specified in the EPA regulations. The thresholds are 700 mature dairy cattle, 1,000 beef cattle, 125,000 chickens, 55,000 turkeys, 2,500 swine, or 10,000 sheep. In some regions, CAFO-size farms are called megafarms, confinements, or advanced farms.

Concentration (economic) — A measure of the degree to which a few large firms dominate total sales, production, or capacity within an industry or market. The concern is that the more concentrated an industry, the greater the likelihood of price and market manipulation. For example, meat packer concentration has long been a concern of cattle producers. It is common to express concentration as a ratio, by stating the share (%) held by the top 4, 8, or 12 firms.

Concessional (export) sale — A sale in which a foreign buyer is allowed loan payment terms that are more favorable than those obtainable in the commercial market. Under P.L. 480, the concessional provisions (compared to the commercial market) may include a lengthy credit period, a grace period before repayment begins, and a low interest rate.

Conditional registration — Under special circumstances, the Federal Insecticide, Fungicide, and Rodenticide Act (FIFRA; 7 U.S.C. 136a) permits registration of pesticide products conditionally upon the submission of additional data. These special circumstances include a finding by the EPA that a new product or use of an existing pesticide will not significantly increase the risk of unreasonable adverse effects. A product containing a new (previously unregistered) active ingredient may be conditionally registered only if the EPA finds that such conditional registration is in the public interest, that a reasonable time for conducting the additional studies has not elapsed, and the use of the pesticide for the period of conditional registration will not present an unreasonable risk.

Conditionality — Most often refers to the set of terms and conditions attached to lending to a country by the International Monetary Fund (IMF), involving agreements and adjustment policies aimed at restoring a sustainable balance-of-payments position within a one-to-three year period. More broadly, any set of conditions attached to bilateral or multilateral aid or used to establish eligibility for trade preferences as, for example, in the case of African beneficiary countries under the African Growth and Opportunity Act (P.L. 106-200).

Conjunctive use — Often used in discussing water supplies and water conservation. This phrase usually is used to describe the practice of storing surface water in a groundwater basin in wet years and withdrawing it from the basin in dry years.

Conservation (cross) compliance — A program authorized by the Food Security Act of 1985 (P.L. 99-198) that requires farmers who operate on highly erodible land to manage this land under an approved conservation system in order to maintain eligibility in specified federal farm programs, including commodity payments and many conservation programs. It has been amended in more recent farm bills. NRCS reviews show that 97% to 98% are meeting compliance requirements, while other reviews, such as a 2004 study by the GAO suggest the rate may be lower, but still is over 90%.

Conservation buffer — Small strips or areas of land in permanent vegetation, designed and strategically placed to intercept pollutants, provide wildlife habitat, and help manage other environmental concerns. Also called buffer strips. Examples of buffers include: riparian buffers separating upland areas from water bodies; filter strips; grassed waterways; shelter belts; windbreaks; and vegetative barriers. Conservation buffers are emphasized in the Buffer Initiative, as well as several other conservation programs.

Conservation Corridor Demonstration Program — This program, enacted in the 2002 farm bill (P.L. 107-171, Sec. 2602-2604) authorizes one or more states on the Delmarva Peninsula (Delaware, Maryland and Virginia) to develop and implement a conservation corridor plan to improve the economic viability of agriculture and environmental integrity of watersheds. Appropriations are authorized at such sums as may be necessary from FY2002 through FY2007, and the federal share will be up to 50% of the total amount.

Conservation district (Soil and Water Conservation District) — A legal subdivision of a state government, with an elected governing body, which develops and implements soil and water conservation programs within a certain area, usually coinciding with county lines. The nearly 3,000 districts in the United States have varying names (i.e., soil conservation district, soil and water conservation district, natural resources district, resource conservation district, resources district, or conservation district).

Conservation easement — Acquisition of rights and interest to a property to protect identified conservation or resource values, using a reserved interest deed. Easements can range from permanent to a multi-year period. A few states prohibit permanent easements. The largest federal programs using conservation easements on private lands are the Farmland Protection Program and the Wetland Reserve Program. Many state governments, especially in the northeast, and local government and nonprofits have been buying conservation easements on agricultural lands. Nonprofit organization have been very active in recent years. More than 1,250 land trusts around the country are using easements to protect about 2.6 million acres of land, according to a 2000 survey by the Land Trust Alliance.

Conservation Effects Assessment Project (CEAP) — A national multi-year assessment of the environmental benefits of conservation practices nationally and in selected watersheds initiated in 2003 by NRCS and FSA. [www.nrcs.usda.gov/technical/nri/ceap].

Conservation Farm Option Program — A provision of the 1996 farm bill (P.L. 104-127) authorized a pilot program for producers who received production flexibility payments to enter into a contract to consolidate payments at rates that were equivalent to payments that would otherwise be received from the Conservation Reserve Program, Wetlands Reserve Program, and/or the Environmental Quality Incentives Program in exchange for implementing practices to protect soil, water, and wildlife.

Conservation Innovation Grants — A new component of EQIP enacted in the 2002 farm bill (P.L. 107-171, Sec. 2301) to provide an unspecified portion of EQIP funds for competitive matching grants of up to 50% for innovative approaches to conservation. Examples of such approaches specified in the statute include market systems for pollution reduction, promoting carbon sequestration in soils, and leveraging these funds with money from other sources to promote environmental accomplishments in connection with agricultural production.

Conservation of Private Grazing Lands — Enacted in the 1996 farm bill (P.L. 104-127) and most recently amended by the 2002 farm bill (P.L. 107-171, Sec. 2502), this program provides coordinated technical, educational, and related assistance to preserve and enhance privately-owned grazing lands. It authorizes the creation of 2 grazing management demonstration districts. Appropriations are authorized at $60 million annually from discretionary funds for FY2002 through FY2007.

Conservation plan — A combination of land uses and farming practices to protect and improve soil productivity and water quality, and to prevent deterioration of natural resources on all or part of a farm. Plans may be prepared by staff working in conservation districts and must meet technical standards. For some purposes, such as conservation compliance, the plan must be approved by the local conservation district. Under the 1996 farm bill (P.L. 104-127, Sec. 315) conservation plans for conservation compliance must be both technically and economically feasible.

Conservation practice — Any technique or measure used to protect soil and water resources for which standards and specifications for installation, operation, or maintenance have been developed. Practices approved by the Natural Resources Conservation Service are compiled at each conservation district in its field office technical guide. Practices can be structural or land management.

Conservation Reporting and Evaluation System (CRES) — The system used by Natural Resources Conservation Service (NRCS) to collect and compile information about staff activities, conservation program implementation, and program accomplishments.

Conservation Reserve Enhancement Program (CREP) — A sub-program of the Conservation Reserve Program, CREP is a state-federal multi-year land retirement program developed by states and targeted to specific state and nationally significant water quality, soil erosion, and wildlife habitat problems. CREP uses the state funds to offer higher payments per acre to participants than the CRP. States may enroll up to 100,000 acres through an approved CREP, and at least three states have more than one CREP. USDA has reserved 4 million acres from the authorized 39.2 million acre total to enroll through either this option or the continuous enrollment option. Currently, 26 states have approved CREPs, and through March 2005, more than 645,000 acres had been enrolled under this option. As of June 2005, proposals are pending for 7 additional states.

Conservation Reserve Program (CRP) — A program created in the Food Security Act of 1985 (P.L. 99-198), to retire from production up to 45 million acres of highly erodible and environmentally sensitive farmland. Landowners who sign contracts agree to keep retired lands in approved conserving uses for 10-15 years. In exchange, the landowner receives an annual rental payment, cost-share payments to establish permanent vegetative cover, and technical assistance. The CRP reportedly has reduced erosion by up to 700 million tons per year. The 2002 farm bill extends authorization to enroll land through FY2007 and caps maximum total CRP acreage at 39.2 million acres. The Act adds a new subprogram to enroll up to 1 million acres of wetlands individually smaller than 10 acres and associated buffers. The Act continues to make the program spending mandatory and finances it through the Commodity Credit Corporation (CCC), adding that the CCC should fund technical assistance in support of the program. As of February 2005, more than 262,000 farms had enrolled 34.8 million acres; states with the greatest participation include Texas, Montana, and North Dakota, each with more than 3 million acres.

Conservation Security Contract — Participants in the Conservation Security Program (CSP) sign a contract stating which of the three levels (also referred to as tiers) of conservation they will maintain on lands in production in return for annual payments. Level I is a 5-year contract of up to $20,000 annually to address at least one conservation problem on a portion of a farm. Level II is a 5- to 10-year contract of up to $35,000 annually to address at least one conservation problem on an entire farm. Level III is a 5- to 10-year contract of up to $45,000 annually to address all conservation problems on an entire farm.

Conservation Security Plan — Producers who wish to participate in the Conservation Security Program (CSP) must prepare a plan that identifies: (1) the land to be enrolled; (2) the level at which the producer will participate and the resources to be protected; (3) the conservation practices to be installed, maintained or improved; and (4) an implementation schedule.

Conservation Security Program (CSP) — The CSP, enacted in the 2002 farm bill (P.L. 107-171, Sec. 2001) and administered by the Natural Resources Conservation Service (NRCS), will enroll land and assist producers in promoting resource conservation on lands producing agricultural crops from FY2003 through FY2007. The law specifies but does not limit the program to 19 allowable conservation practices. The program receives mandatory funding from the Commodity Credit Corporation (CCC). The program currently is capped at $6.0 billion over 10 years. In response to the limited funding, NRCS limits the program to specified watersheds.

Conservation technical assistance (CTA) — CTA has been the central activity of the Natural Resources Conservation Service (NRCS) since it was established in 1936. NRCS field staff help landowners and farm operators plan and implement soil and water conservation and water quality practices. The basis for much of this assistance is the conservation practices described in the field office technical guide. Each year, NRCS issues an accomplishments report. Examples of the accomplishments cited for FY2003 include: planning and applying resource management systems on almost 20 million acres of cropland and grazing land; helping landowners install conservation buffers on almost 500,000 acres; creating, restoring or enhancing 334,000 acres of wetlands; and helping producers apply nutrient management on 3.2 million acres.

Conservation Technology Information Center (CTIC) — A nonprofit center housed at Purdue University that works with public and private partners to promote soil quality by serving as a clearinghouse for information, a facilitator of activities, and a sponsor of meetings. It is associated with the promotion of conservation tillage. [www.ctic.purdue.edu/CTIC/CTIC.html].

Conservation tillage — Any tillage and planting system that leaves at least 30% of the soil surface covered by residue after planting. Conservation tillage maintains a ground cover with less soil disturbance than traditional cultivation, thereby reducing soil loss and energy use while maintaining crop yields and quality. Conservation tillage techniques include minimum tillage, mulch tillage, ridge tillage, and no-till. Conservation tillage varies widely by crop and soil type, but more than 100 million acres is cropped using some form of conservation tillage, according to annual surveys conducted by the Conservation Technology Information Center.

Conservation — The management of natural resources to provide maximum benefits over a sustained period of time (see Sustainable agriculture). In farming, conservation entails matching cropping patterns and the productive potential and physical limitations of agricultural lands to ensure long-term sustainability of profitable production. Conservation practices focus on conserving soil, water, energy, and biological resources. Contour farming, no-till farming, and integrated pest management are typical examples of conservation practices, which can be divided into two categories; land management practices and structural practices.

Conserving use acreage — Farmland diverted from crop production to an approved cultural practice that prevents erosion or other degradation. Though crops are not produced, conserving use is considered an agricultural use of the land.

Considered planted — Under some previous commodity support laws, crop acreage bases were, in general, calculated as a 5-year average of planted and considered planted acreage. Acreage considered planted included acreage idled under production adjustment programs or idled for weather-related reasons or natural disasters; acreage devoted to conservation purposes or planted to certain other allowed commodities; and acreage USDA determined was necessary for fair and equitable treatment.

Consolidated Farm and Rural Development Act of 1972 — P.L. 92-419 authorized a major expansion of USDA lending activities, which at the time were administered by Farmers Home Administration (FmHA). The legislation was originally enacted as the Consolidated Farmers Home Administration Act of 1961 (P.L. 87-128). In 1972, this title was changed to the Consolidated Farm and Rural Development Act, and is often referred to as the Con Act. The Con Act, as amended, currently serves as the authorizing statute for USDA's agricultural and rural development lending programs. The Act includes current authority for the following three major Farm Service Agency (FSA) farm loan programs: farm ownership, farm operating and emergency disaster loans. Also the Act authorizes rural development loans and grants. (7 U.S.C. 1921 *et seq.*).

Consolidated Farm and Rural Development Act of 1961 — P.L. 87-128 authorized a major expansion of USDA lending activities, which at the time were administered by Farmers Home Administration (FmHA), but now through the Farm Service Agency. The legislation was originally enacted as the Consolidated Farmers Home Administration Act of 1961. In 1972, this title was changed to the Consolidated Farm and Rural Development Act, and is often referred to as the Con Act. The Con Act, as amended, currently serves as the authorizing statute for USDA's agricultural and rural development lending programs. Titles in the Act include current authority for the following three major FSA farm loan programs: farm ownership, farm operating and emergency disaster loans. Title III of the ConAct is the Rural Development Act of 1972 (P.L.92-419) authorizing rural development loans and grants.

Consolidation — In agriculture and other economic sectors, consolidation usually is a reference to the trend from numerous smaller-sized operations toward fewer and larger ones. Consolidation can lead to higher concentration. See Industrialization.

Consultative Group on International Agricultural Research (CGIAR) — An informal association of 56 public and private organizations that support 16 international agricultural research centers. All but three of the research centers are located in developing countries. The cosponsors of CGIAR are the United Nations Food and Agriculture Organization (FAO), the United Nations Development Program (UNDP), the United Nations Environment Program (UNEP), and the World Bank (where the CGIAR Secretariat is headquartered). CGIAR's mission is to promote sustainable agriculture in developing countries, primarily by carrying out research and development of agricultural technologies, with the goal of creating food security, alleviating poverty, and preserving natural resources.

Consumer Price Index (CPI-U) — The Bureau of Labor Statistics' general measure of retail prices (for goods and services) paid by urban wage earners and clerical workers. Includes prices of about 400 items, including food, clothing, housing, medical care, and transportation. The CPI-U is commonly used to deflate time series data and is the most widely accepted measure of inflation.

Consumer support estimate (CSE) — An OECD indicator of the annual monetary value of gross transfers to (from) consumers of agricultural commodities, measured at the farm gate level, arising from policy measures which support agriculture, regardless of their nature, objectives or impacts on consumption of farm products. The CSE can be expressed in monetary terms; as a ratio to the value of consumption expenditure valued at farm gate prices, including budgetary support to consumers (percentage CSE); or as a ratio to the value of consumption expenditure valued at world market prices, without budgetary support to consumers. See Producer support estimate (PSE), General services support estimate (GSSE), and Total support estimate (TSE).

Consumer-oriented agricultural products — One of three broad categories of agricultural products used by the Foreign Agricultural Service to report export and import data under its BICO system. The others are bulk commodities and intermediate agricultural products. Consumer-oriented agricultural products are high value products that usually (but not always) are those ready, or easily made ready, for immediate use by consumers. Examples are snack foods, breakfast cereals, bakery mixes, eggs and products, dairy products, fresh or processed red meats and poultry, fresh or processed fruits, vegetables, nuts, pet foods, wine, and beer.

Consumptive water use — Water removed from available supplies without return to a water resources system (e.g., water used in manufacturing, agriculture, and food preparation that is not returned to a stream, river, or water treatment plant). Crop consumptive water use is the amount of water transpired during plant growth plus what evaporates from the soil surface and foliage in the crop area. The portion of water consumed in crop production depends on many factors, especially the irrigation technology.

Continued Dumping and Subsidy Offset Act of 2000 — See Byrd Amendment.

Continuous Coverage — Refers to the ability of farmers to select any level of crop insurance coverage between 50% and 85% of normal crop yield. Current crop insurance law prohibits continuous coverage. Instead, it requires a participating farmer to choose a coverage level in 5% increments to match the premium subsidy structure, which is set in law in 5% increments. The percentage of the crop insurance premium subsidized by the federal government falls as a producer selects higher levels of coverage. Hence, the continuous coverage prohibition is a federal cost-saving measure that prevents producers who would normally choose, for example, a 65% level of coverage, from dropping back to 64% coverage just to receive the higher subsidy level.

Continuous Enrollment — A subprogram of the Conservation Reserve Program (CRP). Continuous enrollment rules allow producers to sign up land at any time without competition if they agree to install and maintain certain conservation practices that will provide greater environmental benefits. Eligible land is automatically accepted. Examples of high priority practices include riparian buffers, filter strips, and grass waterways. In April 2000, USDA announced it would provide enhanced financial incentives for 3 years to attract continuous enrollment participants. USDA has reserved 4 million acres from the authorized CRP total of 39.2 million acres to enroll through either this option or the Conservation Reserve Enhancement Program. Through November 2002, more than 1.8 million acres had been enrolled under this option.

Continuous inspection — USDA's meat and poultry inspection system is often called "continuous" because no animal destined for human food may be slaughtered or dressed unless an inspector is continuously present to examine each one before slaughter (antemortem inspection), and its carcass and parts after slaughter (postmortem inspection). This also is sometimes referred to as "carcass-by-carcass inspection." In processing plants (as opposed to slaughter plants), inspectors need not be present at all times, but they do visit at least once daily. Thus, processing inspection also is considered to be continuous.

Contour farming — Field operations (such as plowing, planting, cultivating, and harvesting) at right angles to the natural slope to reduce soil erosion, protect soil fertility, and limit water runoff. Contour strip farming is a kind of contour farming in which row crops are planted in strips, between alternating strips of close-growing, erosion-resistant forage crops.

Contract — A written or oral agreement spelling out the parties' understanding of how a commodity is to be produced and/or marketed, possibly including specifications for quantity, quality, and price. Marketing contracts are commonly used for crops, while production contracts are more prevalent in the livestock industry. Contracts contrast to cash markets. Cash markets continue to dominate the agriculture sector, accounting for almost 70% of farm commodity sales in 1997. However, contracting likely will continue to grow as a risk management tool for farmers and a coordination tool for processors. Futures contracts provide a way to manage price risk that typically do not involve actual delivery of commodities.

Contract acreage — Base acres enrolled annually in the Direct and Counter-cyclical Program (DCP) authorized by the 2002 farm bill (P.L. 107-171, Sec. 1101-1108) for covered commodities during crop years 2002 through 2007. Previously, the 1996 farm bill (P.L. 104-127) authorized 7-year production flexibility contracts, which guaranteed fixed direct payments but not counter-cyclical target price deficiency payments. The new law uses the term agreement rather than contract, but farmers must sign a Direct and Counter-cyclical Program Contract (Form CCC-509).

Contract commodity — The commodities previously eligible for deficiency payments and subsequently eligible for production flexibility contracts under the 1996 farm bill (P.L. 104- 127)(wheat, corn, sorghum, barley, oats, rice, and upland cotton). Under the Direct and Counter-cyclical Program (DCP) authorized by the 2002 farm bill (P.L. 101-171, Sec. 1101-1108), the phrase "covered commodity" is used and includes wheat, corn, grain sorghum, barley, oats, upland cotton, soybeans, canola, flax, mustard, rapeseed, safflower, sunflowers, and rice.

Contract crops — Crops eligible for production flexibility contract payments (wheat, corn, sorghum, barley, oats, rice, and upland cotton) under provisions of the Agricultural Market Transition Act (AMTA, Title I of the 1996 farm bill (P.L. 104-127)). Sometimes referred to as AMTA payments. The 2002 farm bill (P.L. 101-171, Sec. 1101-1108) used the phrase "covered commodities" to refer to the crops eligible for Direct and Counter-cyclical Program (DCP) payments. Covered commodities include wheat, corn, grain sorghum, barley, oats, upland cotton, soybeans, canola, flax, mustard, rapeseed, safflower, sunflowers, and rice.

Contract for future sale — A sales contract under which a farmer agrees to deliver products of specified quality and quantity to a buyer for a specified price within a prescribed time frame. Contract sales are a growing practice, recently accounting for 86% of poultry, over 80% of tobacco, more than 50% of fruits, and 43% of milk. The benefits to processors are greater uniformity and predictability resulting in lower costs of grading, processing, and packing. The benefits to farmers are more stable income from a guaranteed market and price, and possibly access to a wider range of production inputs and advanced technology. Critics are concerned about lack of accessible price information, and manipulation of markets to the disadvantage of producers.

Contract payments under AMTA — Some $36 billion in Production Flexibility Contract payments made to farmers for contract crops for fiscal years 1996-2002 under Title I of the 1996 farm bill (P.L. 104-127), known as the Agricultural Market Transition Act (AMTA). The total amount made available for each fiscal year was specified in the Act and allocated to commodities each fiscal year using a set of percentages also specified in the Act. These percentages were based on CBO's February 1995 baseline forecast of what deficiency payments would have been if provisions in effect for the 1995 crop had been extended. For example, for fiscal 1997, the total allocation for wheat was 26% of total annual payments of $5.385 billion, or $1.414 billion. The annual payment rate for wheat equaled total spending ($1.414 billion) divided by the sum of all individual wheat payment contract quantities for the year. As with other program commodities, an individual farm's payment quantity equaled the farm's program payment yield multiplied by 85% of the farm's wheat contract acreage. Program yields under the 1996 Act were determined in the same manner as under the 1949 Act for 1995 crops. An individual farmer's PFC payment was the payment quantity times the annual payment rate. The payment was made by September 30 of each of the fiscal years 1996 through 2002. Producers also could choose to receive 50% of the contract payment in December or January of the fiscal year. Farmers had near total planting flexibility on the contract acres (the exception being fruits and vegetables) as well as on the remainder of the farm. PFC contract payments were replaced by Direct and Counter-cyclical Program (DCP) payments beginning with crop year 2002 under the 2002 farm bill (P.L. 101-171, Sec. 1101-1108).

Contract production — A form of vertical integration where a firm commits to purchase a commodity from a producer at a price formula set in advance of the purchase.

Contract sanctity — The concept that U.S. agricultural products already contracted to be exported should not be subject to government cancellation because of short supply, national security, and/or foreign policy reasons. The 1990 farm bill (P.L. 101-624) provides for contract sanctity by prohibiting the President from restricting the export of any agricultural commodity already under contract to be delivered within 270 days from the date an embargo is imposed, except during national emergency or war.

Convention on Biological Diversity — See United Nations Convention on Biological Diversity.

Conventional agriculture — Generally used to contrast common or traditional agricultural practices featuring heavy reliance on chemical and energy inputs typical of large-scale, mechanized farms to alternative agriculture or sustainable agriculture practices. Moldboard plowing to cover stubble, routine pesticide spraying, and use of synthetic fertilizers are examples of conventional practices that contrast to alternative practices such as no-till, integrated pest management, and use of animal and green manures.

Conventional tillage — Tillage operations considered standard for a specific location and crop and that tend to bury the crop residues; usually considered as a base for determining the cost effectiveness of erosion control practices. See Conservation tillage.

Converted wetland — Under the swampbuster program, these are wetlands that were drained or altered to improve agricultural production after December 23, 1985, the date swampbuster was enacted. (Lands converted before December 23, 1985 are called prior converted wetlands, and alterations to these lands are subject to less stringent requirements.) On lands with this designation, no drainage maintenance and no additional drainage are allowed.

Conveyance loss — Water loss in pipes, channels, conduits, ditches by leakage or evaporation.

COOL — Country Of Origin Labeling.

Cooperative — An enterprise or organization owned by and operated for the benefit of those using its services. In agriculture, such an organization is owned and used by farmers mainly to handle the off-farm part of their businesses (i.e., buying farm supplies, marketing their products, furnishing electric and telephone service, and providing business services) at cost. Essential features are democratic control, limited return on capital, and operation at cost, with distribution of financial benefits to individuals in proportion to their use of the services made available by the cooperative (called patronage refunds). In 2002, there were 3,140 farmer cooperatives in the United States. As a variation from the traditional design, so-called "new generation cooperatives" are characterized by limited membership, require substantial investment, and include delivery contracts. Producers are increasingly using this model to create their own value-added business enterprises. The Rural Business-Cooperative Service (RBS) assists in forming new cooperative businesses and improving the operations of existing cooperatives through technical assistance, research, and information products. Cooperatives are afforded certain antitrust exemptions by the Capper-Volstead Act. Many farming-related cooperatives are members of the National Council of Farmer Cooperatives.

Cooperative Extension System — A federal-state-local cooperative education system that provides continuing adult education based on the academic programs of the land grant colleges of agriculture and their affiliated state agricultural experiment stations (on the campuses of major state universities). The system employs approximately 32,000 people located both on campuses and in offices located in virtually every county in the nation. About half of Extension's education programs focus on agriculture and natural resources, one-quarter on youth development (including the vocational 4-H program), and the balance on home economics and community resource development work.

Cooperative Forestry Assistance Act of 1978 — P.L. 95-313 revised authority (of the Clarke-McNary Act of 1924, and other statutes) for the Forest Service to provide financial and technical assistance to states and private landowners on a variety of forestry issues: forest management and stewardship, fire protection, insect and disease control, reforestation and stand improvement, urban forestry, etc., under State & Private Forestry.

Cooperative research and development agreement (CRADA) — The Federal Technology Transfer Act of 1986 (P.L. 99-502). allows industry to enter into research contracts with government laboratories. In exchange for this cooperation, the company involved is entitled to first rights to obtain an exclusive license to any inventions that may emerge as a result of the CRADA. USDA's in-house research agency, the Agricultural Research Service (ARS) has developed more than 1,100 CRADAs with industry since the law was enacted.

Cooperative State Research, Education, and Extension Service (CSREES) — The USDA agency that administers federal funds appropriated for agricultural and forestry research, extension, and education programs at eligible institutions, including the land grant colleges of agriculture in the states, selected veterinary schools, and other institutions with capabilities in the food and agricultural science arena. The agency distributes formula funds to the 1862 land grant colleges under the Hatch Act of 1887 (for agricultural research; 7 U.S.C. 361a *et seq.*), the Smith-Lever Act of 1914 (for Extension programs; 7 U.S.C. 341 *et seq.*), and the McIntire-Stennis Act of 1962 (forestry research; 16 U.S.C. 582a *et seq.*); Evans-Allen funds (7 U.S.C. 3222) for research programs at the 1890 land grant colleges (historically black); the National Research Initiative (NRI)

Competitive Grants program; the Special Grants program; grants for higher education; and the Initiative for Future Agriculture and Food Systems [www.reeusda.gov].

Cooperator Program — Officially known as the Foreign Market Development Cooperator Program (FMDP). One of the agricultural export promotion programs operated by the Foreign Agricultural Service. This program consists of joint government/agri-industry efforts to develop markets by acquainting potential foreign customers with U.S. farm products. Activities under this program include providing technical assistance to prospective foreign buyers, overseas food exhibits, product demonstrations and advertising aimed at foreign consumers. FAS shares the financing of these projects with the cooperators, which are nonprofit commodity trade associations primarily composed of producer-based farm groups.

Coordinated Framework for Regulation of Biotechnology — Proposed in 1984 by the White House Office of Science and Technology Policy and finalized in 1986, this framework spells out the basic federal policy for regulating the development and introduction of products derived from biotechnology. A key principle of the framework is that genetically engineered products would continue to be regulated according to their characteristics and unique features, and not according to their method of production. In other words, for example, if a food product produced through biotechnology is substantially the same as one produced by more conventional means, that food is subject to no additional (or no different) regulatory processes. The framework also maintains that new biotechnology products are regulated under existing federal statutory authorities and regulation. The U.S. policy framework contrasts with that of some of its major trading partners: the EU, Japan, South Korea, China, Australia, and New Zealand either have or are establishing separate mandatory labeling requirements for products containing genetically modified ingredients

Coordinated review effort (CRE) — Federal reviews of the National School Lunch Program conducted in cooperation with state agencies and local school food authorities (LSFAs) to improve the management of the programs, evaluate meal data accuracy, and provide training and technical support to schools to help improve local program accountability.

Cord — 1 cord = 128 cubic feet. This volume measure typically is used to measure pulpwood and cut and spilt firewood. A cord is a stack of logs 4 feet wide by 8 feet high by 4 feet deep. (Because of air spaces between logs, it contains less than 128 cubic feet of solid wood.) Many firewood dealers sell what they call a face cord, a stack 4 feet by 8 feet, but of indeterminate depth (usually about 18 inches).

Corn Belt — That area of the United States where corn is a principal cash crop, including Iowa, Indiana, most of Illinois, and parts of Kansas, Missouri, Nebraska, South Dakota, Minnesota, Ohio and Wisconsin.

Corn gluten — A byproduct of wet milling of corn. Corn gluten is used as a medium-protein (20-24%), medium-fiber (10%) feedstuff. The European Union is the major market for U.S. corn gluten feeds.

Corn/hog ratio — See Hog/corn ratio, and Feed ratio.

Corporate farm — A form of farm ownership that is a separate legal entity from the individual owners of the farm. Tax and other reasons have encouraged conversion of some family farms to Chapter S corporations, where stock is held by family members. The 2002 Census of Agriculture reports 3.5%, or about 74,000, of the 2.129 million farms in the nation were incorporated farms, and 90% of these were family held. By contrast, about 1.910 million (90%) were individual or family-owned operations and about 130,000 (6%) were partnerships.

Cosmetic appearance — The 1990 farm bill (P.L. 101-624, Sec. 1351) defines the term as "the exterior appearance of an agricultural commodity, including changes to that appearance resulting from superficial damage or other alterations that do not significantly affect yield, taste, or nutritional value." The Agricultural Marketing Service sets grades and standards for many agricultural commodities. Some consumer and environmental groups have argued that some of these standards are harmful because they encourage excessive pesticide use merely to make fruits and vegetables attractive. Agricultural interests disagree, countering that consumers prefer blemish-free produce and that cosmetic standards are no less important than other grading factors.

Cost/benefit analysis — See Benefit/cost analysis.

Cost, insurance, and freight (CIF) — In general, the commercial trade term cif means that the seller's price includes the cost of the goods, the marine insurance, and all transportation charges to the named point of destination. Similar terms include CFR, cost and freight; CFI, cost, freight, and insurance; CIF & C, cost, insurance, freight, and commission; CIFC & I, cost, insurance, freight, commission, and interest; and CIFI & E, cost, insurance, freight, interest, and exchange. CAF is the French form of CIF.

Cost of production — The average unit cost (including purchased inputs and other expenses) of producing an agricultural commodity. The Agricultural and Consumer Protection Act of 1973 (P.L. 93-86) requires USDA to make annual estimates of the average cost of producing selected commodities. These cost of production estimates have been used by Congress in considering farm policy options.

Cost-containment (for WIC) — Refers to statutory provisions in the Child Nutrition Act of 1966 (P.L. 89-642, as amended) that require state agencies to control WIC program costs, particularly with respect to the cost of infant formula sold through the program. See Competitive bidding, and Sole source bids.

Cotonou Agreement — A 20 year agreement between the European Union and 77 countries that together are referred to as ACP Countries (African, Caribbean and Pacific countries). The Cotonou Agreement was signed in June 2000 at Cotonou, Benin. In the area of trade cooperation and development aid, the objectives are to promote smooth and gradual integration of ACP economies into the world economy, to enhance production, supply and trading capacities, to create new trade dynamics and foster investment, and to ensure full conformity with WTO provisions. This partnership agreement is the successor to the Lome Convention, first adopted in 1975 and renewed at successive intervals.

Cotton competitiveness provisions — Provisions added by the Food, Agriculture, Conservation, and Trade Act of 1990 to the cotton program designed to keep U.S. cotton prices competitive in domestic and export markets. Sometimes referred to as the "three-step competitiveness" provisions. Step 1 is the discretionary authority for USDA to reduce the adjusted world price (used in the cotton marketing assistance loan program) when world prices are declining to near the adjusted world price, but U.S. prices are higher than world prices. Though rarely used, the Step 1 adjustment is intended to make marketing loans more effective in keeping U.S. cotton globally competitive. Step 2 payments, sometimes referred to as the "user marketing certificate program," are made to U.S. cotton users and exporters when U.S. prices are higher than world prices. Step 2 payments are intended to bridge price gap and keep U.S. cotton competitive. Step 3 mandates the opening of a "special import quota" when the differential between the higher U.S. price for cotton and the lower price for foreign cotton extends for a specified length of time. Its purpose is to allow imports to enter, acting to lower U.S. prices to bring them more in line with world prices. A step 3 quota cannot be established if a limited global quota for upland cotton is in effect, which operates differently and is triggered when other price conditions are met.

Cottonseed Oil Assistance Program (COAP) — Along with the Sunflower Oil Assistance Program (SOAP), COAP was one of two programs that awarded bonuses to exporters to assist in exports of U.S. vegetable oil to targeted markets. Funds for the programs were authorized to be made available under Section 32 of the Agricultural Adjustment Act of 1935 (P.L. 74-320). The provision in the Disaster Assistance Act of 1988 (P.L. 100-387) that authorized the COAP to begin in fiscal year 1989 expired at the end of fiscal year 1995. However, the USDA appropriations act for FY1996 (P.L. 104-37, October 21, 1995) provided authority to operate the program in fiscal year 1996. COAP was not reauthorized by the 1996 farm bill (P.L. 104-127), although export subsidies for cottonseed oil can be financed under the Export Enhancement Program (EEP).

Counter-cyclical payment (CCP) — Under the Direct and Counter-cyclical Program (DCP) created by the 2002 farm bill (P.L. 101-171, Sec. 1101-1108), counter-cyclical payments are made to participating producers when the marketing year average price received by farmers for a covered commodity is less than the target price. The total payment to a producer is the payment acres (85% of base acres) times the payment rate (target price minus average market price, except not more than the difference between the target price and the sum of the national loan rate and the direct payment rate).

Countertrade — A trade transaction of goods and services without the exchange of money. Forms of countertrade include barter, buy-back or compensation, counter-purchase, offset requirements, swap, or triangular trade.

Countervailing duty — A charge levied on an imported article to offset the unfair price advantage it holds due to a subsidy paid to producers or exporters by the government of the exporting country if such imports cause or threaten injury to a domestic industry. The countervailing provisions of the Tariff Act of 1930 (P.L. 71-361), as added by the Trade Agreements Act of 1979 (P.L. 96-39), provide for an assessment equal to the amount of the subsidy, in addition to other duties and fees normally paid on the imported article. Countervailing duties are permitted under the World Trade Organization's Agreement on Subsidies and Countervailing Measures.

Country-of-origin labeling (COOL) — Federal law has long required most imports, including many food items, to bear labels informing the ultimate purchaser of their country of origin. Meats, produce, and several other raw agricultural products generally were exempt. The 2002 farm bill (P.L. 107-171) contained a requirement that many retail establishments provide, starting on September 30, 2004, country-of-origin information on fresh fruits and vegetables, red meats, seafoods, and peanuts. A subsequent appropriation law (P.L. 108-199) postponed mandatory COOL for these products (except seafoods) until September 30, 2006. At issue are whether consumers are more likely to buy the U.S. alternative if such labeling is more prevalent; the business and government costs of implementation and enforcement; and whether foreign countries might challenge the new law as an illegal trade barrier (although many have their own COOL rules).

County committees — Panels of three to five farmers, elected by other farmers, to oversee the local operation of commodity programs, credit, and other programs of the Farm Service Agency. County committees, established by the Soil Conservation and Domestic Allotment Act of 1935 (P.L. 74-46), are so named because they have overseen USDA field offices for farmers that once existed in most rural farm counties throughout the United States. Today, the committees often oversee activities in multi-county areas, due to USDA reorganization and consolidation of its field office structure into a network of about 2,500 field service centers. The committees are responsible for hiring and supervising the County Executive Director (CED), who manages the day-to-day activities of the field service center and its employees. The director and most county office staff legally are employees of the farmer-elected committees rather than the federal government, although their salaries come from federal funds.

County Executive Director (CED) — The supervisor hired by the Farm Service Agency county committee to manage the day-to day activities of a field service center (formerly called the county office).

County loan rate — Nonrecourse loan rates vary from county to county to account for transportation cost differences to the nearest terminal elevator. The weighted average for all county loan rates (the actual loan levels received by farmers) in the United States must equal the national average loan rate, established by USDA according to limits set by Congress.

County office — Usually refers to the local office of the Farm Service Agency, where farmers go to conduct business associated with federal farm commodity and credit programs, and some conservation programs. As a result of reorganization in 1994, local offices are increasingly shared with other USDA agencies having local representatives, such as the Natural Resources Conservation Service. Offices shared by several agencies are called field service centers.

County payments — Forest Service payments of 25% of gross revenues from each national forest to the states for use on road and school programs in the counties where the national forests are located. Technically known as Payments to States, because the states determine which road and school programs can be funded, but the payments are allocated to the counties based on the national forest acreage in each county. Commonly confused with payments in lieu of taxes. Also, modified temporarily by the Secure Rural Schools and Community Self-Determination Act of 2000 (P.L. 106-393).

Cover crop — A close-growing crop, planted primarily as a rotation between regularly planted crops, or between trees and vines in orchards and vineyards, to protect soil from erosion and improve it between periods of regular crops.

Cow-calf operator — A ranch or farm where cows are raised and bred mainly to produce calves usually destined for the beef market. The cows produce a calf crop each year, and the operation keeps some heifer calves from each calf crop for breeding herd replacements. The rest of the calf crop is sold between the ages of 6 and 12 months along with old or nonproductive cows and bulls. Such calves often are sold to producers who raise them as feeder cattle.

Cowardin Classification System — A wetlands classification system used by the U.S. Fish and Wildlife Service to map and inventory wetlands. This system, which was developed in the late 1970s, classifies wetlands using a complex hierarchy that divides wetlands into numerous subcategories of 5 ecological systems.

CP — Contracting party.

CPI — Consumer Price Index.

CRADA — Cooperative Research and Development Agreement.

Crambe — Crambe (Crambe abyssinicia Hochst.) is also referred to as Abyssinan mustard, Abyssinian kale, colewart, or katran. It is a member of the mustard (Cruciferae) family, which includes crops such as rapeseed (canola and industrial oilseed rape) and tame mustard. Crambe is native to the Mediterranean region. It was first introduced into the United States during the 1940s. Commercial production of crambe in North Dakota began in 1990. Crambe seed yields an industrial oil that contains a high level of erucic acid (50 to 60 percent erucic acid by weight). Crambe meal may be used primarily as cattle protein supplement but also can be used for protein isolates and fertilizer. At times, crambe has been designated as eligible for federal oilseed support.

CRBSC — Colorado River Basin Salinity Control Program.

CRC — Crop Revenue Coverage (see Revenue insurance).

CRES — Conservation Reporting and Evaluation System.

Creutzfeldt-Jakob Disease (CJD) — A sporadic, rare, but fatal human disease that usually strikes people over 65. It occurs worldwide at an estimated annual rate of one case per million population. About 10-15% of CJD cases are inherited. A small number of cases have occurred as the result of various medical treatments or procedures that inadvertently transferred the CJD agent. In March 1996, the British government announced a possible link between bovine spongiform encephalopathy (BSE, or mad cow disease) and CJD. The announcement was prompted by the discovery of several atypical cases of CJD in Great Britain.

Critical habitat — Under the Endangered Species Act (P.L. 93-205) critical habitat is an area essential to the conservation of a listed species, though the area need not actually be occupied by the species at the time it is designated. Critical habitat must be designated for all threatened and endangered species under the Act (with certain specified exceptions). If habitat land is nonfederal, there must be a federal connection for the ESA to be triggered; purely private actions are not covered. A federal agency with whom a landowner is dealing must ensure that its actions (which may include giving a loan, increasing irrigation flows, etc.) do not adversely modify these areas.

Crop acreage base — A crop-specific measure equal to the average number of acres planted (or considered planted) to a particular program crop for a specified number of years. The crop-specific nature of this measurement was important prior to the 1996 farm bill (P.L. 104-127), which adopted an inclusive measure of base acreage and allowed planting flexibility among the program crops. The sum of the crop acreage bases for all program crops on a farm could not exceed the farm acreage. The acreage base was used in determining the number of acres a farmer, under an acreage reduction program, had to remove from normal crop production and devote to conserving uses in order to be eligible for USDA price and income supports. See also Base acreage.

Crop insurance — Insurance that protects farmers from crop losses due to natural hazards. A subsidized multi-peril federal insurance program, administered by the Risk Management Agency, is available to most farmers. The program is authorized by the Federal Crop Insurance Act (which is actually title V of the Agricultural Adjustment Act of 1938, P.L. 75-430), as amended. Federal crop insurance is available for more than 100 different crops, although not all insurable crops are covered in every county. With the amendments to the Federal Crop Insurance Act made by the Federal Crop Insurance Reform Act of 1994 (P.L. 103-354, Title I) and the Agriculture Risk Protection Act of 2000 (P.L. 106-224), USDA is authorized to offer basically free catastrophic (CAT) coverage to producers who grow an insurable crop. For a premium, farmers can buy additional coverage beyond the CAT level. Crops for which insurance is not available are protected under the Noninsured Assistance Program (NAP). Federal crop insurance is sold and serviced through private insurance companies. A portion of the premium, as well as the administrative and operating expenses of the private companies, is subsidized by the federal government. The Federal Crop Insurance Corporation reinsures the companies by absorbing some of the losses of the program when indemnities exceed total premiums. Several revenue insurance products are available on major crops as a form of additional coverage.

Crop reports — Reports compiled by the National Agricultural Statistics Service (NASS) on various commodities that are released throughout the year. Information in the reports includes estimates on planted acreage, yield, and expected production, as well as comparison of production from previous years.

Crop residue — That portion of a plant, such as a corn stalk, left in the field after harvest. Crop residues are measured to determine whether farmers are properly implementing conservation tillage if that is a part of their conservation plans to meet conservation compliance requirements. These farmers are required to maintain at least 30% residue cover on the ground to successfully practice conservation tillage.

Crop Revenue Coverage (CRC) — A form of revenue insurance that protects a producer's revenue for an insurable crop whenever low prices, low yields, or a combination of both causes revenue to fall below a guaranteed level selected by the producer. It differs from other revenue insurance programs by allowing producers to use the higher of the planting price or the market price in determining a target level of revenue.

Crop rotation — The growing of different crops, in recurring succession, on the same land in contrast to monoculture cropping. Rotation usually is done to replenish soil fertility and reduce pest populations in order to increase the potential for high levels of production in future years.

Crop scouting — Precise assessments of pest pressure (typically insects) and crop performance to evaluate economic risk from pest infestations and the potential effectiveness of pest control interventions. Scouting is usually sold as a commercial service to farmers.

Crop share rent — In contrast to cash rent, the tenant farmer pays the landlord a share of the crop. This arrangement puts the landlord, like the tenant operator, at risk from variation in yields and prices. For the farm operator, crop share rent is a mechanism for sharing risks with the landlord. In relation to commodity programs for supporting prices and farm incomes, cash rent landlords do not have a beneficial interest in the commodity and are not eligible payments.

Crop year — The calendar year during which a crop is harvested. This contrasts with the marketing year, which is the 12-month marketing period after harvest.

Cropland — Land used primarily for the production of row crops, close-growing crops, and fruit and nut crops. It includes cultivated and noncultivated acreage, but not land enrolled in the Conservation Reserve Program. Approximately 377 million acres are in cropland, and another 32.7 million acres are enrolled into the Conservation Reserve Program, according to the 1997 National Resources Inventory (NRI). Comparing data from NRIs over the past 20 years, the total amount of cropland (including CRP land) has declined by less than 3%. However, increases and decreases are far greater in many states; for example, Alabama's cropland (including CRP land) dropped from 4.5 million acres to 3.5 million acres, while South Dakota's grew from 16.9 million acres to 18.4 million acres. Cropland is 30% of all non-federal rural lands.

Cropping systems — A general term that describes how a producer might grow a crop. A basic distinction is between conventional tillage and conservation tillage.

Cross compliance — A no-longer-used requirement that a farmer who participates in a price support program for one crop must also participate in price support programs for other crops grown on the same farm. See Conservation compliance.

Cross subsidization — The practice of charging higher prices to one group of consumers in order to subsidize lower prices for another group. State trading enterprises with monopoly control over marketing agricultural exports are sometimes alleged to cross subsidize, but lack of transparency in their operations makes it difficult if not impossible to determine if that is the case.

CRP — Conservation Reserve Program.

Crush spread — In the soybean futures market, the simultaneous purchase of soybean futures and the sale of soybean meal and soybean oil futures to establish a processing margin. See Gross processing margin.

CSCE — Coffee, Sugar, and Cocoa Exchange, which merged with the New York Cotton Exchange in 1998, to become the New York Board of Trade. [www.nybot.com].

CSE — Consumer support estimate.

CSFP — Commodity Supplemental Food Program.

CSP — Conservation Security Program

CSPI — Center for Science in the Public Interest. [www.cspinet.org].

CSR — Coalition for Sugar Reform. [www.sugar-reform.org].

CSREES — Cooperative State Research, Education, and Extension Service. [www.reeusda.gov].

CSRS — Cooperative State Research Service (see Cooperative State Research, Education, and Extension Service).

CU — Consumers' Union. [www.consumersunion.org].

Cultural methods — Practices used to enhance crop and livestock health and prevent weed, pest or disease problems without the use of chemical substances. Examples include the selection of appropriate varieties and planting sites; selection of appropriate breeds of livestock; providing livestock facilities designed to meet requirements of species or type of livestock; proper timing and density of plantings; irrigation; and extending a growing season by manipulating the microclimate with green houses, cold frames, or wind breaks.

Custom feeders — Producers who feed animals (e.g., cattle, hogs) they do not own, in return for a fee paid by someone else (such as a packer or a cow-calf operator) who does own the animals. Custom feeding potentially provides packers with more control over supplies and prices of animals. Custom feeding is a form of vertical integration.

Customs union — An agreement between two or more countries to remove trade barriers between each other and to establish common tariff and nontariff policies with respect to other countries. The European Communities (EC) of the European Union (EU) is the best known customs union.

CVD — Countervailing duty.

CVM — Center for Veterinary Medicine. [www.fda.gov/cvm].

CWA — Clean Water Act (Federal Water Pollution Control Act) (33 U.S.C. 1251 *et seq.*).

CWB — Canadian Wheat Board. [www.cwb.ca].

CWD — Chronic wasting disease.

cwt. — Hundredweight, or 100 pounds.

CXT — Common external tariff.

CY — Crop year; calendar year.

Cycle time — The length of time consumed by a freight railcar from one loading to the next.

CYFAR — Children, Youth and Families at Risk Program.

CZMA — Coastal Zone Management Act (see Coastal Zone Management Program).

Dacthal (DCPA) — A selective herbicide, trade name Dacthal, especially used on vegetables. DCPA and its breakdown products are environmentally significant because they may pose unreasonable risks to certain nontarget plants and animals. In addition, EPA is gathering data on occupational exposures due to concerns about chronic cancer risks. EPA has determined that DCPA and its metabolites do not currently pose a significant cancer or chronic non-cancer risk to the general public from most uses.

Dairy and Tobacco Adjustment Act of 1983 — Title I of P.L. 98-180 is known as the Dairy Production Stabilization Act of 1983, authorized a voluntary dairy diversion program, which was operated between January 1984 and March 1985. Producers who elected to participate in the program and reduce their milk marketings by between 5% and 30% below their base production were paid $10 per hundred pounds (cwt.) for these reductions. For a 16-month period (12/1/83- 3/31/85), all dairy farmers were assessed 50¢/cwt. on all milk marketed to help defray the cost of the diversion program. Also, Title I authorized a national dairy promotion program (or, check-off program) for generic dairy product promotion, research and nutrition education. This self-help program is funded through a permanent 15¢/cwt. assessment on all milk production, and is administered by a board of dairy farmers who are appointed by the Secretary of Agriculture. Title II was designated the Tobacco Adjustment Act of 1983. Title II provided for reduced levels of price support for tobacco, the prohibition of lease and transfer of flue-cured quota, the mandatory sale of allotments and quotas by nonfarming entities, the required inspection of imported tobacco, and various other modifications to the tobacco programs.

Dairy Diversion Program — A voluntary supply control program authorized by the Dairy Production Stabilization Act of 1983 (P.L. 98-180, Title I) under which producers in 1984-85 received payments of $10/cwt. of milk, for reducing their milk marketings by between 5% to 30% below an earlier base period.

Dairy Export Incentive Program (DEIP) — A program that offers subsidies to exporters of U.S. dairy products to help them compete with other nations. USDA pays cash to exporters as bonuses to help them sell certain U.S. dairy products at prices below the exporter's cost of acquiring them. The program was originally authorized by the Food Security Act of 1985 (P.L. 99-198) and extended by the 1990 farm bill (P.L. 101-624) and the Uruguay Round Agreements Act of 1994 (P.L. 103-465). The total tonnage and dollar amounts of these and other export subsidies have been limited by the Uruguay Round Agreement on Agriculture. The 2002 farm bill (P.L. 107-171) extended the program through 2007.

Dairy Market Loss Assistance (DMLA) — A series of emergency direct payment programs for dairy farmers funded over three consecutive years (FY1999-2001) by three separate emergency supplemental appropriations measures. The primary purpose of these payments was to supplement dairy farmer income in response to volatile farm milk prices. Dairy farmers received supplemental payments of $200 million provided by the Omnibus Consolidated and Emergency Supplemental Appropriations Act, 1999 (P.L. 105-277) in DMLA-I; $125 million from the FY2000 agriculture appropriations act (P.L. 106-78) in DMLA-II; and $675 million in emergency provisions in the FY2001 agriculture appropriations act (P.L. 106-387) in DMLA-III.

Dairy Price Support Program — The federal program that maintains a minimum farm price for milk used in the manufacture of dairy products. USDA indirectly assures a minimum price for milk by purchasing any cheddar cheese, nonfat dry milk, and butter offered to it by dairy processors at stated prices. These purchase prices are set high enough to enable dairy processors to pay farmers at least the support price for the milk they use in manufacturing these products. The 2002 farm bill (P.L. 107-171, Sec. 1501) mandated a support price of $9.90/cwt, effective through December 31, 2007, when the program by law is scheduled to expire. Also, the farm bill established a Milk Income Loss Contract (MILC) program that makes direct payments to participating dairy farmers whenever the minimum monthly market price for farm milk used for fluid consumption in Boston falls below $16.94 per hundredweight (cwt.). The MILC program expires September 30, 2005.

Dairy Promotion Program — The Dairy Production Stabilization Act of 1983 (P.L. 98-180, Title I) authorized a national producer program for dairy product promotion, research, and nutrition education as part of a comprehensive strategy to increase human consumption of milk and dairy products and to reduce dairy surpluses. Dairy farmers fund this self-help program through a mandatory 15¢/cwt. assessment on all milk produced and marketed commercially in the 48 contiguous states. Dairy farmers can direct up to 10¢ of this assessment for contributions to qualified regional, state or local dairy product promotion, research or nutrition education programs. The national program is administered by the National Dairy Promotion and Research Board (Dairy Board), a group of 36 dairy farmers appointed by the Secretary of Agriculture to staggered 3-year terms. This program should not be confused with a separate, processor-funded National Fluid Milk Promotion Program. [www.ams.usda.gov/dairy/dairyrp.htm].

Dairy Termination Program — Also called the whole herd buy-out, this program was authorized by the Food Security Act of 1985 (P.L. 99-198). Under it, farmers received USDA payments for agreeing to remove their entire dairy herds from production for at least 5 years.

Data call-in — A part of the EPA Office of Pesticide Programs (OPP) process of developing key required test data, especially on the long-term, chronic effects of existing pesticides, in advance of scheduled Registration Standard reviews. Data call-in from manufacturers is an adjunct of the registration standards program intended to expedite reregistration.

DDT — The abbreviated name of a chlorinated hydrocarbon insecticide, dichloro-diphenyl-trichloromethane. It is persistent in the environment and biomagnifies in birds of prey. The EPA canceled U.S. registration of virtually all but emergency uses of DDT in 1972.

***De minimis* exemption** — Under the Uruguay Round Agreement on Agriculture, countries promised to abide by annual limits on their trade-distorting domestic farm subsidies (for the United States, currently $19.1 billion per year). When countries calculate these subsidies, they do not have to count (toward the annual overall limit) those whose total value is equal to or less than 5% of the value of production (the exclusion is 10% for developing countries). If the group of subsidies is commodity-specific, the 5% is applied to the annual production value of that particular commodity; if the subsidies are non-commodity specific, the 5% is applied to the country's annual value of all agricultural production. Negotiators created this *de minimis* exemption based on the assumption that even if a subsidy (or group of subsidies) fits the definition of a trade-distorting one, it is not likely to actually distort trade if its value is relatively low.

Dead zone / hypoxic zone — An area in the Gulf of Mexico off the mouth of the Mississippi River covering about 6,000 square miles where there is not enough oxygen to support fish and shellfish populations. The oxygen depletion is caused by an excessive amount of nutrients that are washed downstream from throughout the Mississippi River watershed. Many of these nutrients are believed to originate from agricultural activities, and the largest portion, over 30%, has been traced to the upper Mississippi drainage, according to research prepared by the U.S. Geological Survey.

Debt-asset ratio — A financial ratio that measures solvency by calculating the percentage of a farm operator's assets that are financed by debt. For example, a ratio of 0.4 means that for every $100 of assets the operator has $40 of debt. The ratio indicates to a lender the degree of security of a loan. Higher values indicate greater risk. Although a safe or acceptable level varies greatly by enterprise, a debt-asset ratio in excess of 0.40 may indicate financial stress. A ratio of 0 means that the operator owes no debt; a ratio greater than 1 means that the borrower's debts exceed the value of assets, indicating the insolvency of the farm business. At the worst point of the credit crisis of the 1980s, the overall farm debt-asset ratio reached 0.237, compared to an average of 0.159 for the decade of the 1990s.

Decouple — The concept of separating federal farm payments from the requirement that farmers produce specific crops and/or divert land from the production of specific crops. A chief goal of decoupling is to remove a seemingly inherent contradiction in traditional policy: asking farmers to reduce production, while implicitly encouraging more output by tying their benefits to each unit produced. The decoupling concept was first introduced during debate over policy options in the 1985 omnibus farm bill, and was effectively implemented by policy changes made by the 1996 farm bill (P.L. 104-127). Under the 2002 farm bill (P.L. 101-171, Title I), direct payments are decoupled from both production and prices, counter-cyclical payments are linked to market prices, and marketing loan program benefits are linked to both production and prices.

Deductions (food stamps) — For the food stamp program, used to calculate a household's monthly food stamp benefit. This calculation is done by subtracting a series of dollar deductions from the household's total cash monthly income, so as to better reflect the amount of money the household has available for food spending. Food stamp deductions include: (1) a standard deduction that is subtracted from income for all recipients, (2) an earned income deduction reflecting taxes and work expenses, (3) a deduction for dependent care expenses related to work or training (up to certain limits), (4) a deduction for child support payments, (5) a deduction for medical expenses above a set amount per month (only available to elderly and disabled recipients), and (6) a deduction for excessively high shelter expenses.

Deferred pricing — A cash forward contract that provides for determining price by formula at a later date. This also may be called "booking the basis," when the formula sets price relative to a futures price.

Deficiency payments — Direct government payments made to farmers who participated in annual commodity programs for wheat, feed grains, rice, or cotton, prior to 1996. The crop-specific deficiency payment rate was based on the difference between the legislatively set target price and the lower national average market price during a specified time. The total payment was equal to the payment rate, multiplied by a farm's eligible payment acreage and the program payment yield established for the particular farm. In the latter years of the program, farmers could receive up to one-half of their projected deficiency payments at program signup. If actual deficiency payments, which were determined after the crop year, were less than advance deficiency payments, the farmer was required to reimburse the government for the difference, except for zero, 50/85-92 payments. The 1996 farm bill (P.L. 107-171) eliminated deficiency payments and replaced them with production flexibility contract payments. The 2002 farm bill (P.L. 101-171, Sec. 1104) reinstituted deficiency payments as counter-cyclical payments with somewhat different payment calculations.

Defoliant — A chemical that removes leaves from trees and growing plants. Defoliants are regulated by EPA as pesticides.

DEIP — Dairy Export Incentive Program.

Delaney Clause — The Delaney Clause in the Federal Food, Drug, and Cosmetic Act (FFDCA; 21 U.S.C. 321 *et seq.*) states that no additive shall be deemed to be safe for human food if it is found to induce cancer in man or animals. It is an example of the zero tolerance concept in food safety policy. The Delaney prohibition appears in three separate parts of the FFDCA: Section 409 on food additives; Section 512, relating to animal drugs in meat and poultry; and Section 721 on color additives. The Section 409 prohibition applied to many pesticide residues until enactment of the Food Quality Protection Act of 1996 (P.L. 104-170, Sec. 404). This legislation removed pesticide residue tolerances from Delaney Clause constraints.

Delayed pricing — A type of deferred pricing that provides for transfer of title before the price is determined and final settlement made. Contracts including this feature are sometimes called price-later contracts.

Delivery month — The specified month within which a futures contract matures and can be settled by delivery. Also referred to as contract month.

Delivery point — A location where a commodity can be delivered to fulfill a futures contract.

Delivery — In settlement of a futures contract, the tender and receipt of the actual commodity, the cash value of the commodity, or of a delivery instrument covering the commodity (e.g., warehouse receipts or shipping certificates). Futures contracts may be settled by delivery, but more often they are settled by offset or cash. Each futures exchange has specific procedures for delivery of a commodity.

Demurrage — The charge that a shipper may be required to pay for detaining a rail car (or water carrier) longer than necessary to load it. What length of time is considered reasonable, and the level of demurrage charges, are frequently points of dispute between agricultural shippers and the railroads, particularly in proceedings before federal or state transportation regulatory bodies.

Department of Agriculture (USDA) — USDA was originally established in 1862 and raised to cabinet status in 1889. In FY2004 it had a workforce equal to 111,501 staff years, working in some 28 separate agencies, carrying out program activities valued at $110.886 billion, with net federal budgetary outlays of $71.769 billion. Forestry, natural resource, and farm activities utilized 50% of the staff time. However, about 63% of USDA expenditures went to domestic food assistance programs. Over 90% of the staff are located in local, state, and regional field offices away from the Washington, DC, headquarters. Approximately 70% of USDA spending is classified as mandatory spending, which by definition is not constrained by the annual appropriations process. Eligibility for mandatory programs is written into law; any individual or entity that meets the eligibility requirements is entitled to a payment as authorized by the law. The vast majority of mandatory spending is in the Food Stamp Program and certain other food and nutrition programs, the farm commodity programs, the crop insurance program, and the Conservation Reserve Program. The other roughly 30% of USDA budget is classified as discretionary and is subject to annual appropriations, including rural development, agricultural research and education, agricultural credit, international food aid, food marketing and inspection, forestry, and certain nutrition programs. All USDA discretionary programs are funded through an annual Agriculture, Rural Development, Food and Drug Administration, and Related Agencies Appropriations Act (except the Forest Service, which is funded through the Department of the Interior appropriations act). Annual appropriations are made to the food stamp and other mandatory nutrition programs based on estimated spending needs. However, supplemental appropriations are generally made if and when these estimates fall short of required spending. An annual appropriation is made to the Commodity Credit Corporation (CCC), which funds the commodity programs and the Conservation Reserve Program, in order to cover its past net realized losses. Most, but not all, USDA programs are under the congressional authorizing jurisdiction of the House Committee on Agriculture and the Senate Committee on Agriculture, Nutrition, and Forestry. [www.usda.gov].

Department of Agriculture Reorganization Act of 1994 — Title II of P.L. 103-354 was designated the Department of Agriculture Reorganization Act of 1994 and gave the Secretary of Agriculture broad authority to reorganize USDA to achieve greater efficiency, effectiveness, and economy. The law called for consolidation of agencies and offices, as well as a reduction in personnel of 7,500 by the end of FY1999.

Department of the Interior (DOI) — This cabinet-level agency, also known as the Home Department, was created on March 3, 1849 by an Act of Congress to oversee and manage the vast national or public domain. Today this Department includes eight bureaus whose functions include managing National Parks and National Wildlife Refuges, collecting revenues from mineral resources, and operating programs developing, protecting, and enhancing resources on the Nation's public lands. [www.doi.gov].

Dermal toxicity — The ability of a pesticide or other chemical to poison people or animals via skin contact. Many organophosphate pesticides exhibit high dermal toxicity.

DES — Diethylstilbestrol.

Desert Terminal Lakes Program — This program, enacted in of the 2002 farm bill (P.L. 107-171, Sec. 2507), authorizes the transfer of $200 million from the Commodity Credit Corporation (CCC) to the Bureau of Reclamation to provide water to "at risk natural desert terminal lakes." The Bureau cannot use these funds to purchase or lease water rights.

Desiccant — A chemical agent that absorbs moisture; desiccants can be used to control certain insect pests or mildew, and also to dry foliage before harvest (as with potatoes). Desiccants are regulated as pesticides under the Federal Insecticide, Fungicide, and Rodenticide Act (FIFRA; 7 U.S.C. 136 *et seq.*).

Designated uses of water — Water uses identified in state water quality standards that must be achieved and maintained as required under the Clean Water Act (P.L. 92-500, as amended; 33 U.S.C. 1251-1387). Uses can include cold water fisheries, public water supply, irrigation, etc.

Desired future condition — Desired future condition of federal forests and rangeland resources is based on ecological, social, and economic considerations during the land and resource management planning process. Desired future condition is usually expressed as ecological status or management status of vegetation and desired soil qualities.

Desired plant community — The plant community that has been determined through a land use or management plan to best meet the plan's objectives for a site. A desired plant community is consistent with the site's capability to produce the required resource attributes through natural succession, management intervention, or a combination of both.

Developed areas — A classification in the natural resources inventory, previously called urban and built-up areas, that includes cities, villages, other built-up areas of more than 10 acres, industrial sites, railroad yards, cemeteries, airports, golf courses, shooting ranges, institutional and public administration sites, and similar areas. The 1997 national resources inventory placed over 98 million acres in this category, an increase of 25 million acres since 1982.

Developed countries — A term to differentiate the more highly industrialized nations, including most of those that are members of the Organization for Economic Cooperation and Development, from developing countries or less developed countries.

Developing countries — Countries with a low per capita income. Terms such as less developed country, least developed country, underdeveloped country, poor, southern or third world have been used to describe developing countries. The Agricultural Trade Development Assistance Act of 1954 (P.L. 83-480) defines developing country, for purposes of receiving U.S. food aid, as "a country that has a shortage of foreign exchange earnings and has difficulty meeting all of its food needs through commercial channels" (Section 402(4), 7. U.S.C. 1732). Under the Uruguay Round Agreement on Agriculture, the World Trade Organization (WTO) accords special and differential treatment to developing countries, i.e., longer periods of time (and for the least developed countries, exemptions) for phasing in required reductions in tariffs, export subsidies, and trade-distorting domestic support. The WTO allows considerable latitude to countries to designate themselves as developing for purposes of meeting WTO obligations. However, a country's status as developed or developing can become an issue if a country is applying for membership in the WTO. China, for example, as a candidate for WTO membership, argued that it should be considered a developing country and should therefore be given special and differential treatment with respect to implementing WTO rules and disciplines. The United States, the European Union, Japan, and other WTO developed country members argued that China was too important a presence in world agriculture to be admitted to the WTO as a developing country.

Development easement — A legal agreement by which a landowner surrenders the right to develop a designated parcel of property. Some local and state governments have programs to acquire development easements from private landowners to prevent conversion of farmland to other uses.

Dietary Guidelines for Americans — Dietary recommendations for healthy Americans age 2 and older about food choices that promote health, specifically with respect to prevention or delay of chronic diseases. These guidelines, which were first published in 1980, are issued every five years by USDA/HHS to reflect the latest nutrition knowledge. Currently, there are seven messages that concern eating a variety of foods, balancing the foods eaten with physical activity, choosing plenty of grains, fruits and vegetables, choosing a diet low in fats, cholesterol, salt and sugar and limiting alcoholic beverages. The food guide pyramid is a practical consumer guide to using the dietary guidelines. [www.health.gov/dietaryguidelines].

Dietary supplements — Defined by law as a vitamin, mineral, herb, botanical, an amino acid, a dietary substance added to supplement the diet to increase the total dietary intake, or concentrate, metabolite, constituent, extract or combination of any ingredients described above. Supplements cannot be represented for use as a conventional food or sole item of a meal or the diet; and they must be labeled as dietary supplements. Regulatory authority rests with the Food and Drug Administration. There are concerns about the degree of regulation by FDA for safety and claims made for the products by manufacturers.

Diethylstibestrol (DES) — A synthetic estrogen hormone. DES was used widely in the United States as a growth promoter in cattle and sheep and as a treatment for estrogen-deficiency disorders in veterinary medicine, and for postcoital contraception. However, because of its carcinogenic properties and other adverse effects, the Food and Drug Administration (FDA) has revoked all use of it in food-producing animals.

Differential (preferential) assessment — Basing the valuation of farmland for property tax purposes on its use for farming rather than on its highest and best use, such as potential development. Every state has some form of differential assessment for agricultural lands. Types of differential assessment include preferential assessment, deferred taxation, and restrictive agreement. Often the lower tax rate is offered for a specified time period. In those instances, if the land is developed within that time period, the beneficiary must repay some portion of the accrued tax reduction. At least 2 states, Wisconsin and Michigan, provide the preferential benefit to farms through state income tax reductions rather than property taxes.

Digital divide — The "digital divide" refers generally to the difference between many rural and urban areas, rich and poor countries, high-income versus low-income families, and racial/ethnic disparities in terms of the availability of advanced telecommunications technologies, e.g., broadband. Advanced telecommunication technologies are increasingly regarded as important infrastructure for economic development in rural areas, much as electricity and telephony were in earlier years. Because of relatively low population densities in many rural areas, private provision of broadband and other advanced telecommunications can be more costly on a per-capita basis. While many rural areas have achieved rough parity in computer use (e.g., Internet services) with urban areas in many telecommunication technologies, rural-urban disparities persist. Rural areas, for example, are now lagging behind central cities and urban areas in broadband penetration at 7.3%, compared to 12.2% and 11.8%, respectively. At almost every income level, those households in rural areas are less likely to own computers than households in urban or central city areas. At every income level, households in rural areas are significantly less likely to have home Internet access than those in urban or central city areas. Black households in rural areas are one-third less likely to own a computer than the average U.S. Black household, and are two-fifths less likely to access the Internet than the average U.S. Black household. For rural areas, the Kindergarten-12th grade school is a popular point of Internet access: 30.0% of rural persons use the school for Internet access outside the home, compared to a national average of 21.8% (data source: U.S. Department of Commerce, National Telecommunications and Information Administration).

Dioxin — Any of a group of toxic chlorinated compounds known chemically as dibenzo-p-dioxins (or the most toxic of these compounds, 2,3,7,8 tetrachloro dibenzo-p-dioxin). They are produced inadvertently as a by-product of chemical production or combustion and are widespread pollutants in the environment.

Direct export credit — See GSM-5, Direct Export Credit Program.

Direct marketing; farmer-to-consumer direct marketing — Any arrangement where the producer (or representative) is selling products directly to the end user. Typical forms of direct marketing are farmers' markets and roadside stands. Other examples include subscription farming or community supported agriculture and regular delivery of fresh farm produce by farmers to homes or restaurants.

Direct payments — Generally, payments (usually in cash but sometimes in commodity certificates) made directly to producers in conjunction with participation in commodity support programs, conservation programs, and disaster assistance programs. Under the 2002 farm bill (P.L. 101-171, Title I), producers of covered commodities and peanuts are eligible to receive fixed, decoupled direct payments (replacing what previously were called contract payments), counter-cyclical payments, and loan deficiency payments (or marketing loan gains). Several other loan commodities also are eligible for loan deficiency payments.

Direct Payments Program (DP or DPP) — Under the Direct and Counter-cyclical Program (DCP) created by the 2002 farm bill for covered commodities and peanuts (P.L. 101-171, Sec. 1101-1108 and 1301-1310), fixed, decoupled direct payments are one element and counter-cyclical payments (CCP) are the other element. Direct payments replaced what previously were called contract payments under the 1996 farm bill. Counter-cyclical payments are analogous to target price deficiency payments (operational after the 1973 farm bill but discontinued by the 1996 farm bill).

Disaster payments — Direct federal payments provided to crop producers when either planting is prevented or crop yields are abnormally low because of adverse weather and related conditions. Between 1988 and 2005, *ad hoc* disaster legislation was enacted for each crop year, providing a total of nearly $20 billion in direct disaster payments to farmers. These payments were made both to producers with crop insurance and those without insurance.

Discharge — In water resources, the term refers to the flow of surface water in a stream or canal or the outflow of ground water from a flowing artisian well, ditch, or spring. In environmental protection, the term is used synonymously with effluent or emission as a term of point source pollution release.

Discretionary spending — The portion of federal spending (other than appropriated entitlements) that is provided in annual appropriations acts. Approximately one-fourth of USDA spending is considered discretionary, and includes agricultural research, rural development, agricultural credit, marketing and regulatory programs, meat and poultry inspection, foreign food aid, and certain conservation and natural resource programs, among other programs.

Disease vectors — Plants or animals that harbor and may transmit pests and diseases to crops or livestock.

Dispute settlement process — As related to trade, this is the process administered by the WTO's Dispute Settlement Body to examine and judge violations of agreed upon trade rules. A dispute arises when one country takes a trade policy measure or some action that one or more other World Trade Organization (WTO) members considers to be breaking the WTO agreements or to be a failure to live up to obligations. Steps and approximate timetables include consultation (60 days); establishment of a dispute panel (45 days); final panel report to parties in the dispute (6 months); final panel report to WTO members (3 weeks); adoption of report by the Dispute Settlement Body (DSB) (60 days). Without appeal, the process would take about one year. With appeal, an additional 3 months would be required. If the country targeted by the complaint loses, it must follow the recommendations of the panel or appeals report and must state its intention to do so at a DSB meeting held within 30 days of the report's adoption. A country can be given a reasonable period of time in which to comply. If the losing country fails to implement the panel report, it has to enter into negotiations with the complaining country(ies) to determine an agreed upon level of compensation, e.g., tariff reductions of interest to the complaining side. If after 20 days, no satisfactory compensation has been agreed upon, the complaining side may ask the DSB for permission to impose limited trade sanctions (suspend concessions or obligations) against the loser. The DSB grants authorization within 30 days after the expiration of the reasonable period of time unless there is a consensus against the request for sanctions.

Dispute Settlement Body (DSB) — A separate entity of the World Trade Organization, composed of all member countries, established to administer dispute settlement rules and procedures. It is responsible for adjudicating disputes arising under various multilateral trade agreements. The DSB has authority to establish panels, adopt panel and Appellate Body reports, maintain surveillance of implementation of rulings and recommendations, and authorize suspension of concessions or other obligations under the various agreements.

Dissolved oxygen (DO) — The oxygen freely available in water, vital to fish and other aquatic life and necessary for the prevention of odors in water. DO levels are a critical indicator of a waterbody's ability to support desirable aquatic life. Secondary and advanced wastewater treatments are generally designed to ensure adequate DO in waste-receiving waters by removing, digesting, or oxidizing oxygen-demanding wastes (see Biological oxygen demand).

Distance Learning and Telemedicine Grant and Loan Program (DLT) — A program authorized by the 1990 farm bill (P.L. 101-624) to provide grants to rural schools and health care providers to help them invest in telecommunications facilities and equipment to bring educational and medical resources to rural areas where the services otherwise might be unavailable. The 1996 farm bill (P.L. 104-127) reauthorized and streamlined the program. The program was also reauthorized in the 2002 farm bill (P.L. 107-171, Sec. 6203). DLT is administered by the Rural Utilities Service. [www.rurdev.usda.gov/rus].

Diversion payments — Payments once but no longer made to farmers who voluntarily reduced their planted acreage of a program crop and devoted the land to a conservation use when a paid acreage diversion was in effect. Also, payments made to dairy producers in the late 1980s under the no longer operating dairy termination program who agreed to reduce their milk marketings below a prescribed level.

DMLA — Dairy Market Loss Assistance Program.

DNA — Deoxyribonucleic acid.

DO — Dissolved oxygen.

Dockage — A factor in the grading of some grains under the official U.S. Grain Standards. Wheat dockage is described as "weed seeds, weed stems, chaff, straw, or grain other than wheat, which can be readily removed from the wheat by the use of appropriate sieves and cleaning devices; also, underdeveloped, shriveled and small pieces of wheat kernels removed in properly separating, properly rescreening, or recleaning." The term also may be used to describe the amount of reduction in price taken because of a deficiency in quality.

Doctrine of prior appropriation — See Prior appropriation.

Doha Development Agenda — The name given to the current round of multilateral trade negotiations under the auspices of the WTO. The name derives from the launch of a new round of multilateral trade negotiations (MTNs) at a WTO ministerial conference held in Doha, Qatar, in November 2001 The name derives also from the stated aim of the negotiations to integrate more fully the developing countries into the WTO.

DOI — Department of the Interior. [www.doi.gov].

Domestic farm labor — Individuals (and the family) who receive a substantial portion of their income from the production or handling of agricultural or aquacultural products. Farm owners and others may be eligible for Section 514 loans to make housing available for domestic farm labor. For purposes of housing loans, the farm laborers must be U.S. citizens or legally admitted for permanent residence in the United States. The term includes retired or disabled persons who were domestic farm labor at the time of retiring or becoming disabled.

Domestic price — The price at which a commodity trades within a country, in contrast to the world price. For those commodities not benefitting from some form of price support, the domestic price is determined by supply and demand. For commodities that receive price support, the domestic price is usually set by the loan rate or some comparable support level that serves as a price floor in the marketplace working in conjunction with any import quota that may be in effect.

Double cropping — The practice of consecutively producing two crops of either like or unlike commodities on the same land within the same year. An example of double cropping might be to harvest a wheat crop by early summer and then plant soybeans on that acreage for harvest in the fall. This practice is only possible in regions with long growing seasons.

Downer (or downed animals) — Commonly used term for animals that are disabled (nonambulatory) due to illness or injury. A longstanding issue is whether these animals are treated humanely or inhumanely by shippers, stockyards, and packers while they are being moved or held for slaughter. Legislation periodically is introduced in Congress to outlaw the sale or transfer of such animals, but livestock producer groups (who generally agree that livestock markets should not accept severely disabled animals) have long contended that their voluntary efforts to end harmful practices have already proven successful. The 2002 farm bill (P.L. 107-171, Sec. 10815) required USDA to investigate and submit a report on nonambulatory livestock, and to issue and enforce regulations, if deemed necessary, to provide for their humane treatment. The downer issue took on another dimension when bovine spongiform encephalopathy (BSE) was found in two North American cows in 2003. Many experts believe that nonambulatory cattle are at higher risk of harboring BSE; as a result, USDA on December 30, 2003, announced an immediate ban on the slaughter of downer cattle for human food use. At issue is how to monitor the hundreds of thousands of U.S. downer cattle (annual estimate) if they are no longer presented for observation and testing at meat packing plants; also at issue is what to do with their remains, and whether they have any non-food value.

DPP — Direct Payments Program

DPSP — Dairy price support program.

Drainage basin — The area of land that drains water, sediment, and dissolved materials to a common outlet or conveyance at some point along a stream channel. Another term describing the same physical setting is watershed.

Drainage wells — Wells drilled to carry excess water off agricultural fields. Because they act as a funnel from the surface to the groundwater below, drainage wells can contribute to groundwater pollution.

Drainage — Improving the productivity of agricultural land by removing excess water from the soil faster or in greater volume than under natural conditions by such means as ditches, drainage wells, or subsurface drainage tiles. See Swampbuster, and Wetlands.

Dryland farming — A system of producing crops in semi-arid regions (usually with less than 20 inches of annual rainfall) without irrigation. Dryland farmers often try to rebuild soil moisture by leaving the land unplanted or mulched in alternate years, called fallow cropland or summer fallowing.

DSB — Dispute Settlement Body.

DTP — Dairy Termination Program.

Dumping — Under World Trade Organization rules, dumping occurs when the price to the importer is less than the normal price of the product charged to the buyer in the country of origin or other 3rd country markets. Countervailing duties are permitted under the World Trade Organization's Agreement on Subsidies and Countervailing Measures. When considering the imposition of an antidumping duty, the U.S. government examines the imported price of a product compared to its domestic price. In addition, before duties are imposed, injury or threat of injury to a U.S. industry must be determined.

Durum wheat — A species of wheat distinct from wheat used to make bread and other bakery products. The hard, flinty kernels of durum wheat are specially ground and refined to obtain semolina, a granular product used in making pasta items such as macaroni and spaghetti. Most durum wheats are grown in Mediterranean countries, the former Soviet Union countries, North America, and Argentina. U.S. durum production is centered in North Dakota, with other producing states being South Dakota, Minnesota, Montana, California, and Arizona.

Duty, import — An import (customs) duty is a charge assessed by a government on an imported item at its point of customs entry into the country, and paid for by the importer;

the term is now used interchangeably with tariff. In terms of assessing duties there are two basic types: an ad valorem duty is assessed in proportion to the value of the imported item, whereas a specific duty is assessed on the basis of a measure other than value, such as the quantity of the product imported. In addition, a compound or mixed duty, which is a combination of an ad valorem and specific duty, is occasionally used in the Harmonized Tariff Schedules of the United States (HTSUS). Special duties such as anti-dumping duties or countervailing duties may also be levied on imports to offset the unfair price advantage of an imported article that is sold below normal value or subsidized by an exporting country.

***E. coli* O157:H7 (*Escherichia coli* O157:H7)** — A particularly virulent and dangerous strain of *E. coli* bacteria that lives harmlessly in the intestines of animals such as cattle, reptiles, and birds. It can be transmitted to humans through fecal contamination on food, causing bloody diarrhea and possibly leading to hemolytic uremic syndrome (HUS), a life threatening disease. *E. coli* O157:H7 has been implicated in several major outbreaks of foodborne illness in recent years. After an outbreak traced to undercooked hamburgers resulted in hundreds of illnesses and several deaths in 1993, USDA began regularly testing samples of ground beef for the pathogen.

EA — Environmental Assessment.

Earmark — Any designation in an annual appropriations bill or its accompanying committee report, which allocates a portion of the appropriation for a specific project, location or institution. In the annual appropriations bill that funds USDA and related agencies, the vast majority of the earmarks are contained within the two largest research agencies within USDA, the Cooperative State, Research, Education and Extension Service (CSREES) and the Agriculture Research Service (ARS). A number of earmarks also can be found in accounts within the Natural Resources Conservation Service (NRCS), the Animal and Plant Health Inspection Service (APHIS), and various rural development programs.

Easement — A landowner sells or surrenders a right to a portion of the property for some purpose, usually in return for a payment or some other benefit. In agriculture, conservation easements receive the most attention, but easements are more common for other purposes, including access and utility strips. Easements may be permanent or temporary (for some specified period of time). Easements are registered as part of the deed for the property. Some local and state governments, and land trusts and other non-governmental organizations, have programs to acquire development easements from landowners to prevent conversion of farmland to other uses.

EBT — Electronic benefit transfer.

EC — European Communities.

ECARP — Environmental Conservation Acreage Reserve Program (replaced by the Comprehensive Conservation Enhancement Program (CCHP).

Economic Research Service (ERS) — USDA's in-house agricultural economics analysis and research agency. It employs about 600 people and has an annual budget of about $53 million. [www.ers.usda.gov].

Economies of scale — See Economies of size.

Economies of size — The concept that the average cost of production per unit declines as the size of the operation grows. One reason farms have been growing in size is to make more economical use of machines capable of covering more ground with less labor, to capture economies of size. Larger sized farms can typically get volume discounts on such inputs as chemicals and seed. They also are more likely to be more vertically coordinated with buyers and processors.

Ecosystem — A functioning community of nature that includes fauna and flora together with the chemical and physical environment with which they interact. Ecosystems vary greatly in size and characteristics; an ecosystem can be a mud puddle, a field or orchard, or a forest. An ecosystem provides a unit of biological study and can be a unit of management.

ECP — Emergency Conservation Program.

EDF — Environmental Defense Fund. [www.environmentaldefense.org].

Edward R. Madigan U.S. Agricultural Export Excellence Award — An award established by the 1996 farm bill (P.L. 104-127) to recognize companies' and other entities' entrepreneurial efforts in the food and agricultural sector for advancing U.S. agricultural exports.

EEP — Export Enhancement Program.

EFAP — Emergency Feed Assistance Program (see Emergency livestock feed programs).

EFAP — Emergency Food Assistance Program (formerly the Temporary Emergency Food Assistance Program, or TEFAP). See Emergency Food Assistance, and Soup Kitchen-food Bank Program.

Effective price — Under the 2002 farm bill (P.L. 101-171, Sec. 1104), for crop years 2002 through 2007, counter-cyclical payments are made when the effective price of a covered commodity is below the target price. The effective price is the sum of: 1) either a) the average price received by farmers over the marketing year, or b) the national average loan rate, whichever is higher; and 2) the payment rate in effect for making fixed direct payments. The difference between the higher target price and the lower effective price is the payment rate. The payment amount to a producer is the product of the payment rate, the payment acres on the farm, and the payment yield of the farm.

Effluent guidelines for feedlots and other point sources — These are national regulations that set numerical or narrative limits for specific pollutants in wastewater from various types of point sources, including concentrated animal feeding operations, which specify "zero discharge except as a result of storms exceeding catastrophic or chronic rainfall." The limits are based on the application of specific processes or treatment technologies within a given industrial category (e.g., food processing, flour mills, etc.), although dischargers may meet their requirements using whatever combination of treatment technologies and process changes they choose. Since guidelines were first issued in 1974, EPA has promulgated limitations and standards for 51 industrial categories. On December 12, 2002, EPA released a final rule on concentrated animal feeding operations, which includes effluent limit guidelines (68 Federal Register 7176-7274).

Effluent limitation — An EPA "standard of performance reflecting the maximum degree of discharge reduction achievable by the best available technology for various categories of sources of water pollution. These categories include feedlots, grain mills, and several kinds of food processing.

Effluent — Waste, usually liquid, released or discharged to the environment. Generally the term refers to point source discharges of sewage or contaminated waste waters into surface waters.

EFNEP — Expanded Food and Nutrition Education Program.

EFP — Emergency Feed Program (see Emergency livestock feed programs).

EI — Erosion index.

EIA — Environmental impact assessment; economic impact assessment.

EIP — Export Incentive Program.

EIS — Environmental impact statement.

Elasticity — See Price elasticity of demand.

Elderly and disabled — For food stamp purposes, elderly persons are age 60 or older; and disabled persons are beneficiaries of disability-based governmental assistance, such as social security disability payments and certain veterans disability payments.

Electronic benefit transfer (EBT) systems — Under an EBT system, food stamp recipients are issued an ATM-like card and a personal identification number (PIN) instead of food stamp coupons. They access their food stamp benefits when purchasing food by using the card at an approved retailer: swiping the card through a point-of-sale terminal and entering their PIN. This electronically debits a food stamp account maintained for them (replenished monthly) and credits the retailer with the purchase amount.

Elevator — A tall warehouse facility that uses vertical conveyors to raise or elevate grain, generally owned privately or by an agricultural cooperative, where grain is stored before being marketed. The term elevator often refers to any grain storage facility, even if the grain is not elevated. The county elevator is where a farmer delivers grain; a terminal elevator is a major transshipment facility; while an export elevator is at a port facility.

ELISA — Enzyme linked immunosorbant assay (test).

ELS cotton — The abbreviation for extra-long staple cotton.

ELS — Extra-long staple (cotton).

EM — Emergency disaster loans.

EMAP — Environmental Monitoring and Assessment Program.

Embargo — A government-ordered prohibition or limitation on trade with another country. Under an embargo, all trade, or selected goods and services, may be restricted. The 1985 farm bill (P.L. 99-198) requires that any prohibition or limitation on such exports should be imposed only when the President declares a national economic emergency under the Act and protects prior contracts to export agricultural commodities and products. The 1990 farm bill (P.L. 101-624) contains contract sanctity provisions that place constraints on the embargo of agricultural commodities from the United States. It also provides for compensation to agricultural producers if the President suspends or restricts exports of a commodity for national security or foreign policy reasons, and requires USDA to develop plans to alleviate the adverse effects of embargoes if imposed. The 1996 farm bill (P.L. 104-127) requires compensation to producers if the U.S. government imposes an export embargo for national security or foreign policy reasons, and if no other country joins the U.S. embargo within 90 days. Under this law, compensation may take the form of payments to producers or as increased funding for agricultural export and food aid programs. The Trade Sanctions Reform and Export Enhancement Act (TSRA) of 2000 (P.L. 106-386) codified the lifting of U.S. sanctions on commercial sales of food, agricultural commodities, and medical products to Iran, Libya, North Korea, and Sudan, and extended this policy to apply to Cuba, but with financing and licensing conditions. Those applicable to Cuba are more restrictive than for the other countries, and are permanent under TSRA , which also gives Congress authority to veto a President's proposal to impose sanctions on the sale of agricultural or medical products to a designated country.

Emergency Conservation Program (ECP) — A program administered by the Farm Service Agency to help farmers to rehabilitate farmland damaged by natural disasters by sharing in the cost of rehabilitation. It is almost always funded in supplemental appropriations that provide federal assistance to deal with a natural disaster.

Emergency Disaster (EM) Loan Program — When a county has been declared a disaster area, by either the President or the Secretary of Agriculture, farmers in that county may become eligible for low-interest emergency disaster (EM) loans available through the Farm Service Agency (formerly Farmers Home Administration). EM loan funds may be used to help producers recover from production losses (when the producer suffers a significant loss of an annual crop) or from physical losses (such as repairing or replacing damaged or destroyed structures or equipment, or for the replanting of permanent crops such as orchards). A qualified producer can then borrow up to 80% of the actual production loss or $500,000, whichever is less, at a subsidized interest rate.

Emergency feeding organization — This refers to an organization serving the food needs of the poor and unemployed that is designated by a state as eligible for commodities and administrative support to distribute commodities or operate a meal service program under the Emergency Food Assistance Program (EFAP).

Emergency Food Assistance Act of 1983 — P.L. 98-92 amended the original Temporary Emergency Food Assistance Act (TEFAA) of 1983 (P.L. 98-8) to authorize multi-year funding and commodity donations from excess Commodity Credit Corporation (CCC) inventories of foodstuffs for food distribution by emergency feeding organizations serving the needy and homeless (7 U.S.C. 7501 *et seq.*). It subsequently was amended in 1985, 1988, 1990, 1996 and 2002 (under the 2002 farm bill (P.L. 107-171, Sec. 4126, Sec. 4204). The law authorizes funding through FY2007 to buy and donate commodities and to provide grants for state and local costs of transporting, storing, and distributing them to emergency feeding organizations, soup kitchens, and food banks serving low-income persons. In addition to discretionary funds authorized to be appropriated by this law, there is a requirement that $100 million of food stamp appropriations be used annually to buy commodities for emergency feeding organizations.

Emergency Food Assistance and Soup Kitchen-Food Bank Program (EFAP-Soup kitchens) — This program provides USDA commodities to emergency feeding organizations to help with the food needs of low-income populations. It also authorizes grants to states to help with the state and local costs of transporting, storing, and distributing the commodities. The program is authorized under the Emergency Food Assistance Act of 1983 (P.L. 98-92, as amended; 7 U.S.C. 7501 *et seq.*). In addition to authorizing funding to buy commodities, the program also requires that $100 million of food stamp funds be used annually for that purpose. Eligible agencies include food banks, food pantries, soup kitchens, and public and private charitable agencies serving the poor. States determine the agencies eligible to participate and set low-income standards for eligibility.

Emergency livestock feed programs — The USDA was given permanent authority by the Disaster Assistance Act of 1988 (P.L. 100-387) to implement an array of emergency livestock feed programs. These programs were designed to assist livestock producers who lost a significant amount of feed grown on the farm due to a natural disaster. The primary livestock feed programs implemented by USDA were: (1) the Emergency Feed Assistance Program (EFAP), which provided farmers who experienced a large loss of feed production with government-owned grain at a subsidized price; and, (2) the Emergency Feed Program (EFP), a cost-share program for farmers affected by a disaster who purchased their needed feed in the marketplace. To meet mandated budget savings requirements, the 1996 farm bill (P.L. 104-127) suspended these programs from the law through 2002. However, *ad hoc* livestock assistance programs have been implemented periodically since 1996 to provide emergency assistance to livestock growers, most notably, the Livestock Assistance Program, Livestock Indemnity Program, and a Livestock Compensation Program.

Emergency Watershed Program (EWP) — A program administered by the Natural Resources Conservation Service to respond to floods, fires, windstorms and other types of natural disasters. Types of work this program funds include: removing debris; reshaping and protecting eroded banks; correcting damaged drainage facilities; repairing levees and other water conveyance structures; and purchasing flood plain easements. For construction activities, it provides up to 75% of the project cost. It is almost always funded in supplemental appropriations that provide federal assistance to deal with a natural disaster.

Emergency Wetlands Reserve Program (EWRP) — Authorized in 1993 under emergency supplemental appropriations to respond to widespread floods in the Midwest, EWRP provided payments to purchase easements and partial financial assistance to landowners who permanently restored wetlands at sites where the restoration costs exceeded the land's fair market value. EWRP was administered by Natural Resources Conservation Service as part of its Emergency Watershed Program and operated in seven midwestern states. Land in this program is considered to be a part of the land enrolled in the Wetland Reserve Program.

Emerging market — For purposes of administering USDA's Emerging Markets Program and Facility Credit Guarantee Program, an emerging market is a country that the Secretary of Agriculture determines: (1) is taking steps toward a market-oriented economy through the food, agriculture, or rural business sectors; and, (2) has the potential to provide a viable and significant market for U.S. agricultural commodities or products.

Emerging Markets Program — A program originally authorized by the 1990 farm bill (P.L. 101-624) and titled the Emerging Democracies Program. The program was authorized to promote U.S. agricultural exports by providing technical assistance and credits or credit guarantees to emerging democracies annually for fiscal years 1991-95. Funds could be used to establish or provide facilities, services, or U.S. products to improve handling, marketing, storage, or distribution of imported agricultural products. The program initially focused on central and eastern Europe and the former Soviet Union. The 1996 farm bill (P.L. 104-127) reauthorized the program through 2002 and renamed it the Emerging Markets Program; the 2002 farm bill (P.L. 107-171) extended it through 2007. The program in 1996 was retargeted to emerging markets (defined as countries that USDA determines have the potential to provide viable and significant markets for U.S. agricultural products). The law requires the Commodity Credit Corporation (CCC) to make available not less than $1 billion annually of direct credit or credit guarantees to emerging markets for fiscal years 1996-2002, in addition to the amounts authorized for GSM-102 and 103.

Emission — Waste released or emitted to the environment. The term is commonly used in referring to discharges of gases and particles to the atmosphere (i.e., air pollutants), and also is used in referring to particles or energy released radioactively. Sometimes the term is used broadly, encompassing any polluting discharge.

Empowerment Zone/Enterprise Communities Program (EZ/EC) — A grant-making initiative to revitalize low-income rural communities (and low-income urban areas) in a way that attracts private sector investment. Selection as an EZ/EC is competitive. Criteria for designation as an EZ/EC are based on physical size of the area, various economic indicators of economic distress (pervasive poverty, unemployment, and general socioeconomic distress), and population. Urban sites can be no larger than 20 square miles, but may include up to 2,000 acres in areas called developable sites located outside the 20 square mile limit. Rural EZ sites can encompass an area up to 1,000 square miles. Maximum population of an urban EZ/EC site is the lesser of (1) 200,000 or (2) the greater of 50,000 or 10% of the population of the most populous city located. In short, nothing larger than 200,000 but at least 50,000. For rural EZ/EC sites, population of towns and cities must be 30,000 or less.

EMU — European Monetary Unit.

END — exotic Newcastle disease.

End-use certificate — The North American Free Trade Agreement Implementation Act (P.L. 103-182) mandates end-use requirements for wheat and barley imported from any country requiring end-use certificates for imports of U.S. produced commodities. Canada is the only nation that requires such certificates, and U.S. wheat is the only commodity subject to the restriction. Regulations implementing the End-Use Certificate Program, implemented February 27, 1995, are found at 7 CFR 782.

Endangered species — Species of animals or plants likely to go extinct in the foreseeable future unless current trends are altered. They are listed by regulation under the Endangered Species Act (P.L. 93-205, as amended) and assigned the Act's highest level of protection. Only scientific factors may be taken into account in deciding whether to list a species as endangered, though economic factors may be taken into account at other stages of the Act. See also Threatened species. For the legal definition, see Section 3 of the Act.

Endangered Species Act (ESA) of 1973 — P.L. 93-205, as amended (16 U.S.C. 1531 *et seq.*), protects species and the ecosystems on which they depend The ESA is administered primarily by the Fish and Wildlife Service (and by NOAA Fisheries, formerly the National Marine Fisheries Service (NMFS), for certain marine species). These agencies list species of domestic plants and animals at risk of extinction as either "endangered" or "threatened" according to the degree of risk. Over 1,000 species have been listed. The ESA has been controversial because: (1) its standards of protection are substantive rather than procedural, and have occasionally prevented activities that would jeopardize a designated species; or (2) because other laws often lack strict substantive provisions the ESA often becomes a battleground by default over larger controversies concerning resource scarcities or altered ecosystems. How the ESA affects farmers and others depends on the listed species, the locale, the nature and health of the ecosystem, the ownership of the land, etc. On private land, ESA prohibits takings of individuals of protected species and requires agencies providing any federal service (such as permitting, increasing irrigation flows, or loans) to ensure the action will not adversely affect critical habitat.

Endocrine disruptor — A chemical agent that interferes with natural hormones in the body. Hormones are secreted by endocrine glands (such as the pituitary, thyroid, pancreas, ovary, and testis), are transported through the body in the bloodstream, and regulate body growth and metabolism, other endocrine organs, and reproductive functions. There is emerging concern that endocrine disruptors may be causing human health or ecological effects, such as abnormal thyroid function, decreased fertility, and alteration of immune and behavioral function. This concern arises from demonstrated instances (an example is the ability of diethylstilbestrol (DES) to disrupt female reproductive function throughout the lifespan in laboratory animals and humans) and the fact that hormones are biologically active at very low concentrations (at parts per billion or less), so low levels of disruptors may similarly be biologically active. In amendments to the Safe Drinking Water Act (P.L. 104-182) and the Federal Insecticide, Fungicide, and Rodenticide Act in 1996 (P.L. 104-170), Congress directed the EPA to study endocrine disruptors. The outcome of this research will be of consequence to agriculture because some pesticides and animal growth stimulants have been hypothesized to act as endocrine disruptors.

Enlargement — The process of adding new countries to membership in th European Union. Ten countries (Cyprus, the Czech Republic, Estonia, Hungary, Latvia, Lithuania, Malta, Poland, Slovakia, and Slovenia) became members on May 1, 2004. Bulgaria and Romania are slated to complete enlargement negotiations and become member countries in 2007. Croatia and Turkey are due to begin enlargement negotiations in 2005.

Entitlement — A legal obligation on the federal government to make payments to a person, business, or unit of government that meets the criteria set in law. Entitlement spending is a subset of mandatory spending. Most entitlement spending is authorized by laws that provide all eligible individuals with financial assistance or other benefits based upon some formula or criteria set out in law. What distinguishes an entitlement from other mandatory spending is that in an entitlement program, eligible individuals retain a legal right to benefits, regardless of the cost. USDA programs such as the commodity price and income support programs and crop insurance would fall into the category of an appropriated entitlement, since participants are legally entitled to benefits, but an appropriation ultimately is required to the Commodity Credit Corporation (the funding source of the commodity programs) and the Federal Crop Insurance Fund (the funding source of crop insurance subsidies).

Entitlement commodities — Food commodities purchased by the USDA for the purpose of meeting legislatively specified rates of commodity assistance for various food assistance programs (e.g., school lunches). These commodities may be in government holdings as a result of agricultural surplus removal or price support activities carried out under a variety of agriculture laws, but more commonly they are items purchased to meet food program needs.

Environment — The totality of the surrounding external conditions (biological, chemical, and physical) within which an organism, community, or object exists. The environment can be defined at any scale. The term is not exclusive in that organisms can be and usually are part of another organism's environment. Thus one can speak of the environment as that within which humankind lives (i.e., separate and external) or, of humankind as a component of the environment.

Environmental Assessment — Under implementing regulations of the National Environmental Policy Act of 1970 (P.L. 91-190) a document used by agencies to determine whether the environmental effects are sufficient to require an Environmental Impact Statement (EIS).

Environmental Benefits Index — An index that has been used by the FSA since 1990 to rank farmers' requests to enroll land into the Conservation Reserve Program during each general sign-up period. The only enrollment mechanisms that do not use the index are the Conservation Reserve Enhancement Program and continuous enrollment, for which a total of 4 million acres has been reserved. The index, as currently structured, assigns points for cost to the government and 6 other factors; 1) wildlife benefits, up to 100 points; 2) water quality benefits, up to 100 points; 3) on-farm erosion control, up to 100 points; 4) enduring benefits, up to 50 points; 5) air quality benefits, up to 35 points; and 6) in a state or national priority area, 25 points. Bids only are accepted if they exceed a threshold level that is determined after the total value of the benefits that each bid would provide are compared. The index continues to be adjusted from general signup to general signup.

Environmental Conservation Acreage Reserve Program (ECARP) — An umbrella program authorized by the 1990 farm bill (P.L. 101-624) that includes the Conservation Reserve Program, and the Wetland Reserve Program. The 1996 farm bill (P.L. 104-127) continued the CRP and WRP and created the Environmental Quality Incentives Program. The goal of the ECARP was to provide long-term protection of environmentally sensitive land. Contracts, easements, and cost-share payments were used to assist landowners and operators of farms and ranches to conserve and enhance soil, water, and related natural resources, including grazing land, wetland, and wildlife habitat. The 2002 farm bill (P.L. 107-171, Sec. 2006) replaced ECARP with the Comprehensive Conservation Enhancement Program (CCHP).

Environmental equity/justice — Equal protection from environmental hazards for individuals, groups, or communities regardless of race, ethnicity, or economic status. This applies to the development, implementation, and enforcement of environmental laws, regulations, and policies, and implies that no population of people should be forced to shoulder a disproportionate share of adverse impacts of pollution.

Environmental impact statement (EIS) — A document required of federal agencies by the National Environmental Policy Act (P.L. 91-190) for major projects or administration-initiated legislative proposals significantly affecting the environment. A tool for decision making, it describes the positive and negative effects of the undertaking and assesses alternative actions.

Environmental Protection Agency (EPA) — An independent federal government agency established in 1970 and charged with coordinating effective governmental action concerning the environment, including setting standards, promulgating and enforcing regulations, and initiating and implementing environmental programs. Two areas of jurisdiction that most directly affect agricultural production are the registration of pesticides required by the Federal Insecticide, Fungicide, and Rodenticide Act (FIFRA; P.L. 80-104, as amended; 7 U.S.C. 136 *et seq.*) and implementation of the Clean Water Act (P.L. 92-500, as amended; 33 U.S.C. 1251-1387). [www.epa.gov].

Environmental Quality Incentives Program (EQIP) — A program created by the 1996 farm bill (P.L. 104-127) to provide primarily cost-sharing assistance, but also technical and educational assistance, aimed at promoting production and environmental quality, and optimizing environmental benefits. The program replaces the Agricultural Conservation Program, the Water Quality Incentives Program, the Great Plains Conservation Program, and the Colorado River Basin Salinity Control Program. EQIP is reauthorized in the 2002 farm bill (P.L. 107-171, Sec. 1241) at $0.4 billion in mandatory spending in FY2002 and rising to $1.3 billion in FY2007. The funding each year is to be divided, with 60% targeted to environmental concerns associated with livestock production and the remainder to crop production. Producers enter into contracts of 1 to 10 years. Participants can receive no more than $450,000 between FY2002 and FY2007. Two new sub programs were created; one provides matching grants for innovative conservation efforts, such as using market systems to reduce pollution and promoting carbon sequestration in soil; and, the second is the Ground and Surface Water Conservation Program (GSWP).

Enzyme linked immunosorbant assay — A tool for the detection of specific proteins and macromolecules in biologic systems.

EPA — Environmental Protection Agency. [www.epa.gov].

Epidemiology — Study of the distribution of disease, or other health-related conditions and events in human or animal populations, in order to identify health problems and possible causes.

EQIP — Environmental Quality Incentives Program.

Equivalence — A term applied by the Uruguay Round Agreement on the Application of Sanitary and Phytosanitary (SPS) Measures. WTO Member countries shall accord acceptance to the SPS measures of other countries (even if those measures differ from their own or from those used by other Member countries trading in the same product) if the exporting country demonstrates to the importing country that its measures achieve the importer's appropriate level of sanitary and phytosanitary protection.

Erosion (erodibility) index (EI) — The erosion (sometimes called erodibility) index is created by dividing potential erosion (from all sources except gully erosion) by the T value, which is the rate of soil erosion above which long term productivity may be adversely affected. The erodibility index is used in conjunction with conservation compliance and the Conservation Reserve Program. For example, one of the eligibility requirements for the CRP is that land have an EI greater than 8.

Erosion — The wearing away of the land surface. Unconsolidated materials, such as soil, erode more rapidly than consolidated materials, such as rock. Rates of erosion vary widely across the landscape, depending on numerous physical factors, so a large portion of all erosion associated with agriculture is concentrated in several relatively small producing areas, such as the Palouse region in eastern Washington. The two most common causes of erosion are wind and moving water. Water causes sheet, rill, and gully erosion. The total annual erosion rate from all cropland and land enrolled in the CRP was 1.9 billion tons in 1997, a significant decline from 3.1 billion tons in 1982, according to the National Resources Inventory.

ERS — Economic Research Service. [www.ers.usda.gov].

ESA — Endangered Species Act of 1973 (P.L. 93-205; 16 U.S.C. 1531 *et seq.*).

ESA — Environmentally sensitive area.

Escape Clause — See Safeguards, import.

Estuary — Regions of interaction between rivers and near-shore ocean waters, where tidal action and river flow mix fresh and salt water. Such areas include bays, mouths of rivers, salt marshes, and lagoons. Estuaries typically include adjoining wetlands. These brackish water ecosystems shelter and feed marine life, birds, and wildlife, but also are sites where commerce and industry are concentrated. The two major federal programs addressing estuarine resource management problems are EPA's National Estuary Program and NOAA's National Estuarine Research Reserve System.

Ethanol ¯ C_2H_5OH; the alcohol product of carbohydrate fermentation used in alcoholic beverages and for industrial purposes (also known as ethyl alcohol or grain alcohol). It is blended with gasoline to make gasohol. When blended with gasoline, ethanol can reduce emissions of carbon monoxide and other ozone-forming pollutants. Ethanol can be produced from many feedstocks, including plants, agricultural residues, and municipal waste. Corn is the primary feedstock for U.S. ethanol production. In the 2002/03 corn marketing year, about 1,400 million bushels (11.9% of the corn crop and 13.3% of domestic corn use) is expected to go into the production of about 2 billion gallons of ethanol (May 2005estimate). One 56-pound bushel of corn yields about 2.7 gallons of ethanol; conversely, 1 gallon of ethanol requires about 0.4 bushels of corn. Brazil uses sugarcane to manufacture what is called sucrose ethanol.

EU — European Union. [www.eurunion.org].

EUP — Experimental use permit.

Euro — The single currency of the 12 countries that are members of the European Monetary Union, one of the institutions of the European Union. The participating countries are Austria, Belgium, Finland, France, Germany, Greece, Ireland, Italy, Luxembourg, Netherlands, Spain, and Portugal. The exchange rates for the currencies of these 11 countries were irrevocably fixed in terms of the euro on December 31, 1998. The transition to the euro occurred in two phases. For financial transactions, the euro replaced national currencies on January 1, 1999, but for notes and coins, it replaced national currencies in 2002. EU members not in the Monetary Union are Denmark, Sweden, the United Kingdom, and the 10 new member states (Cyprus (Greek part), the Czech Republic, Estonia, Hungary, Latvia, Lithuania, Malta, Poland, Slovakia, and Slovenia).

European Economic Community (EEC) — A regional organization created by the Treaty of Rome (1957) that provided for the gradual elimination of customs duties and other interregional trade barriers, a common external tariff, and gradual adoption of other integrating measures, including the Common Agricultural Policy (CAP), and guarantees of free movement of labor and capital. Of the current 25 member countries, the original six were Belgium, France, West Germany, Italy, Luxembourg, and the Netherlands. Membership expanded to include Denmark, Ireland, and the United Kingdom in 1973; Greece in 1981; Spain and Portugal in 1986; Austria, Finland, and Sweden in 1995; and Cyprus (Greek part), the Czech Republic, Estonia, Hungary, Latvia, Lithuania, Malta, Poland, Slovakia, and Slovenia in 2004. In 1993, with establishment of the European Union (EU), the EEC became the customs union component of the EU.

European Food Safety Authority (EFSA) — An agency charged with providing the European Commission of the European Union with independent scientific advice on all matters with a direct or indirect impact on food safety. It's status is that of a separate legal entity independent from other EU institutions. EFSA was established (by regulation 178/2002/EC) in January 2002. Its work covers all stages of food production and supply, from primary production to the safety of animal feed through to the supply of food to consumers. It collects and analyzes information, carries out risk assessments, and can carry out scientific assessments on any matter bearing on the safety of the food supply including matters relating to animal health, animal welfare, and plant health. EFSA also provides scientific advice on non-food and feed GMOs, and on nutrition in relation to EU legislation. In contrast to the United States' FDA, EFSA is strictly an advisory not a regulatory body. It can, however, communicate directly with the public on any issue within its area of responsibility.

European Monetary Union (EMU) — As agreed in the Maastricht Treaty, 11 (later 12) European Union members began participating in the EMU on January 1, 1999. The 12 countries are Austria, Belgium, Finland, France, Germany, Greece, Ireland, Italy, Luxembourg, the Netherlands, Portugal, and Spain. Under the EMU, the 12 countries have a common central bank (the European Central Bank), and thus a common monetary policy, similar to that of the U.S. Federal Reserve System, and a single currency, called the euro.

European Union (EU) — Since 1993, the term used to describe the European Communities and related institutions. The entry into force of the Maastricht Treaty of European Union on November 1, 1993, introduced this change in terminology regarding the EC and many of its institutions. The EU included 15 member countries prior to its recent enlargement (Austria, Belgium,, Denmark, Finland, France, Germany, Greece, Ireland, Italy, Luxembourg, Netherlands, Portugal, Spain, Sweden, United Kingdom of Great Britain and Northern Ireland). Ten new members joined the EU in May 2004 (Cyprus (Greek part), the Czech Republic, Estonia, Hungary, Latvia, Lithuania, Malta, Poland, Slovakia and Slovenia). Bulgaria and Romania are slated to complete enlargement negotiations and become member countries in 2007. Croatia and Turkey are due to begin enlargement negotiations in 2005.

Eutrophication — The process by which a body of water acquires a high concentration of plant nutrients, especially nitrates or phosphates. This nutrification promotes algae growth that, when it dies, can lead to the depletion of dissolved oxygen, killing fish and other aquatic organisms. While eutrophication is a natural, slow-aging process for a body of water, human activities can greatly accelerate the process.

Evans-Allen funds — Federal funds distributed to the 1890 (historically black) land grant colleges of agriculture under a provision in the National Agricultural Research, Extension, and Teaching Policy Act of 1977 (P.L. 95-113, Title XIV), to support research programs. The provision became known by the names of two of its primary proponents in Congress, Representative Frank Evans of Colorado and Senator James Allen of Alabama. These funds are now more commonly referred to in budget documents as Payments to 1890 colleges.

Evapotranspiration — The loss of water from the soil both by evaporation and by transpiration from the plants growing in the soil.

EWG — Environmental Working Group. [www.ewg.org].

EWP — Emergency Watershed Program.

Excess land — Irrigable land, other than exempt land, owned by any landowner in excess of the maximum acreage limitation (ownership entitlement) under the applicable provision of reclamation law.

Exotic Newcastle disease (END) — A contagious and fatal viral disease that can kill birds without their showing any signs of respiratory, digestive, and nervous system problems. Smuggled pet birds, especially Amazon parrots from Latin America, pose a great risk of introducing the disease into U.S. poultry flocks. This disease is also known as Velogenic Viserotropic Newcastle Disease (VVND). The Animal and Plant Health Inspection Service (APHIS) personnel who are now part of the Department of Homeland Security's (DHS) Border and Transportation Security division are responsible for preventing introduction of the disease; APHIS responds to outbreaks, generally by destroying flocks and making indemnity payments to producers for their losses.

Exotic species — A term used in USDA meat and poultry inspection regulations to mean livestock or fowl not covered by the Federal Meat Inspection Act (21 U.S.C. 601 *et seq.*) or the Poultry Products Inspection Act (21 U.S.C. 451 *et seq.*), and for which federal inspection is not mandatory, even though the animals may be processed for human consumption. Deer, elk, reindeer, antelope, buffalo, bison, rabbit, migratory water fowl, game birds, quail, and pheasants currently are defined as exotic species. They are eligible for voluntary, fee-for-service inspection at federally inspected meat and poultry slaughtering facilities (that have agreed to process them), under the authority of the Agricultural Marketing Act. Slaughtering and processing facilities operating under a cooperative state inspection program also can accept exotic species for the voluntary program (and some states require it for certain species), but the finished product can be marketed only within that state. Any of the above-listed exotic species that have not undergone voluntary inspection, but are sold for human consumption, are under the jurisdiction of the Food and Drug Administration (FDA, under the authority of the Federal Food, Drug, and Cosmetics Act (21 U.S.C. 321 *et seq.*).

Expanded Food and Nutrition Education Program (EFNEP) — EFNEP is a program of the Cooperative Extension System that operates in all 50 states and U.S. territories. Started in 1965, its purpose is to provide low-income individuals, particularly youth and families with young children, with the knowledge, skills, and desire to adopt and maintain a nutritious diet. [www.csrees.usda.gov/nea/food/efnep/efnep.html]

Experimental use permit — A permit under the Federal Insecticide, Fungicide, and Rodenticide Act (7 U.S.C. 136c*)* that authorizes the testing of new pesticides or uses thereof in experimental field studies on 10 acres or more of land or one acre or more of water. Such tests provide data to support registration of pesticides.

Export Administration Act (EAA) of 1979 — P.L. 96-72 provided legal authority to the President to control U.S. exports for reasons of national security, foreign policy, and/or short supply. However, the 1990 farm bill (P.L. 101-624) provided for contract sanctity by prohibiting the President from restricting the export of any agricultural commodity already under contract for delivery within 270 days from the date an embargo is imposed under the EAA, except during national emergency or war. With the expiration of EAA in 1994, the President declared a national emergency and exercised authority under the International Emergency Economic Powers Act (P.L. 95-223; 50 U.S.C. 1701 *et seq.*) to continue the EAA export control regulations then in effect by issuing Executive Order 12924 on August 19, 1994. Presidents since then have each year extended these regulations by Presidential Notice.

Export allocations or quotas — Controls applied to exports by an exporting country to limit the amount of goods leaving that country. Such controls usually are applied in time of war or during some other emergency requiring conservation of domestic supplies, as well as to advance foreign policy and national security objectives of the exporting country. The European Union, in 1996, used a licensing system to allocate and restrict exports of wheat because of short supplies and high prices.

Export Enhancement Program (EEP) — A program that USDA initiated in May 1985 under the Commodity Credit Corporation (CCC) Charter Act to help U.S. exporters meet competitors' subsidized prices in targeted markets. The program currently is authorized through 2007 under the 2002 farm bill (P.L. 107-171). Under EEP, exporters are awarded cash payments that enable an exporter to sell certain commodities to specified countries at competitive prices. EEP program activity is constrained by annual dollar and tonnage limits on commodities that can be subsidized, as agreed to under the Uruguay Round Agreement on Agriculture, and these annual limits are incorporated into the EEP authorizing legislation. In practice, the program has been used very little since the mid-1990s.

Export incentive program — See Market Access Program.

Export PIK — A program used in the 1980s that made payment-in-kind to U.S. exporters as export subsidies for surplus commodities.

Export restitutions — A term in the EU for variable export subsidies given to traders to cover the difference between the higher internal Common Agricultural Policy price of a commodity and its lower world price.

Export subsidy — A direct or indirect compensation provided by government to private commercial firms to promote exports of domestic products. Article XVI of the GATT considers that export subsidies are unfair competition and allows countervailing duties to be imposed on subsidized products. Indirect methods of export subsidization include government subsidized financing for exports, export promotion and information activities, tax benefits, or other forms of assistance that may lead to lower than normal costs for exported products. The Uruguay Round Agreement on Agriculture imposes limits on direct agricultural export subsidies.

Exporter Assistance Initiative — Directive, under of the 2002 farm bill (P.L. 107-171 Sec. 3101), that the Secretary of Agriculture maintain an Internet website for U.S. agricultural exporters that will provide a "comprehensive source of information to facilitate exports" of U.S. agricultural commodities.

Exposure assessment — Identifying the pathways by which toxicants may reach individuals, estimating how much of a chemical various individuals are likely to be exposed to, and estimating the number likely to be exposed at each level.

Extension Service — Refers to a nationwide continuing education system that is based on the academic programs of the land grant colleges of agriculture (see Cooperative Extension System). The term also is the former name of the USDA agency that distributes federal funds to the states under the Smith-Lever Act of 1914 (7 U.S.C. 341 *et seq.*) to carry out Extension programs. The 1994 USDA reorganization merged this agency with the Cooperative State Research Service (CSRS) to form the Cooperative State Research, Education, and Extension Service (CSREES).

Extra-long staple (ELS) cotton — ELS cotton, like upland cotton, is eligible for marketing assistance loans and loan deficiency payments (LDPs). The national loan rate for ELS cotton under the 2002 farm bill is $0.7977 per pound. ELS cotton, in contrast to upland cotton, does not qualify for direct payments or counter-cyclical payments. For purposes of federal support, the 2002 farm bill (P.L. 101-171, Sec. 1001) defines ELS cotton. ELS cotton also is called American Pima and was once called American Egyptian cotton, this cotton has a staple length of 1-3/8" or more, is characterized by fineness and high fiber strength, and is used in high-value products such as sewing thread and expensive apparel. The name Pima was applied in honor of the Pima Indians who helped raise the cotton on USDA experimental farms in Arizona in the early 1900s. ELS cotton accounts for less than 5% of U.S. cotton production. It is grown chiefly in California, with small acreages in west Texas, New Mexico and Arizona.

Extra-Long Staple Cotton Act of 1983 — P.L. 98-88 eliminated marketing quotas and allotments for extra-long staple cotton and tied its support to upland cotton through a formula that set the nonrecourse loan rate at not less than 150% of the upland cotton loan level.

Extralabel use in animals — Actual use or intended use in an animal of an FDA approved veterinary-prescribed drug that is not in accordance with the approved labeling, such as for indications (conditions) not listed on the label, use at other than labeled dosage levels, frequencies or routes of administration, and deviation from labeled withdrawal time. This practice is permitted by the Animal Medicinal Drug Use Clarification Act of 1994 (P.L. 103-396). FDA specifically prohibits extralabel use of a number of antibiotics, anti-inflammatory drugs and hormones in food producing animals. FDA also tightly controls the use of certain veterinary-prescribed drugs when administered in the feed of food-producing animals.

FAC — Food and Agriculture Council.

FACA — Federal Advisory Committee Act.

Facility Credit Guarantee Program (FGP) — This is a Commodity Credit Corporation (CCC) credit guarantee program to encourage the construction or improvement of agriculture-related storage, processing, or handling facilities in emerging markets. The CCC provides repayment guarantees (95% of the covered value of the project, and a portion of interest on a variable rate basis) on loans of 1 to 10 years. The longer term goal of the FGP is to expand sales of U.S. agricultural commodities and products to emerging markets. The CCC agrees to pay exporters or their assignee (e.g., financial institution) in the event a foreign bank fails to make payment pursuant to the terms of an irrevocable letter of credit.

FACT Act of 1990 — Food, Agriculture, Conservation, and Trade Act of 1990 (P.L. 101-624); the 1990 farm bill.

FACT — Food Animal Concerns Trust. [www.fact.cc].

Failed acreage — Tracts of properly-planted and managed crops that did not grow or were destroyed due to a natural disaster. Failed acreage is eligible for indemnification if covered by the federal crop insurance program.

Fair Labor Standards Act (FLSA) — P.L. 75-718 (June 25, 1938, as amended) is the primary federal statute governing minimum wages, overtime pay, child labor, and related labor standards. Initially focused upon industrial workers involved in interstate commerce, the statute has gradually been extended to workers in agriculture and immediately related activities, though with many exemptions built into the statute. The Act is administered by the U.S. Department of Labor.

Fair market value (FMV) — The amount in cash, or on terms reasonably equivalent to cash, for which something might be sold by a knowledgeable owner to a knowledgeable purchaser. Several federal statutes specify that the federal government should receive fair market value when exchanging or selling federal lands and resources.

FAIR Act of 1996 — Federal Agriculture Improvement and Reform Act of 1996 (P.L. 104-127); the 1996 farm bill.

FAIR — National Farm Animal Identification and Records.

Fallow cropland — Cropland usually in semi-arid regions that is purposely kept out of production during a cropping season, mainly to conserve moisture for the next season; sometimes called summer fallow. It may be tilled or sprayed to control weeds and conserve moisture in the soil. The 1997 Census of Agriculture reported that 20.9 million acres, almost 5% of the 431 million acres of all cropland, was fallow that year.

FAMC — Federal Agricultural Mortgage Corporation (Farmer Mac). [www.farmermac.com].

Family farm — As defined by USDA regulations related to farm loan programs, a family farm is one that (1) produces agricultural commodities for sale in such quantities so as to be recognized in the community as a farm and not a rural residence; (2) produces enough income (including off-farm employment) to pay family and farm operating expenses, pay debts, and maintain the property; (3) is managed by the operator; (4) has a substantial amount of labor provided by the operator and the operator's family; and (5) may use seasonal labor during peak periods and a reasonable amount of full-time hired labor. (For exact language, see 7 U.S.C. 1941.4,1943.4).

FAO — Food and Agriculture Organization of the United Nations. [www.fao.org].

FAPRI — Food and Agricultural Policy Research Institute. [www.fapri.missouri.edu].

Farm acreage base — The total of the crop acreage bases (wheat, feed grains, cotton, and rice) for a farm for a year, the average acreage planted to soybeans and other non-program crops, and the average acreage devoted to conserving uses (excluding Acreage Reduction Program land). The 1996 farm bill (P.L. 104-127) and the 2002 farm bill (P.L. 107-121) eliminate the need to calculate a farm acreage base.

Farm and ranch risk management (FARRM) accounts — A proposal that would permit farmers and ranchers to put aside money in good (higher-income) years without having to pay taxes on the savings until it is withdrawn at a later time, presumably in lower-income years when taxable income also would be lower. Bills to create these accounts have been introduced in Congress in recent years, but no program has been authorized.

Farm and Ranch Lands Protection Program (FRPP) — The Natural Resources Conservation Service (NRCS) renamed the Farmland Protection Program (FPP) to the Farm and Ranch Lands Protection Program in 2003 to accurately reflect the resources eligible to participate in the program. The program established by the 1996 farm bill (P.L. 104-127) to fund the purchase of conservation easements of 170,000-340,000 acres of land having prime or unique soil or other desirable production qualities that are threatened by urban development. Eligibility depends upon already having a pending offer from a state or local government to protect qualifying land by limiting nonagricultural use. The 2002 farm bill (P.L. 107-171, Sec. 1241) reauthorized the program through FY2007 and provided mandatory funding from the Commodity Credit Corporation (CCC) that was $50 million in FY2002 and will rise to $125 million in FY2004, then slowly decline to $97 million in FY2007. Other changes expanded the definition of eligible land to include cropland, rangeland, grassland, pasture land, incidental forest land, and historic and archeological sites; expanded the list of eligible participants to include Indian tribes and non profit organizations that meet specified qualifications; and directed an unspecified portion of the program funds to carry out a farm viability program. According to the NRCS FY2005 budget request document, more then 306,000 acres in 42 states have or soon will have easement contracts.

Farm bill — A phrase that refers to a multi-year, omnibus law that contains federal commodity and farm support policies, as well as other farm-related provisions. It usually amends some and suspends provisions of permanent law, reauthorizes, amends, or repeals provisions of preceding temporary agricultural acts, and puts forth new policy provisions for a limited time into the future. Beginning in 1973, farm bills have included titles on commodity programs, trade, rural development, farm credit, conservation, agricultural research, food and nutrition programs, marketing, etc. These are referred to as omnibus farm bills. The following is a generally agreed chronological list of farm bills: (1) Food and Agriculture Act of 1965, P.L. 89-321; (2) Agricultural Act of 1970, P.L. 91-524; (3) Agriculture and Consumer Protection Act of 1973, P.L. 93-86; (4) Food and Agriculture Act of 1977, P.L. 95-113; (5) Agriculture and Food Act of 1981, P.L. 97-98; (6) Food Security Act of 1985, P.L. 99-198; (7) Food, Agriculture, Conservation, and Trade Act of 1990, P.L. 101-624; (8) Federal Agriculture Improvement and Reform Act of 1996, P.L. 104-127; (9) Farm Security and Rural Investment Act of 2002, P.L. 107-171.

Farm Bureau — American Farm Bureau Federation (AFBF). [www.fb.com].

Farm Credit Act of 1971 — This Act, as amended, currently serves as the authorizing statute for the Farm Credit System (12 U.S.C. 1200 *et seq.*). P.L. 92-181 (December 10, 1971) recodified all previous acts governing the Farm Credit System (FCS, or System), a cooperatively owned government sponsored enterprise that provides credit primarily to farmers and ranchers. The Act eliminated earlier provisions relating to government capitalization of the System, and expanded the lending authorities of many System institutions. The Agricultural Credit Act of 1987 (P.L. 100-233, January 6, 1988), a major piece of legislation modifying the 1971 Act, authorized up to $4 billion in federal financial assistance to FCS institutions to assist in their recovery from the agricultural credit crisis of the 1980s. The Act created a System entity to issue up to $4 billion in federally guaranteed bonds, required the U.S. Treasury to pay a portion of the interest on these bonds, and also required the FCS to ultimately repay the Treasury for this assistance. The Act also mandated the merger of certain System banks within each farm credit district and expanded other merger authorities, and gave delinquent FCS borrowers certain rights. A separate System institution was established by the Act to insure the timely repayment of principal and interest on consolidated Systemwide debt issues. Farm Credit Banks and Associations Safety and Soundness Act of 1992 (P.L. 102-552) was designed to enhance the financial safety and soundness of FCS banks and associations by establishing new mechanisms to ensure repayment of Farm Credit System debt resulting from federal financial assistance provided to the System under the 1987 Act. The Farm Credit System Reform Act of 1996 (P.L. 104-105) included numerous provisions to provide regulatory relief for the FCS.

Farm Credit Administration (FCA) — The independent federal regulator responsible for examining and insuring the safety and soundness of all Farm Credit System institutions. The FCA is governed by a 3-member, Presidentially-appointed board of directors, one of whom serves as chairman. [www.fca.gov].

Farm Credit Banks — Institutions within the Farm Credit System (FCS) that make direct long-term agricultural loans secured by farm real estate through Federal Land Bank Associations. They provide wholesale loan funds to direct FCS lending associations (Production Credit Associations, Federal Land Credit Associations, and Agricultural Credit Associations).

Farm Credit System (FCS) — A nationwide financial cooperative that lends to agricultural producers, rural homeowners, farm-related businesses, and agricultural, aquatic, and public utility cooperatives. Congress established FCS as a government sponsored enterprise when it enacted the Federal Farm Loan Act in 1916. Current authority is in the Farm Credit Act of 1971 (P.L. 92-181, as amended; 12 U.S.C. 1200 *et seq.*). The fundamental purpose is to provide a permanent, reliable source of credit at competitive interest rates, and related services to agricultural producers, their cooperatives, and related businesses in rural America. FCS is composed of six regional Farm Credit Banks (FCB) and one Agricultural Credit Bank. These banks provide funds and support services to Federal Land Bank Associations (FLBA), Federal Land Credit Associations (FLCA), Production Credit Associations (PCA), and Agricultural Credit Associations (ACA). These associations in turn, provide loans to eligible borrowers. Lending associations are governed by boards of directors elected from FCS borrowers. Funds are raised through the sale of bonds and notes. Federal oversight by the Farm Credit Administration is designed to provide for the safety and soundness of FCS institutions.

Farm Credit System Assistance Board — A temporary board created by the Agricultural Credit Act of 1987 (P.L. 100-233, Title II) and responsible for approving Farm Credit System lender requests for federal financial assistance. Members of the Board consisted of the Secretary of Agriculture, Secretary of Treasury (or their appointees), and an agricultural producer with financial experience.

Farm Credit System Insurance Corporation (FCSIC) — An entity of the Farm Credit System (FCS), established by the Agricultural Credit Act of 1987 (P.L. 100-233, Title III), to insure the timely repayment of principal and interest on FCS debt securities.

Farm equity — The net worth of the farm sector's assets (i.e., farmland, machinery, equipment, facilities, crop and livestock inventories) against which there is no debt. This represents all farm proprietors' residual claims to farm assets. Increases in farm equity in the late 1970s became increasingly important for most agricultural producers as a source of additional collateral against which to obtain credit for operating and expansion purposes. The level of farm equity ranges widely from one farm to another. The overall debt-asset ratio is a measure of the farm sector's financial condition or solvency.

Farm gate price — See Producer price.

Farm income and balance sheet — The income statement measures the profitability of a farm business for a particular period of time, usually one year. The balance sheet measures the wealth or financial position of the business at a particular point in time by reporting the farm's assets, debt, and net worth. The Economic Research Service publishes the income statement and balance sheet of the Nation's farm sector, and the farm sector financial statement for each state.

Farm income — Several measures are used to gauge the earnings of a farming operation over a given period of time. Gross cash income is the sum of all receipts from the sale of crops, livestock, and farm related goods and services as well as all forms of direct payments from the government. Gross farm income is the same as gross cash income with the addition of nonmoney income, such as the value of home consumption of self-produced food and the imputed gross rental value of farm dwellings. Net cash income is gross cash income less all cash expenses such as for feed, seed, fertilizer, property taxes, interest on debt, wages to hired labor, contract labor and rent to nonoperator landlords. Net farm income is gross farm income less cash expenses and noncash expenses, such as capital consumption, perquisites to hired labor, and farm household expenses. Net farm income is a longer term measure of the ability of the farm to survive as a viable income-earning business, while net cash income is a shorter term measure of cash flow.

Farm inputs — The resources that are used in farm production, such as chemicals, equipment, feed, seed, and energy. Most farm inputs are purchased (a change from the days when animals powered most operations), making production costs susceptible to nonfarm economic conditions. Over time, prices of farm inputs have increased relative to commodity prices, creating what farmers describe as a cost-price squeeze. The relationship between prices paid for inputs compared to prices received for output is quantified in the parity ratio.

Farm labor housing grants — Section 516 grants are available through the Rural Housing Service (RHS) to qualified nonprofit organizations to providing housing to farm workers.

Farm labor housing loans — Section 514 loans are available through the Rural Housing Service (RHS) to qualified farm owners for the purpose of providing housing to domestic farm labor.

Farm Labor Contractor Registration Act (FLCRA) — P.L. 88-582 (September 7, 1964, as amended) regulated the activities of farm labor contractors, that is, agents who recruit and are otherwise engaged in the transport, housing, and employment of migratory agricultural workers. Under FLCRA, farm labor contractors were required to secure certification through the U.S. Department of Labor. Strengthened by amendment in 1974, the Act became a target of grower criticism and, in 1983, was repealed and replaced with the Migrant and Seasonal Agricultural Workers Protection Act (P.L. 97-470).

Farm link (land link) — A program concept that would match retiring farmers who want to keep their land in agriculture with beginning farmers, almost always unrelated, who want to buy a farm. Efforts to implement such an effort rest at the state and local level. There is no federal program.

Farm loan programs of the FSA — Loan programs, administered by the Farm Service Agency (replacing FmHA), providing both direct and guaranteed real estate, operating loans, and direct emergency disaster loans to individuals whose primary business is farming and ranching. Loans are targeted to family farms whose operators are unable to obtain sufficient credit from private commercial lenders on reasonable terms. Under the 1996 farm bill (P.L. 104-127), farm lending programs are permanently reauthorized, with new restrictions on the purposes for which loans can be used and on the length of time borrowers are eligible for new credit assistance. Provisions are extended that reserve a portion of loan funds for new and beginning farmers.

Farm operating (OL) loans — The Consolidated Farm and Rural Development Act (P.L. 92-419, Subtitle B, as amended; 7 U.S.C. 1941-1943), authorizes the Farm Service Agency (FSA) (formerly FmHA) to make direct and guaranteed farm operating loans. Applicants must be family-sized farmers, who are denied credit by private and cooperative sources, and have reasonable prospects for success in the farm operation. Operating loans are made to farmers to help them pay their operating expenses for such productions costs as feed, seed, fertilizer, and pesticides, and to meet other essential operating expenses. The scheduled repayment is usually over 1 to 7 years depending on loan purposes. The interest rate on direct loans is determined by the Farm Service Agency and does not exceed the federal cost of borrowing plus 1 percentage point. However, loans to limited resource borrowers can be made at significantly below market rates. The interest rate on guaranteed loans is negotiated between the borrower and the lender. USDA guarantees the timely repayment of 90% of principal and interest on guaranteed loans, and in some cases can subsidize the interest rate on these loans. The amount USDA can directly lend or guarantee each year is determined in the annual congressional appropriations process.

Farm operator — A person who operates a farm, either by doing or supervising the work or by making the day-to-day management decisions. Nationally, farm operators own about 56% of their land and lease or rent the remainder from landlords according to the 2002 Census of Agriculture.

Farm ownership (FO) loans — The Consolidated Farm and Rural Development Act (P.L. 92-419, Subtitle A, as amended, 7 U.S.C. 1922-1925), authorizes the Farm Service Agency (formerly FmHA) to make direct and guaranteed farm ownership loans to eligible family farmers. One of the functions of the FO loan program is to assist farmers, especially beginning farmers, in the purchase and enlargement of farms. An eligible borrower must be unable to obtain sufficient credit from a commercial lender, but must assure reasonable prospects of success in the farm operation. Loans are made for up to 40 years. The interest rate is determined by USDA, and cannot exceed the cost of funds to the Government plus 1 percentage point. However, direct loans to limited resource borrowers can be made at significantly below the federal cost of funds. The interest rate on guaranteed loans is negotiated between the borrower and the lender. USDA guarantees the timely repayment of 90% of principal and interest on guaranteed loans, and in some cases can subsidize the interest rate on these loans. The amount USDA can directly lend or guarantee each year is determined in the annual congressional appropriations process.

Farm price — The price that farmers receive for the commodities they market. Sometimes the term farm-gate price is used to emphasize that the price does not include transportation or processing costs.

Farm programs — This term is generally meant to include the commodity programs administered by the Farm Service Agency, as well as the other USDA programs that directly benefit farmers. Some examples of the other programs include farm loans, federal crop insurance, the noninsured assistance program (NAP), the Conservation Reserve Program (CRP), and conservation cost sharing.

Farm Security and Rural Investment Act of 2002 — Referred to as the 2002 farm bill (and sometimes abreviated FSRI Act or FSRIA), this 407 page omnibus law (P.L.107-171) signed by President Bush on May 13, 2002, revised and extended (in most cases, through 2007) many major government programs and policies related to agriculture. It included titles on commodity programs, conservation, trade, nutrition programs, credit, rural development, research and extension services, forestry, energy, as well as miscellaneous provisions relating to crop insurance, disaster assistance, animal health protection, livestock, specialty crops and general provisions, studies and reports. The total 6-year cost of this law projected by CBO at the time of its enactment was $273.9 billion. This reflected $51.7 billion in new spending, most of which was for payments to farmers under commodity programs (+$37.6 billion) and conservation programs ($+9.2 billion). This law substantially altered the farm commodity policy direction taken by the Agriculture Market Transition Act in the 1996 farm bill (P.L.104-127). Among other things, it increased support for covered commodities (formerly called contract commodities) by raising loan rates and adding counter-cyclical payments to direct payments, and also by expanding the scope to include soybeans, other oilseeds, and pulse crops. Other notable commodity policy changes included: (1) the creation of a new counter-cyclical dairy payment program to replace the Northeast Dairy Compact; (2) the termination of the peanut quota program with compensation for peanut quota holders, and the addition of peanuts to the crops eligible for marketing loan assistance, and direct and counter-cyclical payments; (3) the restoration of federal support for honey, wool, and mohair through marketing assistance loans and loan deficiency payments. Among the noteworthy changes made in the conservation title were significant funding increases for conservation activities, allowance for third-party vendors to carry out technical services, and the creation of a new conservation security program that pays farmers for conserving practices on land that is in production.

Farm Service Agency (FSA) — One result of the 1994 legislative reorganization of USDA was the consolidation of the ASCS, FCIC and FmHA into a single agency, the FSA. This agency is responsible for administering farm income-support programs, conservation cost-sharing programs, noninsured crop assistance (NAP), and the former FmHA farm loan programs. FSA services are provided through field service centers located throughout the agricultural areas of the nation. [www.fsa.usda.gov/pas/default.asp].

Farm size — The most common way to measure farm size is by the value of gross farm sales. Acreage is not used for comparisons across differing kinds of farms because in some cases a farm need not have land (i.e., bee hives may be in constant rotation among parcels not belonging to the beekeeper). There are about 2.1 million farms in the United States. For 2003, according to analysis by the Economic Research Service, commercial farms, which include those having sales of $250,000 or more annually, constituted 9% of all farms and accounted for nearly 72% of production. Intermediate farms constituted 24% of all farms and accounted for 19% of production. The largest number of farms, characterized as rural residence farms, constituted 68% of all farms and accounted for 8% of production.

Farm Storage Facility Loan Program — A loan program for the construction and remodeling of storage facilities on farms producing grains, oilseeds, and pulses. The loan limit for each borrower is $100,000 for up to 7 years. The interest rate is equivalent to the rate on comparable term Treasury securities. The program is administered by the Farm Service Agency (FSA). The 2002 farm bill (P.L. 107-171, Sec. 1402) newly mandated storage facility loans for raw and refined sugar.

Farm to retail price spread — The difference between the farm price and the retail price of food, reflecting charges for processing, shipping, and retailing farm goods (sometimes called the marketing spread). The current spread accounts for about three-fourths of the retail price for a market basket of foods, according to USDA. The farm value varies for each type of food; for example, in 2004 it accounted for about 35% of the retail cost of eggs, compared to about 19% for fresh fruit and vegetables, and about 6% for cereal and bakery products.

Farm typology — The USDA's Economic Research Service (ERS) has developed a farm classification that divides the 2.1 million U.S. farms into 8 mutually exclusive and relatively homogeneous groups: limited resource farms, retirement farms, residential/lifestyle farms, farming occupation/lower sales, farming occupation/high sales, large family farms, very large family farms, and nonfamily farms. Also, the eight categories can be collapsed into 3: rural residence farms, intermediate farms, and commercial farms. Data for 2003 indicate that Commercial farms, those having sales of $250,000 or more annually, constitute 9% of all farms and account for 72% of production. Intermediate farms, constituting 24% of all farms, account for 19% of production. The largest number of farms, characterized as rural residence farms, constitute 68% of all farms and account for 8% of production. See Small farm.

Farm Viability Program — A new program option under the Farmland Protection Program, aimed to increase economic sustainability of farms, enacted in the 2002 farm bill (P.L. 107-171, Sec. 2503), to be funded through appropriations with "such funds as may be necessary" through FY2007.

Farm — Since 1850, when minimum criteria defining a farm for census purposes were first established, the farm definition has changed nine times. A farm currently is defined, for statistical purposes, as any place from which $1,000 or more of agricultural products were produced or sold or would have been sold during the agriculture census year. This was the official U.S. Commerce definition used in the agricultural censuses from 1974 to 1992. It was maintained when the USDA took over the responsibility and cost for the agricultural census of 1997 after the Bureau of Census proposed changing the farm sales minimum to $10,000. In 1995 when this was proposed, the change would have reduced the counted number of farms by nearly half, and was strongly opposed by many agricultural interests. The issue was resolved by permanently transferring the responsibility and cost for agricultural censuses to the USDA, which thus far has maintained the $1,000 threshold.

FARM — Farm Animal Reform Movement. [www.farmusa.org].

Farmable Wetlands Program — First authorized as a pilot program in Title XI of the FY2001 agriculture appropriations legislation (P.L. 106-387) to enroll up to 500,000 acres of farmable wetlands smaller than 5 acres in 6 upper Midwestern states (with no more than 150,000 acres in a single state) into the Conservation Reserve Program. The 2002 farm bill (P.L. 107-171, Sec. 2101) made this a 2 million acre national program (with an enrollment limit of 100,000 acres per state), and made changes in eligibility requirements, such as increasing the maximum size of eligible wetlands from 5 acres to 10 acres.

Farmaceuticals — This term is a melding of the words farm and pharmaceuticals. It refers to medically valuable compounds produced from modified agricultural crops or animals (usually through biotechnology). Proponents believe that using crops and possibly even animals as pharmaceutical factories could be much more cost effective than conventional methods (i.e., in enclosed manufacturing facilities) and also provide agricultural producers with higher earnings. Worldwide, hundreds of genetically modified, agricultural-based pharmaceuticals are being developed and tested, but as of early 2003, none had yet been approved for commercial use. At issue in the United States has been whether the current system for regulating biotechnology is adequate for ensuring the safety (to humans, animals and crops, and the environment) of newly emerging applications, such as farmaceuticals. The terms nutraceuticals and farmaceuticals are sometimes used interchangeably to describe components isolated from foods that can be provided in dosage form for some medical purpose beyond nutrition. The term farmaceuticals is more frequently associated, in agricultural circles, with medical applications of genetically engineered crops or animals. The term nutraceuticals, on the other hand, appears to be associated more with the extraction of naturally occurring components for medical uses.

Farmed wetlands — Under the swampbuster program, these are wetlands that were partially drained or altered to improve crop production before swampbuster was enacted on December 23, 1985. Farmed wetlands have undergone less alteration than prior converted wetlands, and are therefore subject to more stringent rules about further change. Farmed wetlands may be farmed as they were before the 1985 date, and the drainage that was in place before that date can be maintained, but no additional drainage is allowed.

Farmer Mac (Federal Agricultural Mortgage Corporation) — Created by the Agricultural Credit Act of 1987 (P.L. 100-233) as a federally chartered, private corporation responsible for guaranteeing the timely repayment of principal and interest to investors in a new agricultural secondary market. The secondary market allows a lending institution to sell a qualified farm real estate loan to an agricultural mortgage marketing facility, or pooler, which packages these loans, and sells to investors securities that are backed by, or represent interests in, the pooled loans. Farmer Mac guarantees the timely repayment of principal and interest on these securities and, under authorities granted in 1995, can also serve as a loan pooler. [www.farmermac.com].

Farmer to Farmer Program — see John Ogonowski Farmer to Farmer Program.

Farmer-Owned Grain Reserve (FOR) — A program, established under the Food and Agriculture Act of 1977, designed to buffer sharp price movements and to provide reserves against production shortfalls by allowing wheat and feed grain farmers to participate in a subsidized grain storage program. Farmers who placed their grain in storage received an extended nonrecourse loan for at least 3 years. Under certain conditions, interest on the loan could be waived and farmers could receive annual storage payments from the government. The 1996 farm bill (P.L. 104-127) repealed this program.

Farmers Home Administration (FmHA) — Formerly an agency of USDA that provided direct and guaranteed credit to family-sized farmers who were denied credit by a commercial lender. The 1994 USDA reorganization transferred FmHA's farm loan programs to the newly formed Farm Service Agency.

Farmers markets — Farmers markets are venues where agricultural and food producers sell directly to consumers. In the past 20 years, farmers markets have increased significantly in the United States as demand has increased for locally grown products. According to USDA, from 1994 to 2002 the number of farmers' markets increased 79%, to over 3,100 markets across the country. In 2000, 19,000 farmers sold their produce only at farmers markets. There is variation among farmers markets: (1) producers-only markets where vendors sell only produce grown on their own farms; (2) markets where producers sell their own produce along with produce from other local farms; and (3) vendors who sell produce they purchase, locally or not. Some markets have only a few vendors, but the more successful farmers markets may have a large number of venders offering locally and regionally grown produce, meat and poultry, fruits, and value-added agricultural products, e.g., cheese, maple sugar, cut-flowers, breads and pastries, meat and dairy products. The WIC Farmers Market Nutrition Program (FMNP), established by Congress in 1992, provides fresh, unprepared, locally grown fruits and vegetables to WIC recipients, and expands the awareness, use of, and sales at farmers' markets. A similar program, the Senior Farmers Market Nutrition Program (SFMNP), provides locally-grown produce for low-income senior citizens.

Farmers Market Nutrition Program (FMNP) — Authorized under Section 17 of the Child Nutrition Act of 1966 (P.L. 89-642, as amended), this program disburses grant funds to selected states that develop programs promoting the use of farmers markets by WIC recipients.

Farmers Market Promotion Program — A USDA program established by the 2002 farm bill (P.L. 107-171, Sec. 10605) to improve or expand existing farmers' markets, roadside stands, community-supported agriculture programs, and other direct producer-to-consumer market opportunities, and to develop or aid in the development of new farmers' markets, etc. The farm bill authorized unspecified amounts of appropriations for FY2002 through FY2007 for this program.

Farmers Union — National Farmers Union (NFU). [www.nfu.org].

Farming-dependent county — The Economic Research Service (ERS) of USDA categorizes non-metropolitan counties by their dominant economic foundation and by characteristic policy type. The 2004 County Typology Codes were developed for all 3,141 counties, county equivalents, and independent cities in the United States. *Farming-dependent counties* (440 total, 403 nonmetro) are those with either 15 percent or more of average annual labor and proprietors' earnings derived from farming during 1998-2000 or 15 percent or more of employed residents worked in farm occupations in 2000.

Farmland protection — Programs, operated mostly at state and local levels by government agencies or private entities such as land trusts, that are designed to limit conversion of agricultural land to other uses that otherwise might have been more financially attractive to the landowner. Every state provides tax relief through differential (preferential) assessment, and has right-to-farm laws. Less common approaches include establishing agricultural districts, using zoning to protect agricultural land, purchasing development rights, and transferring development rights.

Farmland Protection Program (FPP) — See Farm and Ranch Lands Protection Program (FRPP) for a definition. This became the new name for the program in 2003.

Farmland — Land used for agricultural purposes. The federal government recognizes prime farmland and unique farmland as the most important categories. According to USDA, the United States has had roughly 1 billion acres of farmland. Farmland consists of cropland, pastureland, and grazing land.

FARRM — Farm and ranch risk management accounts.

Farrow-to-finish — Typically, a confinement operation where pigs are bred and raised to their slaughter weight, usually 225-300 pounds. Facilities with a capacity of 2,500 or more swine are considered by the EPA to be concentrated animal feeding operations (CAFOs) subject to point source pollution permit requirements. Other types of hog operations include farrow-to-feeder pig, feeder pig-to-finish, weanling-to-feeder pig, and farrow-to-weanling.

FAS — Foreign Agricultural Service. [www.fas.usda.gov].

FASS — Federation of Animal Science Societies. [www.faseb.org].

FASFAS — Federation of American Societies of Food Animal Sciences. [www.fass.org].

Fast track authority — See Trade promotion authority.

Fat free lean index — One of several measures of hog quality (in this case, leanness) that can be used in determining value. The index was developed by the National Pork Producers Council, an industry trade group.

FATUS — Acronym for Foreign Agricultural Trade System of the United States, FATUS is a system of more than 200 trade codes created and maintained by USDA's Economic Research Service to summarize U.S. agricultural trade in a form accessible to the public. FATUS codes aggregate more than 4,000 import and 2,000 export, 10-digit agricultural trade codes from the Harmonized Tariff Schedule of the U.S. (HTS), under which all U.S. trade data are originally collected by the Census Bureau of the U.S. Department of Commerce. FATUS groupings are similar to, but tend to be more detailed than, those provided to the public through the BICO data system maintained by USDA's Foreign Agricultural Service. See also BICO, and U.S. Trade Internet System.

FB — Farm Bureau (American Farm Bureau Federation). [www.fb.com].

FCA — Farm Credit Administration. [www.fca.gov].

FCB — Farm Credit Bank.

FCC — Farm Credit Council. [www.fccouncil.com].

FCIC — Federal Crop Insurance Corporation. [www.act.fcic.usda.gov].

FCOJ — Frozen concentrated orange juice.

FCS — Farm Credit System. [www.fccouncil.com].

FCS — Food and Consumer Service. [www.fns.usda.gov/fncs].

FCSIC — Farm Credit System Insurance Corporation.

FDA — Food and Drug Administration. [www.fda.gov].

FDPIR — Food Distribution Program on Indian Reservations.

Fecal coliform bacteria — Bacteria found in the intestinal tracts of mammals. Their presence in water or sewage sludge may be an indicator of pollution and possible contamination by pathogens.

Fed cattle — Animals leaving a feedlot, after fattening on a concentrated ration, that are ready to be sold to a packing plant for slaughter. Beef cattle are typically sold to packers at about 1,100 pounds, which yields a carcass weight of about 660 pounds.

Federal Agricultural Mortgage Corporation — An organization more commonly referred to as Farmer Mac, which is a secondary (resale) market for agricultural mortgages. Farmer Mac was authorized by the Agricultural Credit Act of 1987 (P.L. 100-233). [www.farmermac.com].

Federal Agriculture Improvement and Reform (FAIR) Act of 1996 — P.L. 104-127 (April 4, 1996) was the omnibus 1996 farm bill that removed the link between income support payments and farm prices. It authorized 7-year production flexibility contract payments that provided participating producers with fixed government payments independent of current farm prices and production. The law specified the total amount of money to be made available through contract payments under production flexibility contracts for each fiscal year from 1996 through 2002. Payment levels were allocated among contract commodities according to specified percentages, generally derived from each commodity's share of projected deficiency payments for fiscal 1996-2002. The law increased planting flexibility by allowing participants to plant 100% of their total contract acreage to any crop, except with limitations on fruits and vegetables. The authority for acreage reduction programs was eliminated, while nonrecourse loans (with marketing loan repayment provisions) were continued in a modified form. Minimum loan rates generally were calculated each year at 85% of recent past market prices. Authority for the Farmer-Owned Reserve Program was suspended through the 2002 crop year. Authority for the honey program was eliminated. Dairy price support was to be phased down for milk over 4 years and then eliminated, but subsequent legislation continued this program. Had dairy support ended, processors could have obtained recourse loans on dairy products. The peanut program was continued but revised to reduce the likelihood of the federal government incurring loan program costs due to loan forfeitures. The minimum national poundage quota was eliminated. The sugar program also was continued but modified. Trade and food aid programs were reoriented toward greater market development, with increased emphasis on high-value and value-added products. Other provisions established a Commission to conduct a comprehensive review of changes to production agriculture under the 1996 Act, required USDA to conduct research on futures and options contracts through pilot programs, capped expenditures for the Export Enhancement Program, and changed the name of the Market Promotion Program to the Market Access Program. The 1996 Act also reauthorized the Food Stamp Program for 2 years and commodity donation programs for 7 years, and established a Fund for Rural America to augment existing resources for agricultural research and rural development. Other research authorities were revised and extended, some only for 2 years rather than 7 years. The 1996 Act authorized new enrollments in the Conservation Reserve Program to maintain total acreage at up to 36.4 million acres. Other conservation programs were also revised and extended. The Act also contained numerous provisions in the areas of farm credit, rural development, and generic commodity promotion through check-off programs, among others. The 2002 farm bill (P.L. 101-171) superceded many of the 1996 farm bill provisions before they expired.

Federal crop insurance — See Crop insurance

Federal Crop Insurance Corporation (FCIC) — The wholly owned federal corporation within USDA that administers the federal crop insurance program. The 1996 farm bill (P.L. 104-127) created an Office Of Risk Management, which USDA has renamed the Risk Management Agency, and which houses the FCIC [www.act.fciç.usda.gov].

Federal Crop Insurance Fund — The fund within USDA through which all mandatory expenses of the federal crop insurance program (i.e., premium subsidy, program losses, and the reimbursement to participating private insurance companies for their administrative and operating expenses) are funded. Each budget cycle, USDA estimates the amount required to fund the program, but the fund receives an annual appropriation of "such sums as are necessary", since program expenses particularly program losses are contingent on weather and other variables, thus making it difficult to budget in advance.

Federal Crop Insurance Program — See Crop insurance.

Federal Crop Insurance Reform Act of 1994 — This Act is Title I of P.L. 103-354. Beginning with the 1995 crops, it modified the federal crop insurance program by authorizing a new catastrophic (CAT) coverage level available to farmers. The premium on this level of coverage (crop losses in excess of 50% receiving a payment of 60% (now, 55%) of the market price of the insured crop) is 100% subsidized by the government, but requires a farmer to pay a $50 per crop per county administrative fee (since raised to $100 per crop). The Act allows farmers to purchase additional insurance coverage providing higher yield or price protection levels, with the premium on this buy-up coverage partially subsidized by the government. The Act also created the Noninsured Assistance Program (NAP), a permanent disaster payments program for crops not covered by crop insurance. The 1994 Act amended and in many cases suppressed major portions of the Federal Crop Insurance Act of 1980 (P.L. 96-365) which serves as the authorizing statute for the federal crop insurance program. The 1980 Act expanded the scope of the crop insurance program and permitted USDA to subsidize farmer premium payments. The Agriculture Risk Protection Act of 2000 (P.L. 106-224) made subsequent enhancements and reforms to the crop insurance program.

Federal Farm Credit Banks Funding Corporation (FFCBFC) — An entity within the Farm Credit System (FCS) that manages and coordinates the sale of system-wide bonds and notes in the national financial markets. Since the FCS, by law, is not permitted to accept customer deposits, these bonds and notes are the FCS's primary source of loanable funds. [www.farmcredit-ffcb.com/ffcx/owa/ffcweb.home].

Federal Food, Drug, and Cosmetic Act (FFDCA) of 1938 — P.L. 75-717 is the basic authority intended to ensure that foods (other than meat and poultry, which USDA inspects) are pure and wholesome, safe to eat, and produced under sanitary conditions; that human and animal drugs and devices are safe and effective for their intended uses; that cosmetics are safe and made from appropriate ingredients; that animal feeds are safe,and that all labeling and packaging is truthful, informative, and not deceptive (21 U.S.C. 321 *et seq.*). The Food and Drug Administration (FDA) is primarily responsible for enforcing the FFDCA, although USDA also has some enforcement responsibility. The EPA establishes limits for concentrations of pesticide residues on food under this act.

Federal grain inspection program — The grain inspection program administered by the Grain Inspection, Packers and Stockyards Administration. The program establishes official U.S. standards for grain and certain other commodities such as rice, hops, and processed grain products. The program offers a user-financed nationwide inspection and weighing system to certify that grain meets approved standards. By law, all grain exported from the United States must be officially inspected.

Federal Insecticide, Fungicide, and Rodenticide Act (FIFRA) — P.L. 80-104 (June 25, 1947, as amended; 7 U.S.C. 136 *et seq.*), is the basic authority that requires the EPA to regulate the sales and uses of pesticides. The federal government began regulating pesticides in 1910 for the purpose of preventing the exploitation of farmers from adulterated and ineffective products. The original 1947 FIFRA was the first effort to address potential risks to human health. FIFRA was completely revised in 1972 (P.L. 92-516) to become the basis for current federal policy. The law directs EPA to restrict the use of pesticides to prevent unreasonable adverse effects on people and the environment taking into account the costs and benefits of various uses. The sale of any pesticide is prohibited unless it has gone through registration and is labeled to show the approved uses and restrictions.

Federal Land Bank Associations (FLBA) — Institutions within the Farm Credit System that take applications for and service long-term real estate loans for the Farm Credit Banks, but do not have direct lending authority.

Federal Land Credit Associations (FLCA) — Institutions within the Farm Credit System (FCS) that have authority to make long-term real estate loans to eligible retail customers. FLCAs receive their funds directly from the Farm Credit Banks.

Federal Land Policy and Management Act of 1976 (FLPMA) — P.L. 94-579 (October 21, 1976) 1) set out for the Bureau of Land Management standards for managing public land, including land-use plans, sales, withdrawals, acquisitions, and exchanges; 2) authorized local advisory councils to represent major citizens groups interested in land use planning and management; 3) established criteria for review of proposed wilderness areas; and, 4) provided guidelines for other aspects of public land management such as grazing. This law is also known as the BLM organic act.

Federal Meat Inspection Act (FMIA) of 1906 — Enacted June 30, 1906, (Chapter 3913, 34 Stat. 674), and substantially amended by the Wholesome Meat Act 1967 (P.L. 90-201), FMIA requires USDA to inspect all cattle, sheep, swine, goats, and horses when slaughtered and processed into products for human consumption (21 U.S.C. 601 *et seq.*). The primary goals of the law are to prevent adulterated or misbranded livestock and products from being sold as food, and to ensure that meat and meat products (as well as poultry) are slaughtered and processed under sanitary conditions. These requirements also apply to imported meat and poultry products, which must be inspected under equivalent foreign standards. The Food and Drug Administration (FDA) is responsible for all meats not listed in the FMIA, including venison and buffalo, although USDA does offer a voluntary, fee-for-service inspection program for buffalo.

Federal Noxious Weed Act 1975 — P.L. 93-629 was adopted to prevent foreign weeds from entering and becoming established to the detriment of U.S. crops, livestock, and natural resources. Under this act, the Animal and Plant Health Inspection Service (APHIS) inspected incoming international passengers, baggage, and cargo at U.S. ports of entry to intercept noxious weeds (as listed by the Act), and worked with federal, state, and local agencies to confine, eradicate, or control them if any entered the country. Most of the provisions of the Federal Noxious Weed Act were incorporated into the Plant Protection Act of 2000 (Title IV of P.L. 106-224; 7 U.S.C. 7701 *et seq.*). Subsequently, the Homeland Security Act of 2002 transferred APHIS border inspections for noxious weeds (among many other things) to the Department of Homeland Security (DHS); cooperative programs for control or eradication remain in APHIS at USDA. Other agencies' programs related to noxious weeds (e.g. the Fish and Wildlife Service of the Department of the Interior, and the Environmental Protection Agency) now also operate under the authority of the Plant Protection Act of 2000.

Federal Plant Pest Act of 1957 — P.L. 85-36 prohibited the movement of plant pests from a foreign country into or through the United States unless authorized by USDA was superseded by the Plant Protection Act of 2000 (P.L. 106-224, Title IV). Under the new law, Animal and Plant Health Inspection Service (APHIS) retains broad authority to inspect, seize, quarantine, treat, destroy or dispose of imported plant and animal materials that are potentially harmful to U.S. agriculture, horticulture, forestry, and, to a certain degree, natural resources. (7 U.S.C. 7701 *et seq.*).

Federal Register (FR) — This federal document publishes current Presidential orders or directives, agency regulations, proposed agency rules, notices and other documents that are required by statute to be published for wide public distribution and application. USDA publishes its rules, notices and other documents in the Federal Register [www.gpoaccess.gov/fr/index.html]. Final regulations are organized by agency and program in the Code of Federal Regulations [www.gpoaccess.gov/cfr/index.html].

Federal Seed Act — P.L. 76-354 (August 9, 1939) requires accurate labeling and purity standards for seeds in commerce, and prohibits the importation and movement of adulterated or misbranded seeds. The law works in conjunction with the Plant Protection Act of 2000 to authorize the Animal and Plant Health Inspection Service (APHIS) to regulate the importation of field crop, pasture and forage, or vegetable seed that may contain noxious weed seeds. USDA's Agricultural Marketing Service is responsible for enforcing the labeling and purity standard provisions.

Federal Water Pollution Control Act — See Clean Water Act.

Federal-State Marketing Improvement Program (FSMIP)— Sometimes referred to in budget documents as Payments to States and Territories, the program provides matching funds to states for research and innovative projects aimed at identifying new market opportunities for producers or at improving the efficiency of agricultural marketing systems. The program is administered by AMS and recently has been funded at just over $1 million annually. [www.ams.usda.gov/tmd/fsmip.htm].

Feed ban — Usually a reference to the Food and Drug Administration (FDA) regulations that since 1997 have prohibited the feeding of most mammalian-derived proteins to cattle as a method of preventing the spread of bovine spongiform encephalopathy (BSE). Feeding of infected ruminant material back to ruminants is believed to be the most likely means of transmission of the disease. Exceptions to the FDA ban have existed for mammalian blood and blood products; gelatin; inspected cooked meat products for humans; milk products; and products containing pork and equine (and avian) proteins. On January 26, 2004, FDA officials said they would expand the feed ban by prohibiting, from ruminant feeds, ruminant blood and blood products, poultry litter, and restaurant plate waste. At issue are whether recommendations by some scientific experts to ban additional products from feed are necessary, as some foreign countries do where BSE is much more widespread.

Feed grain — Any of several grains most commonly used for livestock feed, including corn, grain sorghum, oats, rye, and barley. These grains and the farms producing them historically have received federal commodity program support. They qualify for marketing assistance loans, direct payments, and counter-cyclical payments under the 2002 farm bill (P.L. 101-171, Title I).

Feed ratio — The relationship of the cost of feeding animals to their market weight, expressed as a ratio to the sale price of animals, such as the hog/corn ratio. This serves as an indicator of the profit margin or lack of profit in feeding animals to market weight.

Feeder cattle — Steers or cows mature enough to be placed in a feedlot where they will be fattened prior to slaughter. Feeder calves are less than 1 year old; feeder yearlings are between 1 and 2 years old.

Feedlot — A confined cattle feeding facility where feeder cattle (usually less than a year old) are put on higher protein rations to prepare them for slaughter as fed cattle at "good" or better grades. Commercial feedlots of 1,000 head or more are considered by the EPA to be concentrated animal feeding operations (CAFOs) and therefore subject to rules requiring permits setting effluent standards for discharges into rivers, streams, lakes.

Feedstuffs — A general term for commercial animal feeds that includes all commercial animal feeds such as silage, hay, feed grains, the many oilseed meals, bone meal, feather meal, beet pulp, grape pomace, etc.

FERM — Forest Ecosystem Restoration and Management.

Fertigation — The application of fertilizer to crops through drip irrigation systems. Both drip irrigation and the application of fertilizer through this system is intended to conserve on the use of these inputs by delivering them to the target crops with minimal loss.

Fertilizer — Any organic or inorganic material, either natural or synthetic, used to supply elements (such as nitrogen (N), phosphate (P_2O_5), and potash (K_2O)) essential for plant growth. If used in excess or attached to eroding soil, fertilizers can become a source of water pollution.

FFA — Future Farmers of America. [www.ffa.org].

FFCBFC — Federal Farm Credit Banks Funding Corporation. [www.farmcredit-ffcb.com/farmcredit].

FFDCA — Federal Food, Drug, and Cosmetic Act (21 U.S.C. 321 *et seq.*).

FFLI — Fat free lean index.

FGP — Facility Credit Guarantee Program.

FIA — Futures Industry Association. [www.fiafii.org].

FICB — Federal Intermediate Credit Bank. (See Federal Farm Credit System)

Field office technical guide — A manual placed in all Natural Resources Conservation Service district offices and field service centers that gives the technical specifications and guidelines for all approved conservation practices. The guide can be viewed electronically at www.nrcs.usda.gov/technical/efotg .

Field service agency — Generally refers to any one of the following USDA agencies that administer programs and provide services to farmers and other rural residents through an extensive network of state and local offices: the Farm Service Agency, Risk Management Agency, Natural Resources Conservation Service, Rural Housing Service, Rural Business-Cooperative Service, and Rural Utilities Service. The Foreign Agricultural Service, because of its overseas offices, also is considered a field service agency under the Administrative Convergence plan being developed by USDA in 1998. Although other USDA agencies and mission areas also have field offices nationwide and overseas, they generally are not considered field service agencies by the Department.

Field service center — A centralized location for a variety of USDA agency field offices. These have been reduced in number from about 3,700 to about 2,600 through closures and consolidations initiated as part of a USDA reorganization and streamlining effort mandated by the Department of Agriculture Reorganization Act of 1994 (P.L. 103-354). The centers are intended to provide one-stop shopping for clients of the Farm Service Agency, the Natural Resources Conservation Service, and USDA's rural development agencies.

FIFRA — Federal Insecticide, Fungicide, and Rodenticide Act (7 U.S.C. 136 *et seq.*).

Filter strip — A type of buffer strip that is an area of vegetation, generally narrow and long, that slows the rate of runoff, allowing sediments, organic matter, and other pollutants that are being conveyed by the water to be removed by settling out. Filter strips reduce erosion and the accompanying stream pollution, and can be a best management practice.

Findley payments — Under the so-called Findley Provision authorized by the Food Security Act of 1985, P.L. 99-198, (and first sponsored by former Illinois Representative Paul Findley), USDA was able to reduce the basic, formula-set nonrecourse loan rate for major crops by up to an additional 20% if that was necessary to keep the United States competitive in international markets. If done, direct compensatory payments were made to producers equal to the amount of the loan rate reduction. These Findley Payments, limited to $200,000 per person, essentially added to the larger direct deficiency payment. The Findley provisions were superseded by the marketing loan repayment provisions of the 1996 farm bill (P.L. 104-127).

FIP — Forestry Incentives Program.

Fire ants — A South American stinging ant that has become established in southern states from North Carolina to Texas. The Animal and Plant Health Inspection Service (APHIS) quarantines nursery products from affected states and conducts research to find promising biological control agents in South America that could be imported to combat the pest in the United States.

Fire protection — Activities that encompass both silvicultural activities to reduce fuels (and thus to reduce the risks and likely damage of wildfires) and fire suppression operations to halt the spread of wildfires. The Forest Service provides fire protection for the national forests and assists with fire protection for non-federal lands through the State Fire Assistance, Volunteer Fire Assistance, and Community Fire Protection Programs.

Fire suppression operations — Activities intended to halt the spread of wildfires, from the outset of a fire (called initial attack) through final control, with the fire being extinguished. Typically, these activities include prescribed burning. Funding, especially in severe fire seasons, may be borrowed from any unobligated funds, which are typically repaid in a supplemental appropriation.

Fish and Wildlife Service (FWS) — FWS, in the Department of the Interior, is the federal agency charged with managing and protecting the nation's wild plants and animals, including migratory birds and endangered and threatened species. It generally works closely with state agencies, which have management primacy for most of these species. It manages the National Wildlife Refuge System, and cooperates with private landowners in habitat conservation. [www.fws.gov].

Fish farming — Usually, freshwater commercial aquaculture; catfish farms are an example.

Flaxseed — Flaxseed is one of the "other oilseed" crops eligible for support from marketing assistance loans and loan deficiency payments, direct payments and counter-cyclical payments under the 2002 farm bill (P.L. 107-171, Title I). North Dakota accounts for about 94% of national production (based on 2003 crop data). Flaxseed oil, like other vegetable oils, contains linoleic acid, an essential fatty acid needed for survival. It also contains significant amounts of alpha linolenic acid (ALA), an omega-3 fatty acid, which has caught the attention of health investigators. Linseed oil, a wood finish and preservative, is made from flaxseed. Another type of flax produces fibers used in paper and fabric (linen). Canada is the largest producer and exporter of flax, and is a major supplier to the U.S. market.

FLB — Federal Land Bank.

FLBA — Federal Land Bank Association.

FLCA — Federal Land Credit Association.

FLCRA — Farm Labor Contractor Registration Act (P.L. 88-582; 7 U.S.C. 2041 *et seq.*).

Flex acreage — The Omnibus Budget Reconciliation Act of 1990 mandated that deficiency payments not be made on 15% of a farm's crop acreage base, called normal flex acres. The acreage could be planted to any program crop (called flexing), but not fruits and vegetables. An additional 10% of the farm's base acreage could be flexed at the option of the operator. Flexing did not diminish the crop acreage base of a farm. The 1996 farm bill (P.L. 104-127)effectively provided total flexibility among all commodities, except for fruits and vegetables, and this policy was continued by the 2002 farm bill (P.L. 101-171, Sec. 1106).

Flood Risk Reduction Program — Though authorized, this program was not implemented. The 1996 farm bill (P.L. 104-127) authorized that contract acreage subject to frequent flooding receive up to 95% of transition payments and projected crop insurance payments in lieu of market transition payments. In return, producers would have complied with swampbuster and conservation compliance provisions and forgone future conservation program payments and disaster payments.

Flow to market — A quantity provision in a fruit or vegetable marketing order that does not change the total quantity that can be marketed during a season, but rather controls the rate or time period that quantities can be shipped to markets by means of shipping holidays and prorates.

FLPMA — Federal Land Policy and Management Act (P.L. 94-579).

Fluoroquinolones — Members of the quinolone group of antibiotics licensed to treat diseases in humans and animals. These are considered to be one of the most valuable antimicrobial drug classes available to treat human infections because of their spectrum of activity, safety, and ease of administration. The use of these types of antibiotics has been approved by the FDA for therapeutic treatment of animals, but is prohibited for extralabel use in food-producing animals because of the possible transmission to humans of resistant microorganisms (21CFR Part 530). In October 2000, the FDA issued a notice proposing to withdraw its approval of the fluoroquinolone antimicrobial enrofloxacin for poultry after determining that its use led to the development of flouroquinolone-resistant *Campylobacter*, a major foodborne pathogen in humans. As of February 2004, this proposal is in the review process.

FLSA — Fair Labor Standards Act of 1938 (P.L. 75-718; 29 U.S.C. 201 *et seq.*).

Flue-cured tobacco — A type of cigarette tobacco that along with burley tobacco account for more than 90% of U.S. tobacco production. Flue-cured farming is centered in North Carolina. Production was limited by national marketing quotas and acreage allotments, and was eligible for nonrecourse price support loans until 2005, when the quota buyout program ended these programs (P.L. 108-357, Title VI).

Fluid differential — In federal milk marketing orders, the fluid differential (or Class I differential) is the amount added to the base price of milk to determine a region's minimum price for milk used for fluid (drinking) purposes.

Fluid Milk Processor Promotion Program — A national program authorized by the Fluid Milk Promotion Act of 1990 (Subtitle H of Title XIX of P.L. 101-624, called the Fluid Act) with the purpose of increasing consumption of milk and dairy products and reducing milk surpluses by developing generic advertising programs. The program is funded by a mandatory 20¢/cwt. assessment on processors for all fluid milk processed in the contiguous 48 states and marketed commercially. The program is administered by the National Fluid Milk Processor Promotion Board. It should not be confused with the dairy farmer funded Dairy Promotion Program. The program originally required periodic reauthorization by Congress. However, the 2002 farm bill (P.L. 107-171, Sec. 1506) gave the program permanent authority. [www.ams.usda.gov/dairy/dairyrp.htm].

Fluid Milk Promotion Act of 1990 — This is the designation given to Subtitle H of Title XIX of the Food, Agriculture, Conservation, and Trade Act of 1990 (P.L. 101-624). Subtitle H authorized the establishment of a national fluid milk processor promotion program, or check-off program for fluid milk promotion. The program is funded through a 20¢/cwt. assessment on all milk processed for fluid consumption. The Act required USDA to conduct a referendum among fluid milk processors to determine if a majority favored implementing the program. The Fluid Milk Order was approved by processors and became effective December 10, 1993. The program originally required periodic congressional reauthorization. However, the 2002 farm bill (P.L. 107-171, Sec. 1506) gave the program permanent authority.

FMD — Foot-and-mouth disease.

FMDP — Foreign Market Development Program; Cooperator Program.

FmHA — Farmers Home Administration.

FMI — Food Marketing Institute. [www.fmi.org].

FMIA — Federal Meat Inspection Act (21 U.S.C. 601 *et seq.*).

FMMO — Federal milk marketing orders.

FMV — Fair market value.

FNS — Food and Nutrition Service. [www.fns.usda.gov/fns].

FO — Farm ownership loans.

FOB — Free on board.

FOIA — Freedom of Information Act (P.L. 89-554; 5 U.S.C. 552).

Foliar nutrient — Any liquid substance applied directly to the foliage of a growing plant for the purpose of delivering an essential nutrient in an immediately available form.

Fonterra Cooperative Group — In 2001, the New Zealand Dairy Board (a state trading agency) was merged with two New Zealand dairy cooperatives to form one of the world's top 10 dairy companies, Fonterra Cooperative Group. Fonterra claims to have a third of all international dairy trade, 95% of all New Zealand dairy exports, and to generate 20% of all (agricultural and nonagricultural) export receipts for the country. Fonterra Cooperative is not a state trading agency. [www.fonterra.com/default.jsp].

Food additives — Any substance or mixture of substances added to the primary food component during any phase of production, processing, packaging, storage, transport or handling. USDA allows food additives in meat, poultry and egg products only after they have received Food and Drug Administration (FDA) safety approval. Food additives are regulated under the authority of the Federal Food Drug and Cosmetic Act (21 U.S.C. 321 *et seq.*).

Food, Agriculture, Conservation, and Trade (FACT) Act of 1990 — P.L. 101-624 (November 28, 1990) was a 5-year omnibus farm bill. This 1990 farm bill continued to move agriculture in a market-oriented direction by freezing target prices and allowing more planting flexibility. New titles included rural development, forestry, organic certification, and commodity promotion programs. The law established a Rural Development Administration (RDA) in the USDA to administer programs relating to rural and small community development. It extended and modified the Food Stamp Program and other domestic nutrition programs and made major changes in the operation of P.L. 480. It revised existing law involving agricultural trade credits and guarantees. The 1990 farm bill was soon altered by the Food, Agriculture, Conservation, and Trade Act Amendments of 1991 (P.L. 102-237) to correct errors and alleviate problems in implementing the law. The amendments allowed the Farm Credit Bank for Cooperatives to make loans for agricultural exports and established a new regulatory scheme and capital standards for the Federal Agricultural Mortgage Corporation (Farmer Mac). The law also established new handling requirements for eggs to help prevent food-borne illness. More policy changes were made by the Omnibus Budget Reconciliation Act (OBRA) of 1993 (P.L. 103-66). This law intended to reduce federal farm spending by $3 billion over 5 years by eliminating USDA's authority to waive minimum acreage set-aside requirements for wheat and corn, reducing deficiency payments to farmers participating in the 0/92 and 50/92 programs from 92% to 85% of the normal payment level, reducing the acreage to be enrolled in the Conservation Reserve Program and Wetlands Reserve Program, and requiring improvement in the actuarial soundness of the federal crop insurance program. The measure also provided for a temporary moratorium on sales of synthetic bovine growth hormone and reduced the loan rate for soybeans. It reduced Market Promotion Program (MPP) funding through fiscal 1997 and provided for a series of significant MPP operational reforms. It also provided, among other provisions, for the designation of a series rural (and urban) empowerment and enterprise zones, eligible for special federal aid and tax credits.

Food Aid Consultative Group — A group created by the 1990 farm bill (P.L. 101-624) (7 U.S.C. 1725) to review and address issues concerning the effectiveness of regulations and procedures that govern U.S. food aid programs. The 2002 farm bill (P.L. 107-171) extended the authority for the Food Aid Consultative Group through 2007.

Food Aid Convention (FAC) — See International Grains Agreement.

Food and Agricultural Act of 1965 — P.L. 89-321 was the first multi-year farm legislation, providing for 4-year commodity programs for wheat, feed grains, and upland cotton, and is generally characterized as the first farm bill. It was extended for 1 more year, through 1970, by enactment of P.L. 90-559. It authorized a Class I milk base plan for the 75 federal milk marketing orders and a long-term acreage diversion under a Cropland Adjustment Program. The law also continued payment and acreage diversion programs for feed grains and cotton, and certificate and diversion programs for wheat.

Food and Agriculture Act of 1977 — P.L. 95-113 was an omnibus farm bill. This 1977 farm bill increased price and income supports and established a farmer-owned reserve for grain. It also established a new two-tiered pricing program for peanuts. Under the peanut program, producers were given an acreage allotment on which a poundage quota was set. Growers could produce in excess of their quota, within their acreage allotment, but would receive the higher of the two price-support levels only for the quota amount. Peanuts in excess of the quota are referred to as "additionals." Title XIII was designated the Food Stamp Act of 1977 and replaced the original 1964 Act with a new law making significant changes, including the elimination of the purchase requirement and simplification of eligibility requirements. Title XIV was designated the National Agricultural Research, Extension, and Teaching Policy Act and made USDA the leading federal agency for agricultural research, extension, and teaching programs. It also consolidated the funding for these programs.

Food and Agriculture Councils (FACs) — These councils were instituted in 1982 by USDA to function as interagency coordinating groups on three levels: national, state, and local. The state FACs are composed of senior level officials of individual USDA agencies within each state, and in the mid-90s the played a major role in managing the reorganization of USDA's field office structure. Local FACs have consisted of USDA representatives at county or area-wide levels; and a national FAC at USDA's Washington headquarters has served as a liaison with the state and local FACs.

Food and Agriculture Organization of the United Nations (FAO) — A UN organization, founded in 1945, that collects and disseminates information about world agriculture. FAO also provides technical assistance to developing countries in agricultural production and distribution, food processing, nutrition, fisheries, and forestry. The FAO's Global Information Early Warning System (GIEWS) monitors famine conditions in regions of risk.

Food and Drug Administration (FDA) — An agency within the Public Health Service of the Department of Health and Human Services. FDA is charged with protecting consumers by enforcing the Federal Food, Drug, and Cosmetic Act (21 U.S.C. 321 *et seq.*) as it pertains to the safety, wholesomeness, and labeling of food. FDA regulates food additives, inspects food manufacturing companies and food imports (all except for meat and poultry), conducts product sampling for pesticide and other residues, investigates outbreaks of foodborne illness, and sets food labeling standards. FDA also regulates animal feeds and veterinary medicines to ensure their safety for use in livestock that will be slaughtered for eventual human consumption. [www.fda.gov].

Food and fiber system — That sector of the U.S. economy that includes agricultural production and all economic activities supporting or utilizing that production, including farm machinery and chemical production, and processing, manufacturing, transportation, and retailing. In 2000, the food and fiber system employed 24.1 million workers, or 17.1% of the U.S. employment, and accounted for $1,264 billion, or 12.8% of the gross domestic product. Farming is one of the smaller components of the system, accounting for 1.2% of U.S. employment and $82 billion in value added to GDP.

Food and Nutrition Service (FNS) — The USDA agency that provides needy people with access to a more nutritious diet, improves the eating habits of the nation's children, and stabilizes farm prices through the distribution of surplus foods. It administers 15 domestic food assistance programs (including the Food Stamp Program, child nutrition programs (e.g., school feeding programs), and the Special Supplemental Nutrition Program for Women, Infants and Children (WIC)). FNS works in partnership with the states and reimburses most of the administrative costs the states incur for carrying out local program administration. [www.fns.usda.gov/fns].

Food Code — Published by the Food and Drug Administration (FDA), the Food Code consists of model requirements for safeguarding public health. Local and state food safety agencies commonly use the Food Code as a guide to judging the performance of restaurants, food services, retail food stores, and food vending operations with regard to food safety and quality.

Food Distribution Program on Indian Reservations (FDPIR) — This program allows Indian Tribal Organizations to operate a food distribution program as an alternative to the Food Stamp Program for those living on or near an Indian reservation. Eligibility for benefits is the same as for the food stamps and funds are drawn from food stamp appropriations. Foods contained in packages include frozen and/or canned meats and poultry, canned fruits and vegetables and juices, dry cereals, cornmeal, flour, butter, macaroni, cheese, evaporated and nonfat dry milk, oats, peanuts and peanut butter, shortening and oils. [www.fns.usda.gov/fdd/programs/fdpir].

Food donations to charitable institutions, soup kitchens, and food banks — Donations of food by the Commodity Credit Corporation (CCC) to help provide meals to needy people. Foods donated are from agricultural surpluses acquired by USDA as part of its price stabilization and surplus removal activities. Eligible charitable groups range from churches operating community kitchens for the homeless to orphanages and homes for the elderly. Other eligible groups include meals-on-wheels programs, soup kitchens, temporary shelters, correctional institutions offering rehabilitative activities, group homes for the mentally retarded, and hospitals that offer general and long-term health care.

Food for Peace Program — A label given to the food donation activities carried out overseas under P.L. 480.

Food for Progress Program (FFP) — A food aid program originally authorized by the Food Security Act of 1985 (P.L. 99-198) to provide commodities on credit terms or on a grant basis to developing countries and emerging democracies to assist in the introduction of elements of free enterprise into the countries' agricultural economies. Commodities may be provided under authority of P.L. 480 (Title I) or Section 416(b). The Commodity Credit Corporation (CCC) may purchase commodities for use in Food for Progress if the commodities are currently not held in CCC inventory. The 2002 farm bill (P.L. 107-171) extends authority for the FFP through 2007. [www.fas.usda.gov/excredits/FoodAid/FFP/ffp.html].

Food guide pyramid — A graphic developed by USDA and first published in 1992 as the consumer guide to implementing the dietary guidelines in their own food choices. The 2005 release, called MyPyramid, translates the principles of the 2005 Dietary Guidelines for Americans and other nutritional standards to assist consumers in making healthier food and physical activity choices. [www.nal.usda.gov/fnic/Fpyr/pyramid.html].

Food insecurity — See Food security.

Food label — A food label is a display of written, printed, or graphic matter upon the immediate container (not including package liners) of any food product, and which bears, in accordance with 21 U.S.C. 343(i)(2), the ingredients in the food contained in the package. Food labels are regulated by the Food and Drug Administration (FDA) under the authority of the Federal Food, Drug and Cosmetic Act (21 U.S.C. 321 *et seq.*)and by the Food Safety and Inspection Service (FSIS) under its statutes, for most meat and poultry products.

Food package — Generally refers to foods contained in the package of specific items provided to those participating in WIC or the Commodity Supplemental Food Program (CSFP). Also, may refer to foods distributed by food banks and pantries, and by Indian Tribal agencies distributing commodities in lieu of food stamps.

Food power — The act of withholding or making available agricultural commodities for export or aid by an exporting nation or group of nations for the purpose of influencing the actions of another country or group of countries. Food power implies a foreign policy motivation rather than a financial or humanitarian motivation to export activities.

Food safety initiative — An interagency initiative begun under the Clinton Administration in 1997 to coordinate the activities of the Food and Drug Administration (FDA), the Centers For Disease Control and Prevention (CDC), EPA, and the USDA to reduce the annual incidence of foodborne illness. [www.foodsafety.gov].

Food Safety and Inspection Service (FSIS) — A 9,300 employee agency within USDA responsible for ensuring food safety in some 6,000 meat and poultry plants throughout the United States. The agency also has a program to ensure that imported meat and poultry has been inspected under a system at least equivalent to that of the United States. Most of the cost of meat and poultry inspection is borne by the government, in contrast to the costs of border inspection of meat product imports by the Department of Homeland Security (DHS), and inspection and grading related to marketing standards by the Agricultural Marketing Service (AMS) and the Grain Inspection Packers and Stockyards Administration (GIPSA), which are fee-based services. [www.usda.gov/agency/fsis/home page.htm].

Food Safety Commission — The 2002 farm bill (P.L. 101-171, Sec. 10807) directs that there be established a Food Safety Commission composed of 15 members appointed by the President. The Commission is to report within a year to the President and the Congress on enhancing the food safety system of the United States.

Food security, international — Access by all people at all times to enough food for an active healthy life. Food security at a minimum includes the ready availability of nutritionally adequate and safe food, and an assured ability to acquire acceptable foods in socially acceptable ways, that is, without having to resort to emergency food supplies, scavenging, stealing, or other coping strategies. The World Food Summit, convened in Rome in November 1996 by the Food and Agriculture Organization of the United Nations, estimated that 800 million people worldwide do not have enough food to meet their basic nutritional needs. Representatives of the more than 180 nations attending the Summit pledged to work to reduce this number by half by no later than 2015. Causes of food insecurity may include poverty, civil conflict, governmental corruption, environmental degradation, and natural disasters. A U.S. position paper on international (world) food security, released in October 1997, argues that food security also requires "...social and economic conditions which empower individuals to gain access to food, either by producing food themselves or earning income to buy food." Food security has taken on an alternate meaning since the terror attacks of 2001, referring to programs and practices designed to protect the food supply from adulteration resulting from acts of terrorism, vandalism, or other crimes. These programs may be enhancements of existing food safety activities, or may be novel programs focused entirely on preventing certain human activities. In this newer context, there are programs for "food safety and security" at FDA and USDA/FSIS. Some of the FDA programs are mandated in the Public Health Security and the Bioterrorism Preparedness and Response Act of 2002 [P.L. 107-188]. See Hunger / Food insecurity, domestic, and Agroterrorism, Bioterrorism.

Food Safety Institute of the Americas (FSIA) — FSIA, based in Miami, Florida, is a cooperative educational and research organization established in 2004 to promote food safety and identify and develop educational programs in the Western Hemisphere. Cooperating organizations include USDA's Food Safety and Inspection Service, Miami-Dade College, the University of Florida, the Pan American Health Organization, and agriculture and food safety agencies in various American countries. [www.fsis.usda.gov/Fact_Sheets/FSIA/index.asp].

Food Security Act of 1985 — P.L. 99-198, a 5-year omnibus farm bill, allowed lower commodity price and income supports and established a dairy herd buyout program. This 1985 farm bill made changes in a variety of other USDA programs. Several enduring conservation program were created, including sodbuster, swampbuster, and the Conservation Reserve Program. Shortly after enactment, the Technical Corrections to Food Security Act of 1985 Amendments (P.L. 99-253 gave USDA discretion to require cross-compliance for wheat and feed grains instead of mandating them, changed acreage base calculations, and specified election procedures for local Agricultural Stabilization and Conservation committees. Technical changes and other modifications were enacted by the Food Security Improvements Act of 1986 (P.L. 99-260), including limiting the non-program crops that could be planted under the 50/92 provision, permitting haying and grazing on diverted wheat and feed grain acreage for a limited period in regions of distress, and increasing deductions taken from the price of milk received by producers to fund the dairy termination program (also called the whole herd buy-out) program. Again in 1986, the Omnibus Budget Reconciliation Act (P.L. 99-509) made changes in the 1985 Act requiring advance deficiency payments to be made to producers of 1987 wheat, feed grains, upland cotton, and rice crops at a minimum of 40% for wheat and feed grains and 30% for rice and upland cotton. The 1985 Act also amended the Farm Credit Act of 1971. Further commodity program changes were made in the FY1987 agricultural appropriations bill (P.L. 99-591). In addition to its funding provisions, P.L. 99-591 set the annual payment limitation at $50,000 per person for deficiency and paid land diversion payments, and included honey, resource adjustment (excluding land diversion), disaster, and Findley payments under a $250,000 aggregate payment limitation. Once again, the Omnibus Budget Reconciliation Act of 1987 (P.L. 100-203) not only set the 1988 fiscal year budget for agriculture and all federal agencies, but also set target prices for 1988 and 1989 program crops, established loan rates for program and non-program crops, and required a voluntary paid land diversion for feed grains. P.L. 100-203 further defined who could receive farm program payments by defining a "person" in terms of payment limitations.

Food Security Commodity Reserve — See Bill Emerson Humanitarian Trust.

Food Security Wheat Reserve (FSWR) — Title III of the Agricultural Act of 1980 (P.L. 96-494) established a reserve of up to 4 million metric tons of wheat for use in meeting emergency food needs in developing countries. This reserve generally was to be used to meet famine or other urgent or extraordinary relief requirements during periods of tight supplies and high prices when commodities are not available under the provisions of P.L. 480. The FSWR was replaced by the Food Security Commodity Reserve under the 1996 farm bill (P.L. 104-127, Sec. 225), which has since been renamed the Bill Emerson Humanitarian Trust. The 2002 farm bill (P.L. 107-171, Sec. 3202) extended the Trust through 2007.

Food Service Management Institute — This institute provides instruction, research, and materials in support of better food service management practices by child nutrition providers receiving federal support (e.g., schools operating school meal programs). It is permanently authorized under Section 21 of the National School Lunch Act (P.L. 79-396, as amended), with an annual entitlement funding level of $3 million. [www.nfsmi.org].

Food Stamp Act of 1964 — P.L. 88-525 provided permanent legislative authority to the Food Stamp Program, which had been administratively implemented on a pilot basis in 1962. It was later replaced and completely rewritten and revised by the food stamp provisions of the Food and Agricultural Act of 1977 (P.L. 95-113, Title XIII; 7 U.S.C. 2011 *et seq.*), which eliminated the purchase requirement and simplified eligibility requirements. Amendments were made to this Act in 1981-82, 1984-85, 1988, 1990, 1994, 1996, 1997, 1998 and 2002 (most recently by Title IV of the 2002 farm bill (P.L. 107-171, Sec. 4101-4126). The current Food Stamp Act (7 U.S.C. 2011 *et seq.*) includes authority through FY2007 for the regular Food Stamp Program, for Nutrition Assistance Grants to Puerto Rico and American Samoa (in lieu of food stamps), for Food

Distribution on Indian Reservations, and for commodity purchases for the Emergency Food Assistance Program.

Food Stamp Program (FSP) — This program supplements the food buying power of eligible low-income households by providing them with monthly benefits through coupons or electronic benefit transfer (EBT) cards. Eligibility is governed primarily by a household's financial resources (e.g., income eligibility generally is limited to those with total cash income below 130% of the federal poverty guidelines). However, applicants and recipients also must meet some non-financial requirements (e.g., the eligibility of noncitizens is limited, work requirements are imposed on most adults). In general, benefits may be used to purchase any food item for home consumption in an approved food store. The regular food stamp program operates in the 50 states, the District of Columbia, Guam, and the Virgin Islands. Variants of the regular program (funded through nutrition assistance grants) operate in Puerto Rico, American Samoa, and the Northern Mariana Islands. This program is administered at the federal level by the Food and Nutrition Service and funded as an entitlement by annual agriculture appropriations acts. [www.fns.usda.gov/fsp].

Food standards — See Standards of identity for food.

Foodborne illnesses — Human illnesses caused by foodborne pathogens.

Foodborne pathogens — Disease-causing microorganisms found in food, usually bacteria, fungi, parasites, protozoans, and viruses. The top ten pathogens are: *Salmonella*; *Staphylococcus aureus*; *Campylobacter jejuni*; *Yersinia enerocolitica*; *Listeria monocytogenes*; *Vibrio cholerae non-01*; *Vibrio Parahemolyticus*; *Bacillus cereus*; *Escherichia coli - enteropathogenic*; and *Shigella*. Meat, poultry, shell eggs, dairy products, seafood, and produce are the foods in which these microorganisms most commonly occur.

Foot-and-mouth disease (FMD) — A major disease of cloven-footed animals (e.g., cattle and pigs) that currently does not exist in the United States. The Animal and Plant Health Inspection Service (APHIS) conducts a surveillance program to track the disease in foreign countries, regulates the importation of animal products from countries where FMD exists, tests imported animals in quarantine, and would lead eradication efforts if FMD were detected in domestic livestock.

FOR — Farmer-Owned Grain Reserve.

Forage value index (FVI) — A derived index of the relative change in the previous year's average monthly rate per head for pasturing cattle on privately owned land in the West. Used in calculating federal grazing fees.

Forage — A plant, other than the separated grain, that is grazed or harvested for use as feed for animals. Generally, the term forage includes pasture vegetation, hay, and silage. Forage production is the largest use of land in farms. All types of forage production involves 60 % of land in farms, with pastureland accounting for the most. Pastureland of all types is a combination of cropland pasture, pasture and rangeland, and woodlands pastured.

Foreign Agricultural Service (FAS) — The USDA agency that administers agricultural export and food aid programs. FAS also is responsible for formulating agricultural trade policy, helping negotiate reductions in foreign agricultural trade barriers, and carrying out programs of international cooperation and technical assistance. The agency maintains a global network of agricultural officers (counselors and attaches) as well as a Washington-based staff to analyze and disseminate information on world agriculture and trade, develop and expand export markets, and represent the agricultural trade policy interests of U.S. producers in multilateral forums.

Foreign Market Development Program (FMDP) — More commonly called the Cooperator Program.

Forest and Rangeland Renewable Resources Planning Act of 1974 (RPA) — P.L. 93-378 directs the Forest Service to prepare and update an assessment every 10 years: to inventory and monitor the status and trends of all forest lands and range lands; to prepare a long-range plan every 5 years to guide Forest Service policies; and, to prepare interdisciplinary forest plans for units of the National Forest System. Since FY1999, provisions in the annual Interior appropriations acts have prevented completion of the 2000 Resource Planning Act (RPA) assessment.

Forest Ecosystem Restoration and Management (FERM) — System designed to facilitate the recovery or re-establishment of native ecosystems to conditions consistent with their evolutionary environments in order to prevent further degradation and conserve native plants and animals. Related to rehabilitation, reclamation, and bioremediation.

Forest health — A term used for a collection of concerns over forest conditions, including both current problems (e.g., insect and disease infestations, wildfires, and related tree mortality) and risks of future problems (e.g., too many small-diameter trees (overstocking), excessive biomass, and an unnatural combination of tree species in mixed stands).

Forest Land Enhancement Program (FLEP) — The Forest Land Enhancement Program (FLEP) was adopted in the 2002 farm bill (P.L. 107-171, Sec. 8002) as an amendment to the Cooperative Forestry Assistance Act of 1978 (P.L. 95-313; 16 U.S.C. 2101 *et seq.*). FLEP replaces the Stewardship Incentives Program (SIP) and the Forestry Incentives Program (FIP). FLEP is optional in each state and is a voluntary program for non-industrial private forest (NIPF) landowners. It provides for technical, educational, and cost-share assistance to promote sustainability of the NIPF forests. The law provided FLEP with $100 million from the CCC through FY07. Half of these funds were diverted to wildfire control in 2003, and $40 million of those funds have not been replenished and the spending authority has been cancelled. [www.fs.fed.us/spf/coop/programs/loa/flep.shtml].

Forest plans — Land and resource management plans for units of the National Forest System under the Forest and Rangeland Renewable Resources Planing Act (P.L. 93-378) and the National Forest Management Act (P.L. 94-588). The Acts specify a detailed process and numerous requirements, including public participation and periodic revision, intended to achieve multiple use and sustained yield of the national forests.

Forest Service (FS) — The largest USDA agency in terms of employees (35,492 staff years, 31% of the total USDA) responsible for administering the National Forest System, for providing financial and technical forestry assistance to states and to private landowners under State & Private Forestry, and for conducting Forestry Research. The gross value of all Forest Service program activities in FY2003 was $4.945 billion. The Forest Service has been funded in the annual Department of the Interior and Related Agencies appropriations acts since 1955. [www.fs.fed.us].

Forest Stewardship Council (FSC) — An international organization that sets standards for sustainable forest management and that accredits other organizations to assess sustainable forest management by forest landowners. [www.fscus.org].

Forestland — A classification of land use in the Natural Resources Inventory (NRI). It includes areas where trees that will be at least 13 feet tall at maturity cover at least 10% of the land and must be at least an acre in size. Forestland was found on 407 million acres, almost 30% of all private lands, in the 1997 NRI. Compare with timberland.

Forestry Incentive Program (FIP) — Initiated in 1975 and administered by the Natural Resources Conservation Service, FIP provided financial assistance for up to 65% of the cost of silvicultural activities on nonindustrial private forest land of generally less than 1,000 acres. The program was terminated in the 2002 farm bill (P.L. 107-171), and replaced with the Forest Land Enhancement Program (FLEP).

Forfeiture penalty (sugar) — This refers to a penalty mandated by the 1996 farm bill (P.L. 104-127) on the forfeiture of sugar pledged as collateral by processors for nonrecourse loans from the Commodity Credit Corporation (see Loan forfeiture). The penalty was 1¢/lb. on raw cane sugar, and 1.072¢/lb. on refined beet sugar. This penalty effectively lowered the price support levels by the penalty amount (i.e., for raw cane sugar, from 18¢/lb. to 17¢/lb.). The 2002 farm bill (P.L. 107-171) repealed this provision.

Formula funds — Federal dollars distributed to the land grant colleges of agriculture through formulas found in the Hatch Act (7 U.S.C. 361a *et seq.*), the Smith-Lever Act (7 U.S.C. 341 *et seq.*), the McIntire-Stennis Act (16 U.S.C. 582a *et seq.*), and the Evans-Allen Act (7 U.S.C. 3222) for (1) agricultural research at the state agricultural experiment stations, (2) Extension Service programs, (3) forestry research at the land grant colleges of agriculture, and (4) research at the 1890 institutions, respectively.

Formula pricing — An arrangement where a buyer and seller agree in advance on the price to be paid for a product delivered in the future, based upon a pre-determined calculation. For example, a packer might agree to pay a hog producer the average cash market price on the day the hogs will be delivered, plus a 2-cent per-pound premium. Such transactions have been used widely in agriculture, particularly for livestock. Users believe that formula pricing brings efficiency and predictability to market transactions. However, as the use of formula pricing expands, fewer animals are sold in cash markets, where prices are more widely reported and understood by producers. Some of these producers believe that formula pricing makes it harder to determine the true value of their animals in the marketplace, and creates greater opportunity for buyers to manipulate and pay lower prices.

Forward contract — A cash transaction common in many industries, including agricultural commodity merchandising, in which a commercial buyer and seller agree upon delivery of a specified quality and quantity of goods at a specified future date. A price may be agreed upon in advance, or there may be agreement that the price will be determined at the time of delivery. Forward contracts, in contrast to futures contracts, are privately negotiated and are not standardized.

Forward market — This refers to informal (non-exchange) trading of commodities to be delivered at a future date. Contracts for forward delivery are personalized (i.e., delivery time and amount are determined between seller and customer).

Forward selling — Forward contracting in which the price is fixed at the time the contract is entered.

Four-firm ratio — It is common to express the degree of concentration within an industry, including agriculture (that is, the degree to which a few firms dominate sales or production) as a ratio, by stating the share (%) held by the top four firms.

FPP — Food for Progress Program. [www.fas.usda.gov/excredits/FoodAid/FFP/ffp.html].

FPP —Farmland Protection Program, renamed to the Farm and Ranch Lands Protection Program (FRPP).

FPPA — Farmland Protection Policy Act (P.L. 97-98; 7 U.S.C. 4201 *et seq.*).

FR — Federal Register. [www.gpoaccess.gov/fr/index.html].

FR — Final rulemaking.

FRAC — Food Research and Action Center. [www.frac.org].

Franc-Zone Africa — Countries in West and Central Africa, all former French colonies, who maintain a common currency, the CFA franc.

Free lunch (or breakfast, snack, or milk) — Refers to a federally subsidized meal (or snacks or half-pints of milk) that is offered under a child nutrition program at no cost to children who apply for and whose family income qualifies them for it. Income eligibility for free meals is set at 130% or less of the federal poverty income level, and substantially higher subsidies generally are provided for these meals than for paid meals, or reduced price meals. The program is federally administered by the FNS of the USDA.

Free market — A system in which the market forces of supply and demand determine prices and allocate available supplies, without government intervention. The concept of a free-market approach in agricultural policy, in its purest form, is no government price and income support programs, supply management programs, export subsidies, or barriers to international trade.

Free on board (FOB or fob) — A commercial trade term indicating that the seller assumes all responsibilities and costs up to the specific point or stage of delivery named (usually a vessel designated by the buyer) including transportation, packing, insuring, etc. A wide variety of fob terms are used, such as fob factor Detroit, fob cars New York, fob ship Norfolk. Free on board vessel, under most P.L. 480 grain contracts, means delivery at the discharge end of the loading spout.

Free rider — In agricultural policy, the term generally refers to a firm or person who benefits from a collectively funded activity (such as a generic advertising and promotion, or check-off, program) without contributing to its costs.

Free stocks — Commodity stocks owned by farmers or others in the trade, rather than by those owned or controlled by the government. (Supplies in the Food Security Commodity Reserve are government-controlled and not considered free stocks.).

Free trade area — A group of countries that have removed trade barriers among the members, but each country may maintain its own trade regime with nonmember countries. The best known current example is the North American Free Trade Agreement (NAFTA).

Freedom-to-farm — A phrase that was used in the congressional arena to characterize the production flexibility contract provisions of the 1996 farm bill (P.L. 104-127).

Freight density — Thousands of gross ton-miles of freight per mile of railroad owned.

FRPP — Farm and Ranch Lands Protection Program, formerly the Farmland Protection Program (FPP).

FS, USFS — U.S. Forest Service. [www.fs.fed.us].

FSA — Farm Service Agency. [www.fsa.usda.gov/pas/default.asp].

FSCR — Food Security Commodity Reserve. See Bill Emerson Humanitarian Trust (7 U.S.C. 1736f-1).

FSIA — Food Safety Institute of the Americas. [www.fsis.usda.gov/Fact_Sheets/FSIA/index.asp].

FSIS — Food Safety and Inspection Service. [www.fsis.usda.gov].

FSMIP — Federal-State Marketing Improvement Program. [www.ams.usda.gov/tmd/fsmip.htm].

FSP — Food Stamp Program. [www.fns.usda.gov/fsp].

FSRIA — Farm Security and Rural Investment Act of 2002, P.L. 107-171; commonly called the 2002 farm bill.

FSU — Former Soviet Union.

FSWR — Food Security Wheat Reserve (P.L. 96-494, Title III). See Bill Emerson Humanitarian Trust (7 U.S.C. 1736f-1).

FTA — Free trade area, or free trade agreement.

FTAA — Free Trade Area of the Americas.

FTE — Full-time equivalent (generally refers to agency staffing levels).

Fuel cell — A device that combines hydrogen (or a hydrogen-bearing compound) with oxygen to produce an electric current. Currently, fuel cells are being studied as a replacement for the internal combustion engine in transportation applications, and as a source of power for stationary applications. Because fuel cells could be used in distributed power systems, there is interest in their use for agricultural applications, as well.

Full-cost water — An annual fee for water delivered from Bureau of Reclamation facilities that includes project construction costs attributed to irrigation, as well as outstanding deficits on operation and maintenance charges, with interest on both accruing from October 12, 1982. The term is defined in Section 202 of the Reclamation Reform Act of 1982 (P.L. 97-293). The Bureau charges full-cost for water delivered to lands above the acreage limitation.

Fumigant — A vaporized pesticide used to control pests in soil, buildings and greenhouses, and chambers holding products such as fruits to be treated. Phosphine is a fumigant often used in stored grain and cargo shipments of grain.

Fumonisin — A mycotoxin that can cause liver and brain damage in horses.

Functional foods — Agribusinesses and consumers increasingly view food not only for its basic nutrition, but also for more specific health benefits. Functional foods and nutraceuticals contain active, naturally occurring components that have shown, or promise to show, physiological benefits or reduced risk of disease, in addition to their traditional nutritional uses. A functional food can be the same as or similar to a conventional food. A nutraceutical is this component isolated from a food and sold in dosage form. Governments, industry, and researchers are keenly interested in nutraceuticals and functional foods both for their potential consumer health benefits and for their ability to help farmers economically. Policy questions revolve around their efficacy, safety and regulation. See Nutraceuticals, and Farmaceuticals.

Fund for Rural America — A fund established by the 1996 farm bill (P.L. 104-127, Sec. 793) to augment existing resources for agricultural research and rural development through an annual transfer of funds from the U.S. Treasury to USDA. The Fund was notable for being the first time that mandatory money (in the form of a mandatory annual transfer to USDA from the U.S. Treasury) was provided for research programs, which traditionally receive discretionary funds as provided annually by Congressional appropriators. One-third of the fund was designated for competitive agricultural research grants, one-third for rural development projects, and one-third for either research or rural development, at the Secretary's discretion. The 1996 farm bill (P.L. 104-127) authorized the U.S. Treasury to transfer $100 million annually to the Fund for 3 years, but a recision reduced that to $80 million. The Agricultural Research, Extension, and Education Reform Act of 1998 (P.L.105-185)extended the authority for the program through FY2003 with an annual transfer to USDA of $60 million. However, the omnibus appropriations law for FY1999 (P.L. 105-277, October 21, 1998) prohibited the expenditure of the $60 million for Fund grants and projects. The program was repealed in 2002 by the 2002 farm bill (P.L. 107-172, Sec. 6043).

Fungibility — The characteristic of interchangeability. Bulk commodities are generally described as fungible, whereas those with special characteristics may be marketed as identity preserved. Futures contracts for the same commodity and delivery month are fungible due to their standardized specifications for quality, quantity, delivery date and delivery locations.

Fungicide — Any pesticide used to control, deter, or destroy fungi, which are forms of plant life (including molds and yeasts) that lack chlorophyll and are unable to make their own food (such as the plant pathogen, powdery mildew).

Furrow irrigation — Small, shallow channels guide water across the surface of a leveled field. Crops are typically grown on a ridge or raised bed between the furrows. This is the major irrigation system that is based on gravity flow.

Futures contract — A standardized agreement calling for deferred delivery of a commodity, or its equivalent, entered through organized futures exchanges. Most agricultural futures contracts call for physical delivery, but feeder cattle futures contracts call for cash settlement at contract maturity. In fact, contracts are usually liquidated before delivery. Traders are classified as hedgers or speculators. The 1996 farm bill (P.L. 104-127)required USDA to conduct research through pilot programs to determine if futures and options contracts can provide producers with reasonable protection from the financial risks of fluctuations in price, yield, and income inherent in the production and marketing of agricultural commodities.

Futures price — (1) Commonly held to mean the price of a commodity for future delivery that is traded on a futures exchange. (2) The price of any futures contract.

FVI — Forage value index.

FW — Farmed wetlands.

FWPCA — Federal Water Pollution and Control Act (Clean Water Act) (33 U.S.C. 1251 *et seq.*).

FWS — Fish and Wildlife Service. [www.fws.gov].

FY — Fiscal year.

GAO — Government Accountability Office, formerly the General Accounting Office. [www.gao.gov].

GAP — Government Accountability Project.

Gas chromatograph / mass spectrometer — An analytical technique for identifying the molecular composition and concentrations of various chemicals in water and soil samples.

Gasohol — Registered trade name for a blend consisting of 90% unleaded gasoline and 10% fermentation ethanol. Gasohol emissions contain less carbon monoxide than those from gasoline.

Gasoline equivalent gallons (GEG) — This measure standardizes fuels on the basis of the Btu value of a gallon of gasoline. One gallon of gasoline is 125,000 Btu. One gallon of ethanol equals 0.65 GEG. One gallon of regular diesel and B20 (20% biodiesel) equals 1.11 GEG.

GATT — General Agreement on Tariffs and Trade (see Uruguay Round and World Trade Organization.).

GDP — Gross domestic product.

GE — Genetic engineering, or genetically engineered.

GEG — Gasoline equivalent gallons.

Gene stacking — Involves using genetic engineering (GE) to combine traits (e.g., herbicide tolerance and insect resistance) in the seed of one crop.

General Agreement on Tariffs and Trade (GATT) — An agreement originally negotiated in Geneva, Switzerland, in 1947 to increase international trade by reducing tariffs and nontariff trade barriers. The agreement provides a code of conduct for international commerce and a framework for periodic multilateral negotiations on trade liberalization and expansion. The original agreement and its elaboration through several rounds of multilateral negotiations is referred to as GATT 1947. The Uruguay Round Agreements (resulting from negotiations that stretched from 1986 through 1993 among over 100 nations) emphasized the elimination of nontariff barriers to GATT's traditional focus on lowering tariffs and also established the World Trade Organization (WTO) to replace the mechanisms created by the GATT. Among the agreements negotiated in the Uruguay Round and administered by the WTO are the Agreement on Agriculture and the Agreement on the Application of Sanitary and Phytosanitary Measures. The WTO officially entered into force on January 1, 1995. The WTO administers the GATT 1947, the various agreements resulting from the Uruguay Round negotiations (known collectively as GATT 1994), and dispute resolution among WTO member countries.

General Sales Manager (GSM) — General Sales Manager of the Foreign Agricultural Service. This office administers the export credit guarantee programs (GSM-102 and GSM-103), the export enhancement program, the P.L. 480 program, and other USDA export assistance programs.

General Services Support Estimate (GSSE) — An OECD indicator of the annual monetary value of gross transfers of general services provided to agriculture collectively, arising from policy measures that support agriculture, regardless of their nature, objectives and impacts on farm production, income, or consumption of farm products. Examples include research and development, education, infrastructure, and marketing and promotion programs. The GSSE can be expressed in monetary terms or as a percentage of the total support to agriculture (percentage GSSE). See Producer support estimate (PSE), Consumer support estimate (CSE), and Total support estimate (TSE).

Generally recognized as safe (GRAS) — A regulatory category created for a group of food additives that were exempted from the more rigorous regulatory requirements for food additives in the 1958 Food Additives Amendment (P.L. 85-929)to the Food Drug and Cosmetics Act of 1938 (21 U.S.C. 321 *et seq.*). A substance is accorded GRAS status if qualified experts generally agree that science has shown it to be safe in food under the conditions of its intended use.

Generic advertising and promotion (check-offs) — The promotion of a particular commodity without reference to a specific producer, brand name, or manufacturer. Because individual producers of nearly homogeneous agricultural commodities cannot easily convince consumers to choose one egg or orange or a single cut of beef over another, they often have joined together in commodity promotion programs to use generic advertising in an effort to expand total demand for the commodity, with the objective of helping their own sales as well. Activities are intended to expand both domestic and export demand; examples include advertising, nutrition education, research to improve product quality and appeal, market research studies, and technical assistance. These activities are often self-funded through assessments on marketings – hence, the name check-off programs. Congress has permitted producer groups to make check-offs mandatory, and this aspect has generated legal challenges by some producers, who contend they must pay taxes for activities they would not underwrite voluntarily. The U.S. Supreme Court [in United States v. United Foods, Inc., 533 U.S. 405, 412 (2001)] ruled that the mushroom check-off violated the Constitutional free speech provisions (First Amendment), creating uncertainty about the future of other check-offs. Since then, separate lower federal courts have ruled that various check-offs also are unconstitutional. However, on May 23, 2005, the Supreme Court ruled that the beef check-off does not violate the First Amendment. In its decision, Johanns v. Livestock Marketing Association and Nebraska Cattlemen v. Livestock Marketing Association (Nos. 03-1164 and 03-1165), a majority of the Court agreed with check-off defenders that the programs are in fact "government speech" (an issue that was not considered in the mushroom decision). Most legal observers predict that this decision will be used to defend other challenges to check-offs.

Generic certificates — Commodity certificates used by the Commodity Credit Corporation (CCC) in the 1980s to meet payment obligations and simultaneously dispose of commodity inventories. Farmers paid with generic certificates could trade them for commodities owned and stored by the CCC.

Genetic engineering — The use of recombinant DNA or other specific molecular gene transfer or exchange techniques to add desirable traits to plants, animals, or other organisms, or to enhance biological processes (7 CFR 340.1). Organisms modified by genetic engineering are sometimes referred to as transgenic, bioengineered, or genetically modified organisms (GMOs). The Animal and Plant Health Inspection Service (APHIS) regulates the release of genetically engineered organisms for field experiments, and the Food and Drug Administration (FDA) regulates their use in food animals and crops.

Genetic use restriction technologies (GURTs) — A group of complex genetic transformations that insert a genetic "off switch" in plants to prevent the unauthorized practice of replanting saved seed and thus benefitting from unique genetic traits without providing any return to the company that created them. (See Terminator seeds.).

Genetically modified organisms (GMO) — A term, currently used most often in international trade discussions, that designates crops carrying new traits that have been inserted through genetic engineering (e.g., Roundup Ready soybeans, Bt cotton, Bt corn). GMO crops are meeting resistence from some trading partners, particularly the European Union, that are responding in turn to consumer concerns over public health and environmental safety aspects of GMOs. GMOs are not permitted in the USDA National Organic Program that became effective in October 2002 (7 CFR 205). The U.S. scientific community maintains that research shows GMOs to be safe and that the regulatory process for their commercial approval, which includes USDA, Food and Drug Administration (FDA), and the EPA, is an adequate safeguard against any potential problems.

Genome — All the genetic material in the chromosomes of a particular organism. USDA's Agricultural Research Service (ARS) has a Plant Genome Mapping Program to identify, characterize, and map the position of agriculturally important genes on the chromosomes of plants grown as crops or trees in order to better use these genes for improving the characteristics of the plant (resistance to disease, higher yields, etc.).

Geographic Information System (GIS) — Computerized systems used to compile, retrieve, analyze, and display spatially referenced data. Farming activities that utilize GIS typically include harvesting, fertilizing, pest control, seeding, and irrigation. Use of GIS is called precision farming.

Geographical indications (GIs) — Geographical indications are place names, or words closely associated with places that are used to identify wines, spirits, and certain other, mainly food and agricultural, products. Examples include Champagne, Idaho potatoes, Roquefort cheese, Scotch whiskey, Swiss watches, or Tequila. The WTO's Agreement on Trade-Related Aspects of Intellectual Property Rights (TRIPS) says geographical indications in general should be protected, at least with a view to avoiding unfair competition and consumers being misled. Currently, TRIPS accords a higher level protection to geographical indications for wines and spirits (even if consumers are not misled or if using the names does not constitute an act of unfair competition). In WTO trade negotiations, some countries have proposed extending the higher level of protection to other food and agricultural products.

Geographically disadvantaged farmers and ranchers — Defined in the 2002 farm bill (P.L. 107-171, Sec. 10906) as a farmer or rancher in an insular area or a state other than 1 of the 48 contiguous states. This law requires the USDA to submit a report to the House and Senate Agriculture Committees describing barriers to agricultural production in such regions, and means of encouraging and assisting those in such areas to own and operate farms and ranches, and participate in USDA agricultural programs.

Germicide — Any compound that kills disease-causing microorganisms. Germicides must be registered as pesticides.

Gilt — See Barrows and gilts.

GIPSA — Grain Inspection, Packers & Stockyards Administration. [www.usda.gov/gipsa].

GIS — Geographic information systems.

Gleaning — Collecting unharvested crops from fields or obtaining unused agricultural products from farmers, processors, or retailers, usually for distribution to food banks and charitable feeding organizations.

Global Food for Education Initiative(GFEI) — A pilot project, launched in FY2000, which used the donation of surplus agricultural commodities under Section 416 of the Agricultural Act of 1949 (P.L. 89-439, as amended) to support a global school feeding program. The GFEI has been superseded by the McGovern-Dole International School Feeding and Child Nutrition Program (IFEP), established by the 2002 farm bill (P.L. 107-

171, Sec. 3107). By decision of the President, the McGovern-Dole Program was to be administered by USDA.

Global Positioning System (GPS) — A network of satellites that can be used by ground-based units to precisely determine their location by latitude and longitude. GPS is part of the infrastructure required to operate geographic information systems that are used to practice precision farming.

Global warming — See Climate change and/or Greenhouse gases.

Gluten — Wheat proteins collectively are called gluten. Generally, higher protein wheat flour has a more sticky dough and makes higher rising yeast breads because the fermentation gasses do not readily escape. Some people are allergic to wheat gluten (gluten intolerance). Concentrated wheat gluten powder is manufactured and sold to bakeries. The import of wheat gluten (in competition with U.S. manufacturers) has created trade disputes in the past. Corn gluten meal and corn gluten feed are coproducts of corn syrup and ethanol production. Corn gluten meal is a higher protein feed for poultry, swine, fish, and pet foods. Corn gluten feed is a medium protein feed ingredient for beef and dairy cattle. Large amounts of corn gluten feed and meal are exported in addition to being used domestically.

GMA — Grocery Manufacturers of America. [www.gmabrands.com].

GMO — Genetically modified organism.

GMP — Good manufacturing practices.

Goldschmidt Thesis — The thesis, by Walter Goldschmidt, that farm scale and other management characteristics (e.g., wage labor) are associated with certain community characteristics. Goldschmidt was a California anthropologist who conducted pioneering rural community research under USDA's Bureau of Agricultural Economics on two California farming communities (Arvin and Dinuba). His 1944 research showed that large-scale, especially industrial, farm structures in one community were associated with adverse community conditions. Smaller-scale, owner-operated farms in the other community, were associated with more vibrant, diverse economies and with higher standards of living. A large body of research has accumulated testing the Goldschmidt Thesis. However, the validity of the thesis that farm structural characteristics dominating an area can produce certain rural community characteristics remains ambiguous. Research results supporting the thesis and other conclusions casting doubt on it have characterized the debate for over 50 years.

Good Faith Provisions — Good faith provisions, enacted in the 2002 farm bill (P.L. 107-171, Sec. 1613), allow the USDA to forgive a participant from the loss of commodity and conservation program benefits when it is determined that the participant either tried but failed to fully comply with program requirements, or relied on faulty (incorrect) advice from the USDA. This provision applies to conservation and commodity programs, but not to credit or crop insurance programs.

Good manufacturing practices (GMPs) — Standards published in the Code of Federal Regulations (CFR) and used by the Food and Drug Administration (FDA) to ensure the quality of marketed products and that products are produced under sanitary conditions. Any FDA-regulated product can be designated adulterated if the manufacturing methods or facilities for processing do not conform with GMPs. GMPs are developed through a consultative process between the FDA and the affected industry. (21 CFR 110 and 210).

Good samaritan laws — With respect to food and agriculture programs, these laws are designed to encourage the donation of food and grocery products to nonprofit organizations serving the needy by minimizing the risks of legal actions against donors and distributors of the foods. The Model Good Samaritan Food Donation Act was amended and revised in 1996 and renamed the Bill Emerson Good Samaritan Act (P.L. 104-210; 42 U.S.C. 1791) in memory of the late Missouri Representative who sponsored and championed good samaritan laws. It excludes from civil or criminal liability a person or nonprofit food organization that, in good faith, donates or distributes donated foods for food relief. Protection does not apply to an injury or death resulting from gross neglect or intentional misconduct and does not supersede state or local health regulations.

Government Performance and Results Act (GPRA) — This 1993 law (P.L. 103-62) requires most federal agencies, including USDA, to develop and adhere to new planning, evaluation, and reporting requirements, such as mission statements, strategic plans, annual performance plans, and annual performance reports. These documents must include explicit goals and objectives, descriptions of how they will be achieved, and establishment of measurable performance indicators to determine success, among other things. Because GPRA requirements must be tied closely to annual budgeting, and because some in Congress have made oversight of the Act a priority, USDA and its agencies have devoted considerable time and resources to implementation.

Government sponsored enterprise (GSE) — A privately owned, federally chartered corporation designed to provide a source of credit nationwide, and limited to servicing one economic sector (e.g., agriculture, education, or housing). Each GSE is given certain features or benefits (such as implicit federal guarantees on its funds) to enhance its ability to borrow money. Such benefits allow GSEs to overcome the barriers that hinder the development of private markets. Examples of agricultural GSEs are the Farm Credit System and Farmer Mac. Other GSEs include the Student Loan Marketing Association (Sallie Mae), the Federal National Mortgage Association (Fannie Mae), the Federal Home Loan Mortgage Association (Freddie Mac), and the Federal Home Loan Bank System (FHLB Banks).

GPCP — Great Plains Conservation Program.

GPRA — Government Performance and Results Act of 1993 (P.L. 103-62).

GPS — Global Positioning System (see Precision farming).

Grade A milk — Milk produced under sufficiently sanitary conditions to qualify for fluid (beverage) consumption. It is also referred to as fluid grade milk. Only Grade A milk is regulated under federal milk marketing orders. Grade B milk (also referred to as manufacturing grade milk) does not meet fluid grade standards and can only be used in cheese, butter and nonfat dry milk. More than 90% of all milk produced nationally is Grade A. Therefore, much of the Grade A milk supply is used in manufactured dairy products.

Grade A Pasteurized Milk Ordinance — Minimum standards and requirements for Grade A milk production and processing are outlined in the Grade A Pasturized Milk Ordinance (PMO) published by the Food and Drug Administration. Grade A standards are recommended by the National Conference on Interstate Milk Shipments (NCIMS), which is comprised of voting representatives from state and local regulatory agencies, and nonvoting representatives of the dairy industry and FDA. As a general rule, FDA accepts the Conference recommendations and incorporates them into the revised PMO. The state regulator (which is usually either the State Department of Agriculture or the State Health Department) adopts the PMO standards as a minimum, and in many cases requires more stringent standards.

Grades and standards — The segregation, or classification, of agricultural commodities into groupings that share common characteristics. Grades provide a common trading language, or common reference, so that buyers and sellers can more easily determine the quality (and therefore value) of those commodities. Two USDA agencies, the Agricultural Marketing Service and Grain Inspection, Packers, and Stockyards Administration, serve as objective sources for this information. These agencies develop common grades and standards and conduct inspection and grading services for most food and farm products, and industry pays for most of the cost through user fees. Adoption and application of official U.S. Grain Standards is authorized by the U.S. Grain Standards Act (USGSA; 7 U.S.C. 71 *et seq*).

Grading certificates — A formal document setting forth the quality of a commodity as determined by authorized inspectors or graders.

Grain Inspection, Packers and Stockyards Administration (GIPSA) — An agency established in 1994 that combined the functions of the Federal Grain Inspection Service (FGIS) and the Packers and Stockyards Administration (P&S). FGIS provides grain marketing standards and an official inspection system. P&S programs are regulatory in nature to protect livestock producers by ensuring open and competitive markets.

Grain reserve — A phrase that might refer to any of several grain reserve programs that have operated in the past (i.e., the farmer-owned grain reserve, the strategic grain reserve, food security wheat reserve, or the food security commodity reserve). The only reserve now in operation is the 4 million metric ton Bill Emerson Humanitarian Trust.

Grain Standards Act of 1916 — See United States Grain Standards Act of 1916 (7 U.S.C. 71 *et seq.*).

Grains Trade Convention (GTC) — See International Grains Agreement.

Grains — Grain crops are categorized various ways, and none of the categories are precisely defined. Cereals are members of the grass family cultivated primarily for their starchy seeds. They may be used as human food or livestock feed depending upon region and/or cultural tradition. Wheat, rice, corn (maize), rye, oats, barley, sorghum (milo), and some of the millets are common cereals. The category small grains typically includes wheat, barley, oats, and rye. Coarse grains or feed grains generally refers to those used primarily for animal feed (i.e., corn, barley, oats, sorghum). Food grains refers to wheat and rice. Sometimes, oilseed crops are included in discussions of grain crops. Wheat, corn, grain sorghum, barley, oats, upland cotton, rice, soybeans, other oilseeds are called covered commodities in the 2002 farm bill and are eligible for marketing loans, direct payments, and counter-cyclical (P.L. 101-171, Title I).

Grange — The Order of the Patrons of Husbandry. [www.grange.org].

Granger-Thye Act of 1950 — P.L. 81-478 established a new direction for some aspects of National Forest System management; authorized the use of grazing fee receipts for rangeland improvement; authorized the Forest Service to issue grazing permits for terms up to 10 years; authorized the Forest Service to participate in funding cooperative forestry and rangeland resource improvements; established grazing advisory boards; and, authorized the Forest Service to assist with work on private forest lands.

GRAS — Generally recognized as safe.

Grass Lands — A general term for land where the dominant vegetation is grasses; examples include prairie, range land, and pastureland.

Grassed waterway — A generally broad and shallow depression planted with erosion-resistant grasses, which is used to convey surface waters off of or across cropland.

Grasslands Reserve Program (GRP) — The 2002 farm bill (P.L. 107-171, Sec 2401), authorizes enrollment of 2 million acres of restored or improved grassland, range land and pastureland under temporary and permanent easements, or contracts of at least 10 years. Enrolled land must be in parcels that exceed 40 acres. Technical assistance is provided to restore grasslands. A total of $254 million in mandatory funding from the Commodity Credit Corporation (CCC) between FY2003 and FY2007 is provided. It also provides cost sharing payments at 75% to restore disturbed grasslands and 90% to protect virgin grasslands. [www.nrcs.usda.gov/programs/GRP].

Grassroots Source Water Protection Program — This program, more commonly called the Source Water Protection Program and enacted in the 2002 farm bill (P.L. 107-171, Sec. 2502), authorizes the appropriation of $5 million annually in discretionary funds from FY2002 through FY2007 to use the technical assistance capabilities of rural water associations that operate wellhead or groundwater protection programs. [www.fsa.usda.gov/pas/publications/facts/html/sourcewater04.htm].

Grazed acreage — Under the 2002 farm bill (P.L. 101-171, Sec. 1206), a producer that would be eligible for loan deficiency payments (LDPs) for wheat, barley, or oats, but agrees to graze livestock and forego harvesting the crops is eligible for a payment equal to the LDP on the date of the agreement. Grazed triticale acreage is eligible for a payment equal to a wheat payment.

Grazing district — An administrative unit of BLM-managed rangelands established by the Secretary of the Interior under Section 3 of the Taylor Grazing Act of 1934.

Grazing fee — A charge, usually on a monthly basis, for grazing a specific kind of livestock. For federal lands, the grazing fee is based on a formula found in the Public Rangelands Improvement Act (PRIA) (P.L. 95-514). The federal grazing fee is equal to a base fee of ($1.23 x the Forage Value Index (FVI)) + (the Beef Price Index (BPI)) - (the Prices Paid Index (PPI)) ÷ (100) and is charged per animal unit month.

Grazing Lands Conservation Initiative (GLCI) — A program started by USDA under its discretionary authority in 1991 and then specifically authorized as the Conservation of Private Grazing Lands Program by the 1996 farm bill (P.L. 104-127) to provide increased technical and educational assistance to conserve and enhance private grazing lands. This program was amended by the 2002 farm bill (P.L. 107-171, Sec. 2502). More than 60% of the country's grazing lands are considered to have serious environmental problems that could lessen their productive capacity if corrective actions are not taken. [www.nrcs.usda.gov/programs/glci].

Grazing Management Demonstration Districts — The Conservation of Private Grazing Lands Program authorizes 2 demonstration districts, to be selected by the Secretary from landowner applications, where sound grazing practices will be promoted. Each district is to be supported by a technical advisory committee.

Grazing permit/license/lease — Official written permission to graze a specific number, kind, and class of livestock for a specified time period on defined federal rangeland.

Grazing preference — The status of qualified holders of grazing permits acquired by grant, prior use, or purchase, that entitles them to special consideration over applicants who have not acquired preference.

Grazing privilege — The benefit or advantage enjoyed by a person or company beyond the common advantage of other citizens to graze livestock on federal lands. Privilege may be created by permit, license, lease, or agreement.

Great Lakes Basin Soil Erosion and Sediment Control Program — The 2002 farm bill (P.L. 107-171, Sec. 2502) authorized $5 million annually in discretionary funds from FY2002 through FY2007 to implement new authority for a soil erosion and sediment control program in this basin that had been established in a number of prior enactments. [www.glc.org/basin].

Great Plains Conservation Program (GPCP) — This program, initiated in 1957, provided cost share and technical assistance to apply conservation on entire farms in 10 Great Plains states from the Dakotas and Montana to Texas and New Mexico. Contracts were limited to $35,000. At the end of 1995, over 6,800 farms in 558 counties with 20 million acres were participating. It was replaced by the Environmental Quality Incentives Program (EQIP) in the 1996 farm bill (P.L. 104-127).

Green box policies — Domestic or trade policies that are deemed to be minimally trade-distorting and that are excluded from reduction commitments in the Uruguay Round Agreement on Agriculture. Examples are domestic policies dealing with research, extension, inspection and grading, environmental and conservation programs, disaster relief, crop insurance, domestic food assistance, food security stocks, structural adjustment programs, and direct payments not linked to production. Trade measures or policies such as export market promotion (but not export subsidies or foreign food aid) are also exempt. See Blue box policies, and Amber box policies.

Green payments — Payments made to producers as compensation for environmental benefits that accrue as a result of or in conjunction with their farming activities. The Conservation Security Program, authorized in the 2002 farm bill (P.L. 107-171, Sec. 2001-2006), generally is viewed as the first green payments program in this country, paying producers to provide conservation benefits on land from which food and fiber are being produced.

Greenhouse effect — The effect of gases, known as greenhouse gases, in the Earth's atmosphere that trap infrared radiation emitted from the Earth's surface, thereby creating the Earth's relatively warm atmospheric temperatures. Over the past century there has been an increasingly rapid increase in the concentration of carbon dioxide and other greenhouse gases in the stratosphere, with predictions by many scientists that the effect of this increase would be a rise in the Earth's temperature. There has been an increase in the Earth's average temperature of about 1 degree Fahrenheit over the past 100 years, but because of a complex array of feedback mechanisms among terrestrial, oceanic and atmospheric systems, and for other reasons, there continues to be debate on specific effects of the increase in greenhouse gases on climate. The climate has varied widely in the past; current predictions for global warming range from very little change associated with the increase in these gases to the most recent range of between 3 to 10 degrees Fahrenheit by the end of this century, posited by the Intergovernmental Panel on Climate Change (IPCC, a large group of scientists from around the world who form an advisory group to the United Nations Framework Convention on Climate Change (UNFCCC)).

GRIP — Gross Revenue Insurance Plan; Group Risk Income Protection.

Gross domestic product (GDP) — Gross domestic product is a measure of the market value of goods and services produced in the United States. The Bureau of Economic Analysis constructs two complementary measures of GDP, one based on income and one based on expenditures. It is measured on the product side by adding up the labor, capital, and tax costs of producing the output. On the expenditure side, GDP is measured by adding up expenditures by households, businesses, government and net foreign purchases. Theoretically, these two measures should be equal. However, due to problems collecting data, there is often a discrepancy between the two measures. The GDP price deflator is used to convert output measured at current prices into constant-dollar GDP.

Gross farm income — The monetary and non-monetary income received by farm operators. Its main components include cash receipts from the sale of farm products, government payments, other farm income (such as income from custom work), value of food and fuel produced and consumed on the same farm, rental value of farm dwellings, and change in value of year-end inventories of crops and livestock.

Gross processing margin (GPM) — This refers to the difference between the cost of a commodity and the combined sales income of the finished products that result from processing the commodity. Various industries have formulas to express the relationship of raw material costs to sales income from finished products.

Gross Revenue Insurance Plan (GRIP) — A form of direct payment, used in Canada, combining a crop insurance component and a revenue protection component. Farmers finance one-third of the premiums paid out under the revenue protection component. The GRIP makes payments when market revenue falls short of a producer's target revenue. Target revenue per acre for an individual crop is based on historical yields, a multi-year moving average of price, and a level of insurance coverage chosen by the producer.

Gross ton-mile — The movement of the combined weight of railcars and their contents a distance of one mile.

Ground and Surface Water Conservation Program (GSWCP) — A new component of EQIP enacted in the 2002 farm bill (P.L. 107-171, Sec. 2301) to improve irrigation and water use efficiency, and reduce water use by agriçulture. Mandatory funding from the Commodity Credit Corporation (CCC) starts at $25 million in FY2002, and increases to $60 million annually between FY2004 and FY2007. In addition, $50 million is to go to the Klamath Basin in Oregon and California to carry out water conservation activities.

Groundwater — The water from wells and underground aquifers. An estimated 95% of the drinking water used in rural areas is from groundwater. Because of its use as drinking water, there is concern over contamination from leaching agricultural and industrial pollutants or leaking underground storage tanks.

Group of Eight (G-8) — See Group of Seven.

Group of Seven (G-7) — An international economic forum, established in 1975, for leaders of the seven largest industrial countries (France, the United States, the United Kingdom, Germany, Japan, Italy, and Canada). The focus of G-7 discussions has been coordination of macroeconomic policies and international trade and monetary policies. Since 1992, the group has been referred to as the Group of 8 (G-8) as Russia has become a regular attendee. In recent years, the European Union has been represented at G-8 deliberations by the President of the European Commission.

Group of Twenty (G-20) — A group of developing countries that coalesced during the 5th Ministerial Conference of the WTO held in Cancun, Mexico, in November 2003. The G-20 (originally called the G-21) developed a counter-proposal to one put forth jointly by the United States and the EU. The group advocated a multilateral agreement on agricultural trade liberalization that would make deeper cuts in farm support than proposed by the United States and EU; eliminate export subsidies; and continue special and differential treatment for all developing countries. The G-20 continued to advocate those positions in on-going WTO agricultural negotiations. The G-20 is composed of Argentina, Bolivia, Brazil, Chile, China, Colombia, Costa Rica, Cuba, Ecuador, El Salvador, Guatemala, India, Mexico, Pakistan, Paraguay, Peru, the Philippines, South Africa, Thailand and Venezuela. Since formation of the group, some, but not all, of the countries negotiating free trade agreements with the United States, e.g., Costa Rica and Guatemala, have distanced themselves from the G-20.

Group Risk Income Protection (GRIP) — A county-based revenue insurance program, that is a variation of Group Risk Protection (GRP). GRIP pays a participating producer when the county revenue per acre for an insured crop falls below a trigger revenue selected by the insured producer, regardless of the actual revenue level of the individual producer. It is available on a limited basis where GRP is currently available.

Group Risk Protection (GRP) — A form of crop insurance available in certain parts of the country that makes an indemnity payment to all participating crop farmers in a particular area when the entire county's crop production is a certain percentage below the normal production level of the county. This differs from the basic crop insurance program that makes payments to participating farmers when the individual farmer's own crop yield is less than the producer's normal yield.

Growing degree day (GDD) — See Growing degree unit (GDU).

Growing degree unit (GDU) — The GDU, sometimes called the growing degree day (GDD), is a commonly used measure of heat accumulation. Unless stressed by other environmental factors like moisture, the development rate from emergence to maturity for many plants depends upon the daily air temperature. Because many developmental events of plants and insects depend on the accumulation of specific quantities of heat, it is possible to predict when these events should occur during a growing season regardless of differences in temperatures from year to year. Growing degrees (GDs) is defined as the number of temperature degrees above a certain threshold base temperature, which varies among crop species. The base temperature is that temperature below which plant growth is zero. GDs are calculated each day as maximum temperature minus the minimum temperature divided by 2 (or the mean temperature), minus the base temperature. GDUs are accumulated by adding each day's GDs contribution as the season progresses. GDUs can be used to: assess the suitability of a region for production of a particular crop; estimate the growth-stages of crops, weeds or even life stages of insects; predict maturity and cutting dates of forage crops; predict best timing of fertilizer or pesticide application; estimate the heat stress on crops; plan spacing of planting dates to produce separate harvest dates. Crop specific indices that that employ separate equations for the influence of the daily minimum (nighttime) and the maximum (daytime) temperatures on growth are call crop heat units (CHUs).

Growing season — The time period, usually measured in days, between the last freeze in the spring and the first frost in the fall. Growing seasons vary depending on local climate and geography. It can also vary by crop, as different plants have different freezing thresholds. It also is an important component in defining wetland areas.

GRP — Grasslands Reserve Program **GSM** — General Sales Manager. **GSP** — Generalized System of Preferences.

GSWP — Ground and Surface Water Conservation Program.

Gully erosion — Also called ephemeral gully erosion, this process occurs when water flows in narrow channels during or immediately after heavy rains or melting snow. A gully is sufficiently deep that it would not be routinely destroyed by tillage operations whereas rill erosion is smoothed by ordinary farm tillage. The narrow channels, or gullies, may be of considerable depth, ranging from 1 to 2 feet to as much as 75 to 100 feet. Gully erosion is not accounted for in the revised universal soil loss equation. In a few states gully erosion is substantial, but in most areas more soil is lost through the other two forms of water erosion, sheet erosion and rill erosion.

GURT — Genetic Use Restriction Technologies.

H-2A program — The H-2A program is a temporary agricultural worker program for foreigners, which was established in its current form in 1986 (P.L. 99-603). It provides for the temporary admission of alien guestworkers to the United States on a nonimmigrant basis to perform agricultural work of a seasonal or temporary nature. The H-2A program takes its name from the section of the Immigration and Nationality Act (INA, Section 101(a)(15)(H)(ii)(A)) that authorizes it. It is administered by the Employment and Training Administration of the U.S. Department of Labor and the Immigration and Naturalization Service of the U.S. Department of Justice. The H-2A program is designed to ensure agricultural employers an adequate supply of labor, while at the same time protecting the job opportunities of U.S. agricultural workers. Under the INA, the Attorney General may not approve an employer's petition to import H-2A workers unless the employer has applied to the Secretary of Labor for a certification that: (1) there are not sufficient U.S. workers who are able, willing, and qualified to perform the labor or services; and, (2) the employment of foreign workers will not adversely affect the wages and working conditions of U.S. workers who are similarly employed (see Adverse effect wage rate). The H-2A program is not subject to numerical limits.

Habitat conservation plans (HCPs) — Plans prepared under the Endangered Species Act (P.L. 93-205) by nonfederal parties wishing to obtain permits for incidental taking of threatened and endangered species. The number of HCPs has expanded enough in the last 5 years that there are concerns over cost, effectiveness, contributions to recovery, monitoring, and other issues.

Habitat suitability index (HSI) — A wildlife habitat suitability model providing habitat information for evaluating impacts on fish and wildlife resulting from water or land use changes.

Habitat — The place where a population (e.g., human, animal, plant, microorganism) lives, characterized by physical features (e.g., desert) and/or dominant plants (e.g., deciduous forest).

HACCP — Hazard analysis and critical control point. [www.cfsan.fda.gov/~lrd/haccp.html and www.fsis.usda.gov/Science/Hazard_Analysis_&_Pathogen_Reduction/index.asp].

Handler — Generally, the first buyer of a farmer's commodity, often destined for fresh market use (in contrast to processing). Under marketing orders, handlers are defined as anyone who receives the commodity from producers, grades and packs it, and sells it to someone else for further marketing. Usually, the requirements spelled out in marketing orders technically apply to handlers, although producers absorb their effects.

Hard white wheat incentive payments — Payments authorized by the 2002 farm bill (P.L. 101-171, Sec. 1616) to encourage the production of hard white wheat. This wheat largely is produced by Australia.

Hardiness zones — See Plant hardiness zones.

Hardwood Trees — One major botanical group of trees, also known as deciduous. Members of this group have broad leaves, such as oaks, and heavy close-grained wood. The other major botanical group of trees is conifers, which have needle leaves. The CRP has provisions which encourage planting enrolled lands to hardwood trees.

Harmonization — The process of establishment, recognition, and application of internationally recognized measures or standards. Used most often in reference to tariffs (as in the Harmonized Tariff Schedule of the United States (HTSUS)), technical standards, or sanitary and phytosanitary measures applied to imported food products.

Harmonized system — The international classification system for goods, implemented by most countries on January 1, 1998, which is used for tariff classification, trade statistics, and ultimately, transport documentation. Officially known as the Harmonized Commodity Description and Coding System, conversion was begun by the Customs Cooperation Council in 1970 as a replacement for the Customs Cooperation Council Nomenclature also known as the Brussels Tariff Nomenclature.

Harmonized Tariff Schedule of the United States (HTS, HTSUS) — This document, published and maintained by the U.S. International Trade Commission, provides the applicable tariff rates and statistical categories for all merchandise imported into the United States. It is based on the international Harmonized System, the global classification system that is used to describe most world trade in goods. [www.usitc.gov/taffairs.htm].

Hart-Scott-Rodino — The Hart-Scott-Rodino Antitrust Improvements Act of 1976 (P.L. 94-435) established a notification and review process that precedes mergers and acquisitions. Generally, large companies planning a merger or acquisition of another firm first must notify both the Department of Justice and Federal Trade Commission, and provide the two agencies with information (which must be kept confidential) that enables them to assess the likely competitive impacts. The agencies usually have 30 days to make this assessment. During this period, USDA (which has no authority to approve mergers) might be asked, or decide on its own, to contribute information if the proposed merger affects agriculture.

Harvested acres — The cropland actually harvested for a particular crop, usually somewhat smaller at the national level than planted acres due to abandonment caused by weather damage or low market prices, or for livestock grazing.

Hatch Act of 1887 — The permanent statute (24 Stat. 440) authorizing federal funds to state agricultural experiment stations affiliated with the land grant colleges of agriculture. Congress amended the act in 1955 to add a formula that uses rural and farm population factors to allocate the annual appropriation among the states. Under the 2002 farm bill (P.L. 107-171, Sec. 7212), states will continue to be required to provide at least 100% matching funds (traditionally, most states have provided more). On average, Hatch Act formula funds constitute 10% of total funding for each experiment station. (7 U.S.C. 361a *et seq.*).

Haulage rights — Rights obtained by one railroad to have its cars or trains operated by another railroad over that railroads's tracks, using the other's crews and usually its locomotives.

Haying and grazing rules — In response to natural disaster conditions (usually a drought) CRP land can be used for haying and grazing, but only after the Secretary approves it for the entire county. CRP participants who then choose to hay or graze receive reduced financial payments, reflecting the decreased environmental benefits being provided on enrolled lands. See Emergency livestock feed programs.

Hazard analysis and critical control point (HACCP) — A production quality control system intended to minimize the entry of pathogens into the food supply. Under a HACCP system, food manufacturers identify the points in processing where contamination could occur (hazard analysis), and establish critical control points where control steps are implemented. Plant employees are responsible for monitoring the control steps and taking corrective actions when necessary. Record keeping and verification are used to ensure that the system is working. USDA's meat and poultry regulatory agency, the Food Safety and Inspection Service (FSIS), implemented HACCP (to complement the agency's traditional organoleptic inspections) as one of the major elements of new inspection regulations that became fully effective in January 2000. FSIS inspectors are responsible for checking plants' records to assure that the HACCP system is working effectively. Under separate rules that became effective in December 1997, the Food and Drug Administration (FDA) requires seafood processors and importers also to implement the HACCP system. [www.cfsan.fda.gov/~lrd/haccp.html and www.fsis.usda.gov/Science/Hazard_Analysis_&_Pathogen_Reduction/index.asp].

HAZMAT — Hazardous materials.

HCP — Habitat conservation plan.

Head month — A month's use and occupancy of rangeland by a single animal or equivalent. A full head month's fee is charged for each month of grazing by adult animals if the grazing animal (1) is weaned, (2) is 6 months old or older when entering National Forest System land, or (3) will become 12 months old during the period of use. For fee purposes, a head month is equivalent to one animal unit month.

Headage payments — The EU Common Agricultural Policy provides for subsidies to be paid to producers based on the number of head of a specific type of livestock. Such payments may be contingent upon participation in a supply control program (e.g., limiting the number of livestock on a farm).

Headwaters — The source, or upper part, of a stream or watershed. Often used in discussing water rights related to wilderness or other federal land designations.

Healthy Forests Restoration Act of 2003 — P.L. 108-148 provides an expedited planning and review process for fuel reduction treatments on federal lands, with priority for protecting communities in the wildland-urban interface. Also, the law authorizes: woody biomass research and development, and grants for woody biomass use; watershed forestry assistance; insect and disease surveys and treatments; payments for private forest reserves; and forest surveys and monitoring. [www.healthyforests.gov/initiative/legislation.html].

Healthy Meals for Healthy Americans Act of 1994 — P.L. 103-448 reauthorized several expiring programs under the National School Lunch Act (P.L. 79-396, as amended; 42 U.S.C. 1751 *et seq.*) and Child Nutrition Amendments of 1966 (P.L. 89-642) through

FY1998. Required that federally subsidized meal programs conform their meal requirements to the Dietary Guidelines for Americans, permanently authorized the nutrition education and training (NET) program and made it an entitlement, and expanded the outreach and coordination of WIC. Subsequently, P.L. 104-193 restored NET to discretionary status and to a multi-year authorization rather than a permanent authorization. This law was superceded by William F. Goodling Child Nutrition Reauthorization Act of 1998 (P.L. 105-336).

Hectare (ha) — A metric measure of area equal to 10,000 square meters. One hectare = 2.47 acres. One acre = 0.405 hectares. See Acre.

Hedging — Taking a position in a futures market opposite to a position held in the cash market to minimize the risk of financial loss from an adverse price change; a purchase or sale of a futures contract as a temporary substitute for a cash transaction that will occur later (i.e., long hedge and short hedge). Hedgers use the futures markets to protect their business from adverse price changes. This contrasts with speculative buyers and sellers in the futures market who may be on the other end of a hedgers contract.

HEL — Highly-erodible land.

Herbicide-tolerant crops — Crops developed to survive certain herbicides that previously would have destroyed the crop along with the targeted weeds. The most common herbicide-tolerant crops (cotton, corn, soybeans, and canola) are Roundup Ready (RR) crops resistant to glyphosate, a herbicide effective on many species of grasses, broadleaf weeds, and sedges. Other genetically modified herbicide-tolerant crops include Liberty Link (LL) corn resistant to glufosinate, and BXN cotton resistant to bromoxynil.

Herbicide — Any pesticide used to destroy or inhibit plant growth; a weed killer.

HFCS — High-fructose corn syrup.

HFI — Healthy Forests Initiative. [www.fs.fed.us/projects/hfi] [www.healthyforests.gov].

HHS — Department of Health and Human Services. [www.hhs.gov].

High moisture feed grains — Corn and grain sorghum must have moisture content below Commodity Credit Corporation (CCC) standards in order to qualify for marketing assistance loans. Higher moisture feed grains do not serve as suitable collateral for nonrecourse loans. However, the 1996 farm bill (P.L. 104-127)first made, and the 2002 farm bill (P.L. 101-171, Sec. 1209) continued, the policy of making recourse loans available to producers of high moisture corn and grain sorghum.

High value products (HVP) — Consumer- oriented agricultural products that are high in value, often but not necessarily due to processing. HVPs can be divided into three groups: 1) semi-processed products, such as fresh and frozen meats, flour, vegetable oils, roasted coffee, refined sugar; 2) highly processed products that are ready for the consumer, such as milk, cheese, wine, breakfast cereals; and 3) high-value unprocessed products that are also often consumer-ready, such as fresh and dried fruits and vegetables, eggs, and nuts. In recent years HVPs have accounted for a greater percentage than bulk commodities of the total value of U.S. agricultural exports.

High-fructose corn syrup (HFCS) — HFCS is a natural sweetener created by converting glucose in corn starch to fructose . HFCS production expanded during the 1980s as a substitute for higher-cost beet and cane sugar used in soft drinks. HFCS-55 (55% fructose), which is as sweet as sugar, has almost completely replaced liquid sugar in beverages. HFCS-42 (42% fructose) is roughly 90% as sweet as sugar, and is mainly used in cereal, baking, dairy, and processed foods. HFCS and two other corn-derived sweeteners (glucose syrup and dextrose) accounted for 55% of total U.S. natural (caloric) sweetener use in recent years compared to less than one-third in 1980.

Highly erodible land (HEL) — Land that is very susceptible to erosion, including fields that have at least 1/3 or 50 acres of soils with a natural erosion potential of at least 8 times their T value. About 104 million acres of cropland meet this definition of HEL, according to the 1997 National Resources Inventory. Farms cropping highly erodible land and under production flexibility contracts must be in compliance with a conservation plan that protects this cropland.

Historic peanut producers — Those producers who were actively involved in planting and harvesting peanuts in the 1998-2001 period. Under the 2002 farm bill (P.L. 107-171, Sec. 1301-1310), only these historic producers are eligible to receive fixed direct payments and counter-cyclical payments under the new peanut program, irrespective of whether or not they continue to produce peanuts. Payments made to these producers are based on past production on historical acreage.

HNI — Human Nutrition Institute. [www.hni.ilsi.org].

Hog cholera (swine fever) — A highly contagious viral disease of swine that was eradicated from the U.S. in 1978. The acute form causes death within 5 to 14 days. The mild form is characterized by short periods of illness followed by periods of recovery, but animals die after about 3 months. Animal and Plant Health Inspection Service (APHIS) is responsible for preventing the disease from reappearing in the U.S.

Hog/corn ratio (corn-hog ratio) — Number of bushels of corn equal in value to 100 pounds of live hogs (feed ratio). Put another way, the price of hogs, per hundredweight, divided by the price of corn per bushel. Since corn is a major input cost to hog producers, the higher the price of hogs relative to corn, the more profit there is in feeding hogs.

Holding pond — A pond or reservoir, usually made of earth, built to store wastewater or polluted runoff.

Homeland Security Act of 2002 — This law (P.L. 107-296) created the Department of Homeland Security (DHS). In doing so, it transferred approximately 2,680 agriculture border inspectors from the Animal and Plant Health Inspection Service (APHIS) to the new Department of Homeland Security (DHS) Bureau of Customs and Border Inspection (CBP). USDA, however, retains a significant presence in border inspection because of its scientific expertise. APHIS continues to set agricultural inspection policies, pre-clear certain commodities, and inspect passengers and cargo entering the mainland from Hawaii and Puerto Rico. The Act also transferred the land and buildings of the Plum Island Animal Disease Center in New York to DHS.

Homeland Security Presidential Directive 7 (HSPD-7) — This directive added agriculture to the list of industries for critical infrastructure protection in December 2003. HSPD-7 replaces the 1998 Presidential Decision Directive 63 (PDD-63) that omitted agriculture and food. These directives designate the physical systems that are vulnerable to terrorist attack and are essential for the minimal operation of the economy and the government. Federal agencies are to develop plans to prepare for and counter the threat. In terms of combating terrorism and weapons of mass destruction (WMD), agriculture was included in the National Security Council's WMD preparedness group formed by Presidential Decision Directive 62 (PDD-62) in 1998. Agroterrorism preparedness is more directly addressed by HSPD-9. [www.nimsonline.com/presidential_directives/hspd_9.htm].

Homeland Security Presidential Directive 9 (HSPD-9) — This directive, Defense of United States Agriculture and Food (January 30, 2004), establishes a national policy to protect against terrorist attacks on agriculture and food systems. It directs federal departments and agencies to coordinate their efforts, assess vulnerabilities, prepare response plans, create public awareness, and conduct necessary research. HSPD-9 builds upon HSPD-7, which added agriculture to the list of industries for critical infrastructure protection. HSPD-5 (regarding the national response plan) and HSPD-8 (regarding preparedness) are being used by the USDA Homeland Security Staff in implementing HSPD-9.

Homestead Act — This federal law, adopted in 1862, essentially provided that heads of families or persons 21 years old, who were U.S. citizens or had filed for citizenship, could obtain up to 160 acres of unappropriated public land. Five years' residence on and cultivation of the land were required in order for the individual to receive title to the land. The land was free, except for small transaction fees. The Act was repealed generally by the Federal Land Policy and Management Act of 1976 (P.L. 94-579); it remained in effect in Alaska until 1986. Approximately 287,000,000 acres were provided to homesteaders during the life of the act.

Homestead protection — When a USDA farm loan borrower lacks the financial resources to make payments on a delinquent loan, is ineligible for a restructured loan, and is unable to buy out the loan at the net recovery value of the collateral property, the borrower can convey the property to USDA in lieu of loan payments. The 1996 farm bill (P.L. 104-127) reduces the time period in which the borrower may apply to lease and/or purchase the residence and up to 10 acres of adjoining land, to within 30 days from acquisition by USDA.

Honey program — Non-recourse marketing loans had long been available to support honey prices until FY1994, when the funding was suspended by provisions in annual appropriations legislation. The 1990 farm bill (P.L. 101-624) had set honey loan rates at $0.538 per pound and permitted deficiency payments. The 1996 farm bill (P.L. 104-127) repealed the statutory authority for the honey program. A Honey Recourse Loan Program was made available for the 1998 crop only through broader emergency spending authority in the FY1999 agriculture appropriations act (P.L. 105-277). The 2002 farm bill (P.L. 107-171, Sec. 1201) made honey eligible for marketing assistance loans (and loan deficiency payments, LDPs) from 2002 through 2007.

Honey Recourse Loan Program — A program authorized by the emergency provisions of the FY1999 USDA appropriations act (P.L. 105-277) that made recourse loans based on a national average rate of $0.56 per pound on 1998-crop honey. Final date to obtain a loan was May 7, 1999. The producer-owned honey was required to be merchantable and stored in acceptable containers. Loans carried an administrative fee of $0.009 per pound, bore an interest rate 1% higher than the Commodity Credit Corporation (CCC) borrowing interest rate, and matured not later than 9 months following disbursement.

Horse Protection Act of 1970 — P.L. 91-540, as amended (15 U.S.C. 1821 *et seq.*), makes it a crime to exhibit, or transport for the purpose of exhibiting, any sored horse (one whose feet have been injured deliberately to accentuate the animal's gait). The Animal and Plant Health Inspection Service (APHIS) enforces the law, which covers all breeds, although Tennessee Walking Horses are the most frequent victims of this procedure.

Horticultural specialty crops — The Census of Agriculture includes as horticultural specialties bedding plants, florists' greens, flower and vegetable seeds, flowers, foliage, fruit stocks, nursery and ornamental plants, shrubbery, sod, mushrooms, and vegetables grown under cover (e.g., in greenhouses).

Household (foodstamp) — A food stamp household is composed of all those who purchase food and prepare meals in common. All related co-residents must apply as a single food stamp household, no matter how they purchase and prepare food (except for elderly persons who are medically certified as unable to purchase and prepare meals separately). Other co-residents may apply separately if they purchase and prepare food separately, and residents in certain eligible institutional settings (e.g., shelters for battered women, residential drug treatment programs) may apply as separate households no matter how they purchase and prepare food.

Housing Act of 1949 — Title V of P.L. 81-171 authorized USDA to make loans to farmers to construct, improve, repair, or replace dwellings and other farm buildings to provide decent, safe, and sanitary living conditions for themselves, their tenants, lessees, sharecroppers, and laborers. The USDA was authorized to make grants or combinations of loans and grants to farmers who could not qualify to repay the full amount of a loan, but who needed the funds to make the dwellings sanitary or to remove health hazards to the occupants or the community. Over time, the Act has been amended to authorize housing loans and grants to rural residents in general and these are administered by the Rural Housing Service (RHS). The rural housing programs are generally referred to by the authorizing section number in the Housing Act of 1949, as amended (42 U.S.C. 1441 *et seq.*). See Section 502 loans, Section 504 loans and grants, Section 514 loans, Section 515 loans, Section 516 grants, Section 521 rental assistance, Section 523 loans, Section 524 loans, Section 533 grants.

HSI — Habitat suitability index.

HSUS — Humane Society of the United States. [www.hsus.org].

HTS, HTSUS — Harmonized Tariff Schedule of the United States. [www.usitc.gov/t affairs.htm].

Humane Methods of Slaughter Act — The Food Safety and Inspection Service (FSIS) enforces this Humane Methods of Slaughter Act (P.L. 85-765; 7 U.S.C. 1901 *et seq.*) under a provision in the Federal Meat Inspection Act (21 U.S.C. 603(b), 610(b), 620(a)). The law requires that cattle, calves, horses, mules, sheep, and swine be rendered entirely insensible to pain before slaughter. This is accomplished with powerful stun guns (see Stunning). Poultry currently is not subject to humane slaughter requirements. FSIS inspectors at slaughtering plants are responsible for overseeing compliance, and have the authority to stop slaughter lines and order plant employees to take corrective actions. Following news reports in early 2002 alleging significant non-compliance, FSIS assigned additional veterinarians to its district offices specifically to monitor humane slaughter and handling procedures and to report to headquarters on compliance. The 2002 farm bill requests an annual compliance report to Congress, and in the FY2003 agricultural appropriations act, Congress designated $5 million of FSIS funding for hiring 50 additional compliance inspectors. Language in the FY2004 consolidated appropriations act directs FSIS to continue fulfilling that mandate, and the FY2005 budget request calls for another $5 million to be allocated for enforcement activities. A January 2004 GAO report states that compliance problems persist (GAO-04-247). Earlier concerns about humane treatment of non-ambulatory (downer) cattle at slaughter houses became irrelevant when FSIS issued regulations in January 2004 (69 FR 1892) prohibiting them from being slaughtered and inspected for use as human food.

Hundredweight — One hundred pounds (abbreviated as cwt.). A standard unit of measure for milk, rice, and some meat livestock.

Hunger / food insecurity (domestic) — This is defined as the lack of food due to the limited or uncertain availability of nutritionally adequate and safe foods, or limited or uncertain ability to acquire acceptable foods in socially acceptable ways. In the 3-year period from 1998 through 2000, 11.3 million households reported food insecurity and 3.4 million of these reported hunger, according to a Study of Hunger and Food Insecurity released in August 2002 by the Food Security Institute, Center on Hunger and Poverty at Brandeis University. See also Food Security (international).

Hunger Prevention Act of 1988 — P.L. 100-435 amended the Temporary Emergency Food Assistance Act of 1983 (P.L. 98-8) to require the USDA to make additional types of commodities available for the Temporary Emergency Food Assistance Program (TEFAP), to improve the child nutrition and food stamp programs, and to provide other hunger relief.

HVP — High value products.

Hybrid — Plants with desirable attributes that have been developed by fertilizing one variety of the plant with pollen from another that also has desirable traits. Corn and horticultural plants are the most commonly hybridized crops in the United States. Hybrids provide a form of patent protection to the seed company that produces them because second generation plants grown from first generation seeds do not produce a second crop that uniformly has all the same attributes as the first generation.

Hydric soil — Soil that, in its undrained state, is flooded long enough during a growing season to develop anaerobic conditions that support the growth and regeneration of hydrophytic vegetation (plants specialized to grow in water or in soil too waterlogged for most plants to survive). This term is part of the legal definition of a wetland included in the Food Security Act of 1985 (P.L. 99-198). The Natural Resources Conservation Service maintains the official list of hydric soils.

Hydrogen — The lightest and most abundant element. It is usually produced from the reformation of methane (natural gas), although it can be produced as a renewable fuel using solar energy to electrolyze water. Hydrogen fuel is seen as an ultimate replacement for petroleum-based transportation fuels, especially in fuel cell vehicles. A fuel cell running on pure hydrogen produces electricity, heat, and water exclusively; pollutant emissions from hydrogen fuel cells are zero. Currently, there are very few hydrogen vehicles, and most of these are test prototypes.

Hydroponics — The growing of plants without soil by using an inert medium such as sand, peat, or vermiculite and adding a nutrient solution containing all the essential elements needed by the plant for its normal growth and development. Water culture, when plant roots are suspended in a liquid medium containing the nutrient solution while their crowns are supported in a thin layer of inert medium, is true hydroponics. Often called soilless culture, it also includes aeroponics where plant roots are suspended in a dark chamber and sprayed with the nutrient solution.

Hypoxia — A low oxygen condition in the water that may occur where a nutrient-laden free-flowing body of water (like a river) enters a lake or ocean. The high nutrient content promotes rapid growth of plankton/phytoplankton that subsequently die and, in the process of decomposition, consume large amounts of oxygen (see Biochemical oxygen demand). While fish and shrimp can migrate away from a hypoxic area, less mobile bottom-dwelling organisms are unable to escape. These areas expand and contract with the seasons; they are largest in the summer after maximum spring runoff when the water is warmest and conditions are calm. The size of these areas also varies greatly from year to year because of changing physical conditions. Hypoxia can be found in many water bodies, such as the deep portions of the Chesapeake Bay in the summer, but the best-known instance is an area in the Gulf of Mexico called the dead zone.

IBA — Independent Bakers Association. [www.independentbaker.org].

ICAC — International Cotton Advisory Committee. [www.icac.org].

ICBA — Independent Community Bankers Association. [www.ibaa.org].

ICM — Integrated crop management.

ICO — International Coffee Organization. [www.ico.org].

Identity Preservation (IP) — A general term used to describe the technical and managerial techniques used to track and document the paths that agricultural products move in the production process. A fully integrated IP system might track and document a commodity's seed characteristics, initial planting, growing conditions, harvesting, shipping, storage, processing, packaging, and ultimate sale to the consumer. Separating organic products from conventionally raised ones is a one type of IP system. IP systems are a central component of value-chains.

Identity preserved (IP) — This is the designation given to bulk commodities marketed in a manner that isolates and preserves the identity of a shipment, presumably because of unique characteristics that have value otherwise lost through commingling during normal storage, handling and shipping procedures. The concept of IP has been accorded greater importance with the introduction of genetically modified (GM) organisms into agriculture. Although the U.S. scientific community maintains that GM crops are safe, critics want them segregated from non-GM commodities out of concerns about their potential environmental and food safety implications.

IDFA — International Dairy Foods Association. [www.idfa.org].

IFAD — International Fund for Agricultural Development. [www.ifad.org].

IFDA — International Foodservice Distributors Association Food Distributors International. [www.ifdaonline.org].

IFIC — International Food Information Council. [ific.org].

IFIS — International Food Information Service. [www.ifis.org].

IFSA — International Farming Systems Association.

IFT — Institute of Food Technologists. [www.ift.org].

IGC — International Grains Council. [www.igc.org.uk].

IJO — International Jute Organization.

Immunohistochemistry (IHC) test — A laboratory method that detects antibodies of prions (mis-shapen proteins thought to transmit bovine spongiform encephalopathy, BSE or mad cow disease) by exposing a brain sample to a stain that appears as a specific color under a microscope. The IHC test is used by USDA researchers in their BSE surveillance program because they consider it the gold standard, providing a high level of confidence about the results. However, IHC tests are expensive and time-consuming. More rapid and less expensive testing alternatives ("rapid tests") have been used in some other countries, but until recently USDA has viewed them as less reliable because they can deliver more false positive and/or false negative results than the IHC. However, in June 2004 USDA embarked on a greatly expanded BSE testing program to test more than 200,000 cattle over a 12-18 month period (compared with 20,000 in each of 2002 and 2003). It is now using rapid test kits at regional laboratories to conduct initial screening; any samples that test "positive" for BSE (which USDA terms "inconclusive") must be subjected to an IHC test for confirmation.

Import Administration — The branch of the International Trade Administration that administers U.S. antidumping and countervailing duty laws.

Import fee — Generally, a charge assessed for a service rendered. For example, when an import stamp or import license is issued, the government assesses a fee for this service. Within the context of Section 22 of the Agricultural Adjustment Act of 1935, fees were imposed on imports of agricultural products when deemed necessary to protect domestic farm programs. Then, under the North American Free Trade Agreement (starting in 1994) and the Uruguay Round Agreement on Agriculture (starting in 1995), Section 22 import fees and quotas were converted into tariff-rate quotas.

Import license — A document required and issued by some national governments authorizing the importation of specified goods into their respective countries. When used in a discriminatory manner, these licenses can become nontariff trade barriers.

Import quota — A trade barrier that sets the maximum quantity (quantitative restriction) or value of a commodity allowed to enter a country during a specified time period. The Uruguay Round Agreement on Agriculture requires the conversion of import quotas and other quantitative restrictions to tariff-rate quotas and/or bound tariff rates.

Import relief — See Safeguards, and Trade relief.

Import sensitive product — A product that is particularly susceptible to competition from imports from other country suppliers. Import sensitive products generally receive longer phase-in periods for tariff reduction or elimination in trade agreements. There is no definitive list of import sensitive agricultural products in the United States, but such products as sugar, dairy, beef, citrus, winter fruits and vegetables, among others have been identified as import sensitive. In some cases, import sensitive products have been excluded altogether from negotiated trade agreements as was sugar in the U.S.-Australia Free Trade Agreement.

Incentive payments — Direct payments made under the National Wool Act (P.L. 83-690, Title VII) to producers of wool and mohair, which were similar to deficiency payments made to producers of grains and cotton. The incentive payment rate was the percentage needed to bring the national average return to producers (the market price plus the incentive payment) up to the annually set national support price. Each producer's direct payment was the payment rate times the market receipts. Producers with higher market receipts got larger support payments. This created an incentive to increase output and to improve quality. The wool and mohair commodity programs ended after the 1995 marketing year as required by P.L. 103-130. The 2002 farm bill (P.L. 107-171, Sec. 1201) made wool and mohair eligible for marketing assistance loans and loan deficiency payments. For conservation programs, incentive payments refer to cost-sharing payments that producers may receive to attract participation. In some programs, the federal portion of these payments can vary; for these programs, the higher share is provided for the more desirable conservation practices that will provide greater benefits.

Income insurance — A concept, similar to revenue insurance, that envisions an insurance program that would insure farm families a specified minimum annual income.

Income Protection (IP) — A form of revenue insurance that protects a grower of an insurable crop whenever low prices, low yields, or a combination of both causes revenue to fall below a guaranteed level selected by the producer.

Income support — Generally, programs providing direct, income-supplementing payments to farmers. Intended to protect farm income without affecting market prices. The Direct and Counter-cyclical Program (DCP) payments under the 2002 farm bill (P.L. 101-171, Sec. 1101-1108) provide income support, not price support. The phrase also is used to characterize the nature of support provided to low-income families by various food assistance programs.

Indemnity — The payment that eligible producers receive if they realize a qualifying loss under crop insurance, revenue insurance, or other insurance programs. Also refers to the compensation farmers may receive when the Animal and Plant Health Inspection Service (APHIS) exercises its authority to destroy animals or crops to control a contagious disease, such as avian influenza and citrus canker.

Industrial crops — Crops that primarily have industrial applications in contrast to food or livestock feed uses. Industrial uses account for a relatively small share of the current market for farm commodities. Some of the industrial and experimental crops include: cuphea (soap, surfactants); guar (food stiffeners, drilling muds, paper manufacturing); guayule (natural rubber and hypoallergenic latex products); hesperaloe (specialty pulp paper); kenaf (twine, fiberboard, carpet padding, newsprint); lesquerella (lubricants, cosmetics); meadowfoam (cosmetics, lubricants, water repellents); milkweed (insulated clothing, filler for comforters, nonwoven textiles) and plantago ovato (high fiber additive to laxatives). While corn is the primary feedstock for ethanol, it is not considered an industrial crop because about 80% of utilization is livestock feed and export. Authority for a government-sponsored corporation that Congress created in 1990 to identify and develop new industrial uses for crops, the Alternative Agricultural Research and Commercialization Corporation (AARC), was repealed in the 2002 farm bill (P.L. 107-171, Sec. 6201).

Industrial hemp — A variety of marijuana containing trace amounts of the psychoactive chemical found in abundance in the varieties used as drugs. Industrial hemp fiber can be used to manufacture clothing, paper, carpeting, construction materials, auto parts, animal bedding, cosmetics, pharmaceuticals, and food products (using hemp seed oil). Approximately 40 countries allow legal cultivation of industrial hemp. In the United States it is considered a controlled substance under the Controlled Substances Act (P.L. 91-513; 21 U.S.C. 801 *et seq.*). It is illegal to grow for any purposes, and any imported hemp products must meet a zero tolerance level. Several states have passed or attempted to pass laws to permit growing research plots of hemp for investigation as a high-value alternative crop, particularly in tobacco-growing regions, but at the federal level no legislation to legalize the production of industrial hemp has ever been introduced.

Industrialization — When used in agriculture, this term generally refers to the consolidation of farms into very large production units that are more vertically coordinated with suppliers or markets or both.

Inert ingredient — A pesticide component such as a solvent, carrier, dispersant, or surfactant that is not active against target pests. Inert ingredients may be toxic and may be subject to testing under the Federal Insecticide, Fungicide, and Rodenticide Act (FIFRA) (7 U.S.C. 136 *et seq.*).

Infant formula cost-containment — Refers to statutory provisions in the Child Nutrition Act of 1966 (P.L. 89-642, as amended) that require state WIC agencies to solicit bids from infant formula companies for the sale of infant formula used in WIC food packages.

Infiltration — The downward entry of water into soil. Also called percolation. A high rate of infiltration means that soil moisture for crops will be higher. Many conservation practices, such as conservation tillage, reduce rates of runoff and increase infiltration rates.

Infoshare — USDA established this program in 1993 to merge and coordinate the business management and information technology (computer) activities of its agencies, particularly in the field, in order to support consolidation of field offices into one-stop field service centers for farmers and other USDA clients. However, the program, which initially had been budgeted at nearly $3 billion, was terminated by early 1996 in the wake of critical reviews by USDA's Office of Inspector General, the General Accounting Office, and others, which found, among other things, that despite Infoshare, individual USDA agencies were continuing to buy their own computers, were not sharing information technology with each other, and were still not operating in a common computing environment. Infoshare was replaced by another computer modernization initiative designed and coordinated by the Farm Service Agency.

Initiative for Future Agriculture and Food Systems — A competitive research grants program authorized in the Agricultural Research, Extension, and Education Reform Act of 1998 (P.L. 105-185) and reauthorized in the 2002 farm bill (P.L. 107-171, Sec. 7205). The law allows mandatory funds available from savings in food stamp program administration to be used to support the program. Grants were awarded in FY2000 and FY2001; in all other years congressional appropriators have prohibited USDA from spending funds on personnel to operate the program except for oversight of existing grants.

INRO — International Natural Rubber Organization. [www3.jaring.my/inro].

Insecticide — A pesticide used to kill, deter, or control insects.

Instream use — Water use taking place within the stream channel. Examples are hydroelectric power generation, navigation, fish propagation and use, and recreational activities. Some instream uses, usually associated with fish populations and navigation, require a minimum amount of water to be viable. Often used in discussions concerning water allocation and/or water rights.

Insular area — Insular area is defined by the National Agricultural Research, Extension, and Teaching Policy Act of 1977 (P.L. 95-113), as amended, to include the Commonwealth of Puerto Rico, Guam, American Samoa, the Commonwealth of the Northern Mariana Islands, the Federated States of Micronesia, the Republic of the Marshall Islands, the Republic of Palau, the Virgin Islands of the United States.

Integrated Farm Management Program (IFMP) — A program authorized by the 1990 farm bill (P.L. 101-624) to assist producers in adopting resource-conserving crop rotations by protecting participants' base acreage, payment yields, and program payments. The program's goal was to enroll 3 to 5 million acres over 5 years. The 1996 farm bill (P.L. 104-127) replaced the IFMP with production flexibility contracts and a pilot conservation farm option program.

Integrated Pest Management (IPM) — A pest control strategy based on the determination of an economic threshold that indicates when a pest population is approaching the level at which control measures are necessary to prevent a decline in net returns. In principle, IPM is an ecologically based strategy that relies on natural mortality factors, such as natural enemies, weather, and crop management, and seeks control tactics that disrupt these factors as little as possible. Also, a USDA/EPA program that aims to decrease pesticide applications by teaching farmers to use a variety of alternative control techniques to minimize pesticide use. These techniques include biological controls, genetic resistance, tillage, pruning, and others.

Integration — The combination (under the management of one firm) of two or more identical (horizontal) or successive (vertical) stages in the production or marketing process of a particular product. Generally the stages are capable of being operated as separate businesses. The firm that has management responsibility is called the integrator. The poultry industry, for example, is vertically integrated, from production through processing and distribution. Diversification, on the other hand, is the production of two or more products by one firm or farmer.

Integrator — See Integration.

Interagency Coordinating Committee on the Validation of Alternative Methods (ICCVAM) — A permanent coordinating committee established by Congress in 2000 with representation from 15 federal agencies. It replaced a standing committee at the National Institute for Environmental Health Sciences. ICCVAM's purpose is to evaluate the scientific validity of new and alternative testing methods that may reduce the use of animals in testing chemicals for toxicity. The committee develops test recommendations based on its technical evaluations, which are forwarded to the federal agencies for final acceptance decisions.

Intermediate agricultural products — Generally refers to agricultural products that have a higher per-unit value than bulk commodities. They are often partly processed but not necessarily ready for the consumers. Examples might include soybean meal, wheat flour, vegetable oils, feeds and fodders, animal fats, hides and skins, live animals, and sweeteners such as sugars. Applied to trade policy, intermediate products are one of three categories of agricultural products used by the Foreign Agricultural Service to report export and import data under its BICO system (the others are bulk and consumer-oriented agricultural products).

Intermediate Export Credit Guarantee Program (GSM-103) — One of the Commodity Credit Corporation (CCC)'s export credit guarantee programs. See GSM-103.

Intermodal service — Freight moving via at least two different modes of transport. Intermodal service generally involves the shipment of containers and trailers by rail, truck, barge, or ship.

International commodity agreement — An undertaking by a group of countries to stabilize trade, supplies, and prices of a commodity for the benefit of participating countries. An agreement usually involves a consensus on quantities traded, prices, and stock management. A number of international commodity agreements serve solely as forums for information exchange, analysis, and policy discussion.

International Dairy Agreement (IDA) — The IDA replaced the International Dairy Arrangement, which had been established in 1980. Its primary function was to expand and liberalize world trade in dairy products through international cooperation. On January

1, 1995, the IDA was placed under the aegis of the World Trade Organization. Its members included Argentina, Bulgaria, the European Union, Japan, New Zealand, Romania, Switzerland, and Uruguay. The United States, which had been one of the original members, withdrew from the organization in 1985 to protest sales by the European Union of butter and other basic dairy products at prices below the minimum export prices established by the Committee on Certain Milk Products that, along with the International Dairy Council, administered the Arrangement. The International Dairy Council suspended minimum prices for dairy products in 1995. The IDA was terminated, by decision of the International Dairy Council, as of January 1, 1998.

International Epizootics Organization — See Office International des Epizooties. [www.oie.int].

International Grains Agreement (IGA) — Replaced the International Wheat Agreement in 1995. The IGA comprises a Grains Trade Convention (GTC) and a Food Aid Convention (FAC). The IGA is administered by the International Grains Council (IGC), an intergovernmental forum for cooperation on wheat and coarse grain matters. The Grains Trade Convention provides for information-sharing, analysis and consultations on grain market and policy developments. Under the Food Aid Convention, donor countries pledge to provide annually specified amounts of food aid to developing countries in the form of grain suitable for human consumption, or cash to buy suitable grains in recipient countries. The International Grains Agreement does not contain any mechanisms for stabilizing supplies, prices, or trade.

International Grains Council (IGC) — An intergovernmental forum responsible for administering the International Grains Agreement (IGA). The United States is a member of the IGC. [www.igc.org.uk/index.htm].

International Monetary Fund (IMF) — A multilateral financial institution established in 1945 to help member countries with international payments problems and to maintain orderly exchange rate policies. U.S. agricultural exports benefit indirectly from activities of the IMF that maintain the global trade in commodities and food. [www.imf.org].

International Plant Protection Convention (IPPC) — An international treaty for cooperation in plant protection, operated under the Food and Agriculture Organization of the United Nations. Launched in the early 1950s, the IPPC provides for international cooperation to protect plant resources from harmful pests. Currently, more than 100 countries are active. With the coming into force in 1994 of the Uruguay Round multilateral trade agreements, and particularly the Agreement on the Application of Sanitary and Phytosanitary (SPS) Measures, the IPPC was seen as an important arena for encouraging international harmonization in phytosanitary matters affecting trade, and for helping to ensure that phytosanitary measures were not used as unjustified barriers to trade. [www.fao.org/legal/treaties/004t-e.htm].

International Sugar Organization (ISO) — The body established by the International Sugar Agreement, 1992 whose objectives are: "(a) To ensure enhanced international cooperation in connection with world sugar matters and related issues; (b) To provide a forum for intergovernmental consultations on sugar and on ways to improve the world sugar economy; (c) To facilitate trade by collecting and providing information on the world sugar market and other sweeteners; (d) To encourage increased demand for sugar, particularly for non-traditional uses." [www.sugaronline.com/iso].

International Trade Administration (ITA) — An agency of the U.S. Department of Commerce. Among its responsibilities are the promotion of U.S. exports (excluding agricultural exports for which promotion is a function of the Foreign Agricultural Service (FAS) of USDA) and U.S. companies, and the implementation of U.S. antidumping and countervailing duties laws. [www.ita.doc.gov].

International Trade Commission (ITC or USITC) — An independent, quasi-judicial U.S. federal agency that provides objective trade expertise to both the legislative and executive branches of government and determines the impact of imports on U.S. industries. It makes recommendations concerning countervailing duty and antidumping petitions submitted by U.S. industries seeking relief from imports that benefit from unfair trade practices. The agency also updates and publishes the Harmonized Tariff Schedule of the United States. Established by Congress in 1916 as the U.S. Tariff Commission, the Trade Act of 1974 (P.L. 93-618) changed its name to the U.S. International Trade Commission. [www.usitc.gov].

International Wheat Agreement (IWA) — Replaced in 1995 by the International Grains Agreement. [www.igc.org.uk].

Interregional Project 4 — See IR-4. [pmep.cce.cornell.edu/ir4/index.html].

Interstate Shellfish Sanitation Conference (ISOC) — The federal-state-industry cooperative body which manages the National Shellfish Sanitation Program. [www.issc.org].

Intervention price — The price at which national intervention agencies in the EU are obliged to purchase any amount of a commodity offered to them regardless of the level of market prices (assuming that these commodities meet designated specifications and quality standards). Thus, the intervention price serves as a floor for market prices. Intervention purchases have constituted one of the principal policy mechanisms regulating EU markets in sugar, cereal grains, butter and skimmed milk powder, and (until 2002) beef.

Intervention stocks — Stocks held by national intervention agencies in the EU as a result of intervention buying of commodities subject to market price support. Intervention stocks may be released onto internal markets if internal prices exceed intervention prices. Otherwise, they may be sold on the world market with the aid of export restitutions under the regulation of commodity-specific Management Committees.

Invasive species — Alien (non-native or exotic) species of plants, animals, and pests whose introduction causes or is likely to cause economic or environmental harm or harm to human health. Executive Order 13112, issued February 8, 1999, seeks to prevent the introduction and minimize the impacts of invasive species through better federal agency coordination under a National Invasive Species Management Plan. Examples of invasive species receiving recent attention include the Asian long-horned beetle, Africanized honeybees, zebra mussels, and the Formosan termite. Inspection and quarantine programs at U.S. ports of entry (jointly carried out by Department of Homeland Security (DHS) and Animal and Plant Health Inspection Service (APHIS) inspectors) prevent entry of invasive species. A number of laws are aimed at prevention and control, including among others the Plant Protection Act of 2000 (P.L. 106-224, Title IV; 7 U.S.C. 7701 *et seq.*), the Federal Seed Act (7 U.S.C. 1551 *et seq.*), the Animal Damage Control Act (7 U.S.C. 426-426c), the Nonindigenous Aquatic Nuisance Prevention and Control Act of 1990

(P.L. 101-646, Title I; 16 U.S.C. 4701 *et seq.*), and the Alien Species Prevention and Enforcement Act (P.L. 102-393, Title VI, Sec. 631).

IOOC — International Olive Oil Council. [www.internationaloliveoil.org].

IP — Income protection program (see Revenue insurance).

IPM — Integrated pest management.

IPPC — International Plant Protection Convention

IR-4 — Inter-regional Project 4, also known as the Minor Crop Pest Management Program, is funded by the Cooperative State Research, Education, and Extension Service (CSREES) to generate data to register pesticides and biological pest control agents for minor crops where there is no economic incentive to do so on the part of the pesticide manufacturing industry. IR-4 provides coordination, funding, and scientific guidance for both field and laboratory research at land-grant universities to develop data for the registration by the EPA of pest control products on a wide variety of commodities. The program has been responsible for data to support over 500 food use clearances, 7,000 clearances on ornamental crops, and over 100 biopesticide clearances to date. About 40% of all EPA newly registered pesticides are the result of IR-4 activities. [pmep.cce.cornell.edu/ir4/index.html].

IRIS — Instructional resources information system; integrated risk information system.

IRM — Integrated resource management.

Irradiation — The process of exposing food or other items to ionizing radiation (e.g., from cobalt-60, cesium-137, x-ray machines, or electron accelerators) in order to destroy insect pests or microorganisms. In December 1999, the Food Safety and Inspection Service (FSIS) published the final rule to guide the red meat industry in the use of the technology and in labeling irradiated products (irradiation of packaged poultry has been permitted since 1992). (The Food and Drug Administration (FDA), which considers irradiation a food additive and thus under its jurisdiction, approved irradiation of red meat in 1997.) The 2002 farm bill (P.L. 107-171, Sec. 10808) contains a provision that, among other things, requires the FDA to begin rulemaking to permit irradiated meat and poultry to be labeled as pasteurized.

Irrigation return flow — Part of artificially applied water that is not consumed by plants or evaporation, and that eventually returns to an aquifer or surface water body, such as a lake or stream. Commonly used when discussing water conservation techniques and measurement.

Irrigation water management — Managing irrigation applications based on the water-holding capacity of the soil and the need of the crop. The water is applied at a rate and in such a manner that the crop can use it efficiently and resource losses are minimized. Irrigation efficiency is the ratio of the amount of water stored in the crop root zone compared to the amount of water applied. Water conservation has become more important as costs have risen and demands have grown for wildlife and urban uses.

Irrigation — Applying water (or wastewater) to land areas to supply the water (and sometimes nutrient) needs of plants. Producers who irrigate in arid areas are more likely to use it throughout the growing process (full irrigation), while producers in more humid areas may use irrigation to supplement rainfall and soil moisture under drought conditions. Techniques for irrigating include furrow irrigation, sprinkler irrigation, trickle (or drip) irrigation, and flooding. More than 55 million acres of land are irrigated in the United States, according to the 1997 Census of Agriculture's irrigation survey. More acres of corn are irrigated than any other crop, but that amounts to only about 15% of the harvested corn acres. In contrast, irrigation is used for 100% of rice, 81% of orchards, 64% of vegetables, and 36% of cotton. About 40% of freshwater withdrawals, and about 80% of the water consumed, are for irrigation, making agriculture the single largest user of water. The West had about 78% of the irrigated land in 1997; this percentage has been declining as irrigated acreage in the East has been increasing by approximately the same number of acres as in the West since 1969.

ISAAA — International Service for the Acquisition of Agri-Biotech Applications. [www.isaaa.org].

ISO — International Sugar Organization. [www.sugaronline.com/iso]. International Standards Organization. [www.iso.org/iso/en/ISOOnline.frontpage].

ISOC — Interstate Shellfish Sanitation Conference. [www.issc.org].

Isolated Wetlands (Geographically) — Wetlands that have no apparent surface water connection or outlet to some form of drainage, including perennial streams and rivers, and estuaries. A Supreme Court decision in January 2001 exempted some isolated wetlands from the federal dredge and fill permit program (section 404) under the Clean Water Act (P.L. 92-500; 33 U.S.C. 1344). The 2002 farm bill (P.L. 107-171, Sec. 2101) made some isolated wetlands eligible for protection under the Farmable Wetlands Program.

ITA— International Trade Administration. [www.ita.doc.gov].

ITC — International Trade Commission. [www.usitc.gov].

ITTO — International Tropical Timber Organization. [www.itto.or.jp].

IWA — International Wheat Agreement. See IGC, International Grains Council.

John Ogonowski Farmer to Farmer Program — Initiated as a P.L. 480-funded pilot project and authorized first under the Food Security Act of 1985 (P.L. 99-198), the program taps U.S. agricultural expertise to provide technical assistance to farmers in developing, middle income, and emerging market countries. The 2002 farm bill (P.L. 107-171) extends funding authority through FY2007 and requires that no less than 0.5% of P.L. 480 funds be used for the program. John Ogonowski was the pilot on one of the planes that crashed into the World Trade Center on September 11, 2001. The program was renamed to honor his extensive work with immigrant Southeast Asian farmers using his land in rural Massachusetts.

Joint Agricultural Weather Facility (JAWF) — Created in 1978, the facility is a cooperative effort between USDA's World Agricultural Outlook Board and the National Oceanic and Atmospheric Administration of the U.S. Department of Commerce to collect, on an ongoing basis, global weather data and agricultural information to determine the impact of weather conditions on crop and livestock production. JAWF reports are followed closely not only by producers but also by commodity traders.

Joint Subcommittee on Aquaculture (JSA) — Created by the National Aquaculture Act of 1980 (P.L. 96-362, Sec. 6; 16 USC 2805), the JSA has served as a federal interagency coordinating group to increase the overall effectiveness and productivity of federal aquaculture research, transfer, and assistance programs. The JSA's Working Group on Quality Assurance in Aquaculture Production is co-chaired by representatives from the Food and Drug Administration's Center for Veterinary Medicine (FDA-CVM) and USDA's Cooperative State Research, Education and Extension Service (USDA-CSREES). This Working Group serves as an open national public-private sector forum on issues associated with new animal drug approvals, biologics, licensing, producer quality assurance programs, international harmonization, etc.

Jones Act — The common reference for Section 27 of the Merchant Marine Act of 1920 (41 Stat. 988), which requires that all water transportation of goods between U.S. ports be on U.S.-built, -owned, -crewed, and -operated ships. The purpose of the law is to support the U.S. merchant marine industry, but agricultural interests generally oppose it because, they contend, it raises the cost of shipping their goods, making them less competitive with foreign sources.

JSA — Joint Subcommittee on Aquaculture (federal).

Just-in-time delivery — An inventory control system that replenishes and delivers products to a retailer just as a current supply is depleted. This practice, which the agriculture and food sector have, as well as industry, has required increased vertical coordination.

Karnal bunt — A fungus disease of wheat that reduces yields and causes an unpalatable but harmless flavor in flour milled from infected kernels. Appearance of the disease in the United States in early 1996 resulted in the Animal and Plant Health Inspection Service (APHIS) implementing an emergency quarantine, inspection, and certification program for wheat moving out of the infested areas, along with regulations on sanitizing machinery and storage facilities. Many foreign countries have a zero tolerance for karnal bunt in import shipments.

KCBT — Kansas City Board of Trade. [www.kcbt.com/index.asp].

Killer gene — See Terminator seeds.

Kyoto Protocol — The Kyoto Protocol to the United Nations Framework Convention on Climate Change (UNFCCC), agreed to by parties to the convention in 1997, establishes an agreed schedule for specified, legally binding reductions of greenhouse gas emissions by 38 developed countries. President Bush made the decision in 2001 that the United States would reject the Protocol and rely upon voluntary domestic measures. Since that time, the United States has declined to participate in negotiations related to the Protocol. The Protocol will take effect when not less than 55 countries, accounting for at least 55% of developed country greenhouse gas emissions for 1990, have ratified it. Until then, the terms of the Protocol are not binding on any country, and after ratification it will be binding only upon those who have ratified it. [unfccc.int/essential_background/ convention/items/2627.php].

LAA — Local administrative area.

Lagoon — For purposes of livestock waste management, a lagoon is designed to store manure while it is decomposing. The Natural Resources Conservation Service (NRCS) has developed specifications and provides technical assistance for the proper construction of such lagoons. Lagoons are the most widely used technology to manage manure among cattle and hog producers, and are used by Concentrated Animal Feeding Operations (CAFOs) to meet EPA standards for point source pollution prevention. However, they often generate health and waste management concerns, especially associated with concentrated odors.

Lamb Meat Adjustment Assistance Program — A 4-year program initiated in 1999-2000 to help producers deal with import competition and help stabilize the lamb market. Through 2002, the USDA-administered program provided some $50 million in incentive payments to help producers increase the supply of domestic lamb meat.

Land capability (classification) — The quality of soil resources for agricultural use is commonly expressed as land capability classes and subclasses, which show, in a general way, the suitability of soils for most kinds of field crops. Soils are grouped according to their limitations when they are used to grow field crops, the risk of damage when they are used, and the way they respond to treatment. Capability classes, the broadest groups, are designated by Roman numerals I through VIII, with I being the best soils and VIII being the poorest.

Land evaluation and site assessment system (LESA) — A technique that can be used at the local level to determine the quality of land or agricultural uses and to assess sites or areas of land for their agricultural viability. It was first used in the early 1980s.

Land grant colleges of agriculture — The Morrill Act of 1862 (7 U.S.C. 301 *et seq.*) granted federal land to states to sell, and instructed each state to use the proceeds to endow a college to teach agriculture and the mechanical arts. States not having any federal land within their borders were given land in scrip, permitting them to sell federal land located in other (usually western) states for revenue to establish an agricultural college. The original schools are called the 1862 Institutions. Subsequently, the Morrill Act of 1890 (7 U.S.C. 321 *et seq.*) created the black colleges of agriculture, called the 1890 Institutions. The Equity in Educational Land-Grant Status Act of 1994 (P.L. 103-382, Sec. 531-535) gave land grant status to more than 25 Native American colleges, called the 1994 Institutions.

Land grant university — The term used to identify a public university in each state that was originally established as a land grant college of agriculture pursuant to the Morrill Act of 1862 (7 U.S.C. 301 *et seq.*). In most states the original agricultural colleges grew over time into full-fledged public universities by adding other colleges (e.g., arts and sciences, medicine, law, etc.); in states where a public university existed prior to 1862, this first Morrill Act resulted in a college of agriculture being added to the university. USDA funds go only to the original land grant colleges of agriculture within the so-called land grant universities. [www.nasulgc.org].

Land management (conservation) practices — Conservation practices that are carried out to protect soil water and related resources from degradation through programs such as the Environmental Quality Incentives Program. Conservation tillage is a land management practice.

Land management services contracts — A proposed national forest timber sale contract where purchasers would be required to perform activities, other than those directly related to timber cutting and removal, in or near the sale area, in exchange for a reduction in the stumpage price. Pilot tests of this contract arrangement have been conducted, but its general use is not authorized.

Land treatment — Any activity or project to improve conservation of soil, water, or other resources and to improve production.

Land trust — A private nonprofit organization exempted from federal taxes, if it conforms to Section 501 (c)(3) of the federal tax code, that may receive donations of money, property or development rights, and may use its assets to purchase property or development rights. According to a 2000 survey by the Land Trust Alliance, there are more than 1,250 land trusts that have protected about 6.2 million acres, and many of them have used easements to protect about 2.6 million acres of all types of land. A portion of the protected lands are used for agriculture.

Land use and development controls — Ordinances, resolutions, and controls enacted by local government under the authority of state enabling legislation to protect public health, safety, or welfare. Many controls can affect agriculture either directly or indirectly. How agriculture is affected depends on the design of the controls.

Land use plan — A coordinated collection of data, programs, and activities related to existing and potential uses of land and resources within a defined area. Commonly associated with local units of government trying to anticipate and organize uses of space so as to meet defined goals. For producers, conservation plans are a type of land use plan.

LATS — Long-Term Agricultural Trade Strategy.

LD_{50}/lethal dose / LC_{50}/lethal concentration — The dose (median concentration) of a toxicant that will kill 50% of the test animals within a designated period. The lower the LD_{50}/LC_{50}, the more toxic the compound. Testing to determine lethal dosages is used to characterize the acute toxicity of pesticides and other toxic chemicals.

LDCs — Less developed countries. See also Developing countries.

LDP — Loan deficiency payment.

Leaching — The process by which chemicals are dissolved and transported through the soil by percolating water. Pesticides and nutrients from fertilizers or manures may leach from fields, areas of spills, or feedlots and thereby enter surface water, groundwater, or soil. Leaching from concentrated sources such as waste sites and loading areas vulnerable to spills can be prevented by paving or containment with a liner of relatively impermeable material designed to keep leachate inside a treatment pond, landfill, or a tailings disposal area. Liner materials include plastic and dense clay.

Legumes — A family of plants, including many valuable food, forage and cover species, such as peas, beans, soybeans, peanuts, clovers, alfalfas, sweet clovers, lespedezas, vetches, and kudzu. Sometimes referred to as nitrogen-fixing plants, they can convert nitrogen from the air to build up nitrogen in the soil. Legumes are an important rotation crop because of their nitrogen-fixing property.

LESA — Land evaluation and site assessment.

Letters of credit — Letters of credit are commonly used in agricultural export trade. A letter of credit from a bank (or other financial institution) is the bank's guarantee that a buyer of goods will pay the seller. A letter of credit gives the buyer of goods the financial backing of a bank, thereby becoming a financially reliable customer. The buyer can use a letter of credit to help assure that proper shipment is made prior to actual payment. Also, the buyer may obtain a lower price for the purchased product due to the seller being exposed to a lower risk. The seller is assured prompt payment and is protected against cancellation of the sales contract. An *advising bank* to the seller will notify the seller that a letter of credit has been issued on behalf of the buyer. This advising bank is responsible only for authenticating and forwarding the letter of credit, but makes no commitment to pay unless it agrees to act as a confirming bank. A *confirming bank* adds its commitment to pay along with the commitment of the issuing bank.

Levy — The USDA defines levy as an import charge assessed by a country or group of countries not in accordance with a definite tariff schedule. The variable import levy of the European Communities was an example. The EC's levy on grains varied from day to day, depending on the offering price of third-country suppliers. In USDA's view the variable import levy is a nontariff trade barrier because, unlike a moderate customs duty or even a quota, it can completely bar imports. The Uruguay Round Agreement on Agriculture resulted in the replacement of variable levies by fixed tariffs.

Liability — When used as a term in the federal crop insurance program, it represents the maximum amount a participating farmer could collect in the program if a total crop loss were experienced.

LIBOR — London interbank offered rate.

Limited global quota for upland cotton — A provision of the Food and Agriculture Act of 1977 (P.L. 95-113) that authorized the President to proclaim an import quota whenever the USDA determines that the spot market average price in any one month exceeds 130% of the previous 36-month average. If triggered by such a determination, the established quota allows for imports of up to 21 days of mill consumption during a 90-day period. A limited global quota cannot overlap with the step 3 quota, one of the cotton competitiveness provisions.

Limited Resource Farmers and Ranchers — The Agricultural Credit Act of 1987 (P.L.100-233, Section 622) requires the Secretary of Agriculture to maintain a Limited Resources Farmer Initiative in carrying out USDA programs. Natural Resources Conservation Service (NRCS) provides conservation technical assistance and other resources to these farmers and ranchers who, are defined as being those who, when compared to other farmers, ranchers and farm operators in a given geographic area (such as a state, county, or project area), have distinct disadvantages in obtaining USDA program assistance. The NRCS determines this by low income and farm sales: (1) total household income must be at or below the national poverty level for a family of four, or not more than 50% of the county median income in each of the previous two years, and (2) gross farm sales may not exceed $100,000 in each of the previous two years, with this amount adjusted for inflation beginning in FY2004. The limited resource phrase often is used along with small, disadvantaged, under served, and minority farmers in describing outreach, education, loan programs and support activities operated by other USDA agencies. For example, farm ownership and operating loans are available to farmers and ranchers unable to get conventional credit. These limited resource applicants are described as those who operate small or family farms and meet the general requirements for a loan but are unable to pay the regular interest rate because of low income. See also Socially disadvantaged farmers and ranchers,and Beginning farmer and rancher.

Line capacity — The maximum number of trains that can operate safely and reliably over a given segment of track during a given period of time.

Line-haul service — The movement over the tracks of a carrier from one city to another, not including the switching service.

Linola — A new form of linseed known by the generic crop name Solin, which produces a high-quality edible polyunsaturated oil similar in composition to sunflower oil. It was developed and released in Australia in 1992 and first commercially grown in 1994. Linola is being produced in Australia, Canada, the U.K. and in the states of Washington and Idaho. Linola substitutes for flax in cropping rotations, is claimed to have lower production costs than canola, but brings prices comparable to canola or other edible oils. Linola is Generally Recognized as Safe (GRAS) by the Food and Drug Administration.

Linters — The short fibers that remain on cottonseed after ginning. They are used mainly for batting, mattress stuffing, and as a source of cellulose.

Listeria — *Listeria monocytogenes*, a pathogenic bacterium found widely in nature, can be carried in a variety of foods such as dairy products, red meat, poultry, seafood, and vegetables. *Listeria* is a major cause of foodborne illness (listeriosis) in the U.S.

Live weight — The weight of live animals purchased or sold by a producer.

Livestock Assistance Program (LAP) — An emergency livestock assistance periodically authorized and funded by Congress in response to natural disasters. The most recent version of LAP provides direct payments to eligible livestock producers who suffered grazing losses due to natural disasters during either calendar year 2001 or 2002 (not both). For an individual producer to be eligible, the producer's county must have suffered a minimum 40% loss of available grazing for at least 3 consecutive months due to a disaster during the year. The county also had to be declared a disaster area by either the President or the Secretary of Agriculture in 2001 or 2002. Once the county qualified for assistance, a producer had to suffer a minimum loss of 40% in order to qualify for a payment to partially compensate for purchases of off-farm feed. Producers with more than $2.5 million of gross revenue are ineligible. The maximum payment is $40,000 per person. [www.fsa.usda.gov/pas/publications/facts/html/lap01.htm].

Livestock Compensation Program — A program administratively authorized by USDA in 2002 to compensate certain livestock producers for feed and pasture losses caused by a natural disaster declared in 2001 and 2002. Under the program, estimated total direct payments of just over $1 billion were made to all producers of beef, dairy, sheep and goats in any county that was declared a disaster area by the Secretary between January 1, 2001, and February 20, 2003, regardless of the individual producer's loss experience. The payment rates under the LCP were $31.50 per adult dairy cattle, $18 per adult beef cattle, $13.50 for certain livestock over 500 lbs., and $4.50 per sheep or goat. Payments were limited to $40,000 per person, and were not made to any person with qualifying gross revenue over $2.5 million. The program was not specifically authorized by Congress but was initially implemented by USDA under existing authorities. The Livestock Compensation Program was initiated because funding was not available to implement the Livestock Assistance Program. [www.fsa.usda.gov/pas/publications/facts/html/LCP03.htm].

Livestock Indemnity Program (LIP) — A program periodically authorized and funded on an emergency basis by Congress to compensate livestock producers for losses caused by a natural disaster. Under the program, a payment is made to help producers defray the cost of replenishing their herds when livestock are killed by a natural disaster. [disaster.fsa.usda.gov/lip.htm].

Livestock Mandatory Reporting Act of 1999 — Title IX of the FY2000 USDA appropriations act (P.L. 106-78), requires large packers and importers to report to USDA the details of all transactions involving purchases of livestock and imported boxed lamb cuts, and the details of all transactions involving domestic and export sales of boxed beef cuts, sales of domestic and imported boxed lamb cuts, and sales of lamb carcasses. Additional provisions impose, in turn, new data reporting requirements on USDA, including more frequent price reports along with new monthly information on retail prices for meat and poultry products. Policy issues include the ability of USDA to effectively implement the mandatory program, which has a 5-year authorization that expires October 22, 2004, and whether mandatory reporting is more or less helpful to producers than the longstanding voluntary reporting system. See Mandatory price reporting.

Livestock — When used in agricultural policy discussions, the term broadly refers to farm animals, often, but not always, exclusive of poultry and seafood. However, federal legislation sometimes more narrowly defines the term in order to make specified agricultural commodities either eligible, or ineligible, for a program or activity. For example, the Livestock Mandatory Reporting Act of 1999 (P.L. 106-78, Title IX) defines livestock only as cattle, swine, and lambs. However, 1988 disaster assistance legislation defined the term as "cattle, sheep, goats, swine, poultry (including egg-producing poultry), equine animals used for food or in the production of food, fish used for food, and other animals designated by the Secretary."

LMA — Livestock Marketing Association. [www.lmaweb.com].

LOAEL — Lowest-observed-adverse-effect-level.

Loan commodity — Under the 2002 farm bill (P.L. 101-171, Sec. 1201-1205), the following commodities are eligible for marketing assistance loans and are called loan commodities: wheat, corn, grain sorghum, barley oats, upland cotton, extra long staple cotton, rice, soybeans, other oilseeds, wool, mohair, honey, dry peas, lentils, and small chickpeas. With the exception of extra long staple cotton, farmers agreeing to forgo the loans are eligible for loan deficiency payments (LDPs) on actual production of loan commodities.

Loan deficiency payments — A farm income support program first authorized by the Food Security Act of 1985 (P.L. 99-198) that makes direct payments, equivalent to marketing loan gains, to producers who agree not to obtain nonrecourse loans, even though they are eligible. Loan deficiency payments are available under the 2002 farm bill (P.L. 101-171, Sec. 1205) for wheat, corn, grain sorghum, barley, oats, upland cotton, rice, soybeans, other oilseeds, wool, mohair, honey, dry peas, lentils, and small chickpeas.

Loan forfeiture level (sugar) — The mandatory nonrecourse loan rates for raw cane sugar and refined beet sugar are 18¢/lb. and 22.9¢/lb. However, borrowers are likely to forfeit loan collateral when raw can prices are below 20¢-22¢/lb. and refined beet prices are below 23¢-26¢/lb, depending upon geographic location. These higher loan forfeiture levels result from the added costs of interest and transportation associated with repayment of loans. Therefore, USDA seeks to keep market prices above the loan forfeiture levels through the administration of sugar import quotas, marketing allotments, and payment-in-kind authority. Accordingly, the loan forfeiture level serves as the effective price support level.

Loan forfeiture — Under commodity program rules, a producer or processor who pledges a stored commodity as collateral to the Commodity Credit Corporation (CCC) to obtain a nonrecourse loan can settle the repayment obligation by forfeiting the commodity without any penalty. This happens, by design, if forfeiture is more profitable than selling the commodity in the marketplace. The difference between the loan price and the market price at the time of forfeiture is called the marketing loan gain to the farmer and is taxed as income. The incentives for loan forfeiture largely were eliminated by the introduction of loan deficiency payments (LDPs). Per person payment limits on LDPs and marketing loan gains can be avoided when producers purchase commodity certificates at posted county prices for use in settling loans. Until the introduction of marketing loans, loan forfeiture was used to reduce the available market supply of commodities and raise market prices.

Loan rate — The price per unit (bushel (bu.), bale, pound (lb.), or hundredweight (cwt.), depending on the commodity) at which the government will provide nonrecourse or recourse loans to farmers (or associations acting on their behalf) by the Commodity Credit Corporation (CCC). This short term financing at below market interest rates enables farmers to hold their commodities for later sale. The introduction of marketing assistance loan provisions (repayment at the posted county price and loan deficiency payments) largely eliminates the forfeiture of nonrecourse loan commodities when market prices fall below loan rates.

Local school food authority (LSFA) — Administering unit for the operation of a school feeding program, it receives federal meal reimbursements for meal programs and is responsible for ensuring that meal counts and eligibility criteria are met. This may be a school district, several school districts, or individual schools.

London interbank offered rate — Interest rate offered by a specific group of London banks for U.S. dollar deposits of a stated maturity. Used as base index for setting some adjustable rate financial instruments, including adjustable rate mortgages. Used by Fannie Mae.

Long ton — A measure of weight equal to 2,240 pounds. By contrast, a short ton is 2,000 pounds; a metric ton (MT) equals 2,204.62 pounds.

Long — (1) One who has bought a futures contract or option to establish a market position; (2) a market position that obligates the holder to take delivery; (3) one who owns an inventory of commodities. The opposite of short.

Long-term agreements — Providing benefits through conservation programs for periods of longer than a single year. Many of the newer conservation programs, such as the Conservation Reserve Program (CRP) and the Farmland Protection Program FPP), use multi-year agreements with land owners.

Loss Ratio — A term used to measure the financial performance of the federal crop insurance program. It is expressed as the ratio of indemnity, or loss payments, to total premiums. For example, a loss ratio of 1.3 to 1.0 means that for every $1.30 of indemnity or loss payments, premiums amounted to $1.00.

Low-flow irrigation systems — These systems (drip, trickle, and micro sprinklers) provide water in small volumes and generally target it to plants with less waste than furrow irrigation. Drip and trickle systems apply water through small holes in small diameter tubes placed on or below the surface of the field. Another type of system, micro sprinklers, supplies water from low-volume sprinkler heads located above the surface. Low flow systems are expensive and their use is generally limited to high-value crops such as vegetables, fruits, and vineyards.

Lowest-observed-adverse-effect-level — Lowest concentration or amount of a substance found by experiment or observation which causes an adverse alteration of morphology, function, capacity, growth, development or life span of a target organism distinguished from normal organisms of the same species under defined conditions of exposure. Federal agencies use set approval standards below this level.

Lump-sum sales — A common term for tree measurement sales.

M-W price — Minnesota-Wisconsin price.

Mad cow disease — The common term used for bovine spongiform encephalopathy (BSE).

Main line — Primary rail line over which trains operate between terminals. It excludes sidings, and yard and industry tracks.

Maintenance fees — The annual fees paid by pesticide manufacturers and formulators to continue registration of pesticide active ingredients and products with EPA. The fees supplement funds appropriated from general U.S. revenues, which cover most administrative costs of the EPA pesticide program under the Federal Insecticide, Fungicide, and Rodenticide Act (FIFRA; 7 U.S.C. 136a-1). Fees are deposited into a separate Reregistration and Expedited Processing Fund to offset costs associated with EPA reregistration activities and expedited processing of pesticide registrations that are substantially similar to registrations already in effect or which are for public health pesticides, as defined in FIFRA Section 2(nn). Congress mandated collection of an annual maintenance fee from each pesticide registrant in 1988 amendments to(P.L. 100-532). EPA has authority to cancel a registration if a registrant fails to pay the maintenance fee.

Maize — Term used in many parts of the world for corn.

Major land resource area (MLRA) — Major land resource areas are geographically associated land resource units delineated by the Natural Resources Conservation Service and characterized by a particular pattern that combines soils, water, climate, vegetation, land use, and type of farming. There are 204 MLRAs in the United States, ranging in size from less than 500,000 acres to more than 60 million acres.

Make allowance (or milk manufacturing marketing adjustment) — The margin between the government support price for milk and the Commodity Credit Corporation (CCC) purchase price for butter, nonfat dry milk, and cheese. This margin is administratively set to cover the costs of "making" milk into butter, nonfat dry milk, or cheese to reach the desired level of prices for milk in manufacturing uses.

Mandatory price reporting — In the past, packers and processors were not required to report the prices they paid for the animals or the terms of sale. Rather, daily sales and price information was collected by AMS from companies on a voluntary basis. AMS reporters also attended auctions to collect price information. However, as more and more animals were sold under formula pricing, other contract, or captive supply arrangements, the open cash markets became less helpful as benchmarks. On the argument that such arrangements also enabled packers to more easily conceal potential anti-competitive practices, Congress passed the Livestock Mandatory Reporting Act of 1999 (P.L. 106-78, Title IX). This law requires large packers and importers to report prices and other transaction details to the Agricultural Marketing Service.

Mandatory spending — Budget authority and ensuing outlays provided in laws other than appropriations acts, including annually appropriated entitlements. Nearly three-fourths of USDA spending is classified as mandatory (or appropriated entitlement) spending, including the farm commodity price and income support programs, crop insurance, food stamps and child nutrition programs.

Manure management — Capturing, holding, treating, and utilizing animal wastes in ways that minimize pollution. Animal waste can occur in a liquid, slurry, or solid form. It can be retained in various holding facilities. It is disposed by distribution on fields in amounts that enrich soils without causing water pollution or unacceptably high levels of nutrient enrichment. Manure management is a component of nutrient management.

MAP — Market Access Program. [www.fas.usda.gov/mos/programs/mapprog.html].

MARAD — Maritime Administration. [www.marad.dot.gov].

Margin call — (1) A request from a brokerage firm to a customer to bring margin deposits up to initial levels; (2) a request by the clearinghouse to a clearing member to make a deposit of original margin, or a daily or intra-day variation payment, because of adverse price movement, based on positions carried by the clearing member.

Margin — The amount of money or collateral deposited by a customer with a broker, by a broker with a clearing member, or by a clearing member with the clearinghouse, for the purpose of insuring against loss on open futures contracts. The margin is not partial payment on a purchase. (1) Initial margin is the total amount of margin per contract required by the broker when a futures position is opened. (2) Maintenance margin is a sum that must be maintained on deposit at all times. If the equity in a customer's account drops to, or under, the level because of adverse price movement, the broker must issue a margin call to restore the customer's equity. Sometimes called a performance bond.

Mariculture — The form of aquaculture where fish, shellfish, other invertebrates, or aquatic plants are cultured in a saltwater environment.

Market access — The extent to which a country allows imports. A variety of tariff and nontariff trade barriers can be used to limit the entry of foreign products. Governed by provisions of the Uruguay Round Agreement on Agriculture, which refer to concessions contained in the country schedules with respect to bindings and reduction of tariffs and to tariffication, the only allowable exceptions to the process are those described under the Special Safeguard Provisions and the special and differential treatment accorded developing countries. Free trade agreements (i.e, NAFTA, Chile) contain similar concessions.

Market Access Program (MAP) — MAP, previously called the Market Promotion Program, is administered by the Foreign Agricultural Service and uses funds from the Commodity Credit Corporation (CCC). It helps producers, exporters, private companies, and other trade organizations finance promotional activities for U.S. agricultural products. MAP is designed to encourage development, maintenance, and expansion of commercial agricultural export markets. Activities financed include consumer promotions, market research, technical assistance, and trade servicing. The Export Incentive Program, which is part of MAP, helps U.S. commercial entities conduct brand promotion activities including advertising, trade shows, in-store demonstrations, and trade seminars. MAP is authorized through 2007 by the 2002 farm bill (P.L. 107-171). The program promotes exports of specific U.S. commodities or products in specific markets. Under MAP, program participants are reimbursed for their expenses in carrying out approved promotional activities. Participating organizations include nonprofit trade associations, state regional trade groups, and private companies. Funding authority was limited to $100 million in fiscal year 2002, rising gradually to $200 million each in fiscal years 2006 and 2007. [www.fas.usda.gov/mos/programs/mapprog.html].

Market allocation — A quantity provision in a fruit or vegetable marketing order specifying the maximum amount of the regulated commodity that can be sold for a given use or market (such as the domestic fresh market).

Market basket — Average quantities of consumables, including U.S. farm foods, purchased per household for a given base period, used to compute an index of retail prices.

Market loss assistance — See Market loss payments.

Market loss payments — A designation first used in the Omnibus Consolidated and Emergency Appropriations Act, FY1999 (P.L. 105-277) to describe the $3.1 billion in emergency income support payments authorized for eligible grain, cotton, and dairy farmers. The Act stated that such funds were to compensate farmers for the loss of 1998 income caused by "regional economic dislocation, unilateral trade sanctions, and the failure of the government to pursue trade opportunities aggressively." Similar economic emergency support payments for selected commodities were subsequently enacted in P.L. 106-78 ($6.5 billion), in P.L. 106-224 ($6.5 billion), in P.L. 106-387 ($0.9 billion), and in P.L. 107-25 ($5.5 billion). Market loss assistance to grain and cotton producers were distributed in parallel manner to the contract payments authorized by the Agricultural Market Transition Act.

Market price support (MPS) — An indicator, developed by the OECD, used in the calculation of Producer and Consumer Subsidy Equivalents (PSE/CSE). MPS is the annual monetary value of gross transfers from consumers and taxpayers to agricultural producers arising from policy measures creating a gap between domestic market prices and border prices of a specific agricultural commodity measured at the farm gate level. Conditional on the production of a specific commodity, MPS includes the transfer to producers associated with both production for domestic use and exports and is measured by the price gap applied to current production. The MPS is net of producer levies on sales of the specific commodity or penalties for not respecting regulations such as production quotas (price levies). In the case of livestock production MPS is net of the market price support on domestically produced coarse grains and oilseeds used as animal feed (excess feed cost).

Market price — The price per bushel (or pound or hundredweight) of an agricultural commodity paid in the private sector. It can sometimes refer to the price paid at domestic seaports or large inland terminal markets (such as daily cash prices listed in newspapers) and sometimes refers to the price received by farmers, or farm price.

Market Promotion Program (MPP) — An export promotion program authorized by the 1990 farm bill (P.L. 101-624) that replaced the Targeted Export Assistance (TEA) program authorized by the Food Security Act of 1985 (P.L. 99-198). The MPP was renamed the Market Access Program (MAP) under the 1996 farm bill (P.L. 104-127). [www.fas.usda.gov/mos/programs/mapprog.html].

Market sharing quota (MSQ) — In Canada, the MSQ for industrial milk is determined by estimating the domestic demand for dairy products on a butterfat basis, adding about 3% to cover exports and subtracting the volume of approved imports. Provincial shares of the national quota are adjusted in line with changes in the total and each province allocates its share to its producers according to its own quota policies. The Canadian Dairy Commission sets a target price for industrial milk based on production costs, including a return to labor, capital and management. Dairy farmers receive direct government payments (which are part of the target price) on in-quota deliveries of industrial milk and cream. Farmers who produce in excess of their quota do not receive direct government payments and face an over-quota levy. Each province maintains and administers its own quota scheme for fluid milk.

Market structure — Characteristics of an industry that relate to its economic performance, such as the number of buyers and sellers, product differentiation among firms, barriers to entry, costs, degree of integration, and diversification.

Market transition payments — Referred to variously as AMTA payments, contract payments, or production flexibility contract payments made to farmers under Title I (the

Agriculture Market Transition Act (AMTA)) of the 1996 farm bill (P.L. 104-127). These payments were replaced by direct payments in the Direct and Counter-cyclical Program (DCP) of the 2002 farm bill (P.L. 101-171, Sec. 1103).

Marketing agency (or board) — Generally, a statutory body possessing certain legislated regulatory powers over prices, quality standards, foreign trade, etc. In the United States, marketing orders and agreements exercise control over quality standards and flow to market. The Canadian Wheat Board and the Australian Wheat Board handle domestic and export marketing in those countries. See also State trade enterprises (STEs).

Marketing agreements — Marketing agreements (and orders) are authorized by the Agricultural Marketing Agreement Act of 1937 (50 Stat. 246), as amended). They may be designed to (1) maintain the high quality of produce that is on the market; (2) standardize packages and containers; (3) regulate the flow of product to market; (4) establish reserve pools for storable commodities; and (5) authorize production research, marketing research and development, and advertising. In contrast to marketing orders, agreements are enforceable only against those handlers who enter into the agreement. Federal oversight is provided by AMS. See Marketing orders and agreements. [www.ams.usda.gov/fv/moab.html].

Marketing assessments — At times, producers and first purchasers of some supported commodities are required to pay assessments as a contribution toward achieving budget deficit reduction targets. Under the 1996 farm bill (P.L. 104-127), assessments were imposed on sugar processors and on producers and first buyers of peanuts. However, the 1996 farm bill (P.L. 104-127) eliminated a milk marketing assessment. The 2002 farm bill (P.L. 107-171) eliminated the assessments for peanuts and sugar. Tobacco was subject to a no-net-cost assessment on all marketings to offset Commodity Credit Corporation (CCC) losses on price support loan operations until support was ended in 2005 under the quota buyout provision (P.L. 108-357, Title VI).

Marketing assistance loans — Nonrecourse loans made available to producers of loan commodities (wheat, corn, grain sorghum, barley oats, upland and ELS cotton, rice, soybeans, other oilseeds, honey, wool, mohair, dry peas, lentils, and small chickpeas) under the 2002 farm bill (P.L. 101-171, Sec. 1201-1205). The new law largely continued the commodity loan programs as they were under previous law. Loan rate caps are specified in the law. Marketing loan repayment provisions apply when market prices drop below the loan rates. For farmers who forego the use of marketing assistance loans, loan deficiency payment (LDP) rules apply (but not for ELS cotton).

Marketing certificate — A certificate that may be redeemed for a specified amount of Commodity Credit Corporation (CCC)-owned commodities. The certificates may be generic or for a specific commodity.

Marketing contract — Prices (or pricing mechanisms) are established for a commodity before harvest or before the commodity is ready for marketing. Most management decisions remain with the grower, who retains ownership of both production inputs and output until delivery. The farmer assumes the risks of production but shares price risks with the contractor. Marketing contracts are commonly used for crops and not livestock. According to the USDA, about 40% of the value of all fruits and vegetables produced in 1997 were under marketing contracts. Marketing contract shares for selected other commodities were: sugar beets, 82%; milk, 60%; cotton, 33%; cattle, 10%; soybeans, 9.4%; corn, 8%. See Production contract.

Marketing loan repayment provisions — A loan settlement provision, first authorized by the Food Security Act of 1985 (P.L. 99-198), that allowed producers to repay nonrecourse loans at less than the announced loan rates whenever the world price or loan repayment rate for the commodity were less than the loan rate. Marketing loan provisions became mandatory for soybeans and other oilseeds, upland cotton, and rice and were permitted for wheat, corn, grain sorghum, barley, oats, and honey under amendments made by the 1990 farm bill (P.L. 101-624). The 1996 farm bill (P.L. 104-127)retained the marketing loan provisions for wheat, feed grains, rice, upland cotton, and oilseeds. The 2002 farm bill (P.L. 101-171, Sec. 1201-1205) continued marketing assistance loans and expanded their application to wool, mohair, dry peas, lentils, and small chickpeas.

Marketing orders and agreements — Orders and agreements (authorized by the Agricultural Marketing Agreement Act of 1937 (50 Stat. 246), as amended) allow producers to promote orderly marketing through collectively influencing the supply, demand, or price of a particular commodity so as to create orderly marketing. Research and promotion can be financed with pooled funds. Once approved by a required number of producers (usually two-thirds) the marketing order is binding on all handlers of the commodity within the geographic area of regulation. It may limit the quantity of goods marketed, or establish the grade, size, maturity, or quality of the goods. Marketing orders have been established for milk, fruits, vegetables, and other commodities. Marketing agreements may contain more diversified provisions, but are enforceable only against those handlers who enter into the agreement. An order can be terminated when a majority of all producers favor its termination or when USDA determines that the order no longer serves its intended purpose. Federal oversight is provided by USDA's Agricultural Marketing Service (AMS). See Market allocation, Orderly marketing, Prorate, Reserve pool, Shipping holiday, and Specialty crops. [www.ams.usda.gov/fv/moab.html].

Marketing quotas (or allotments) — Authorized by the Agricultural Adjustment Act of 1938, these quotas (sometimes called poundage quotas) limit marketings of certain commodities. The marketing quota, which must be approved by at least two-thirds of the eligible producers voting in a referendum, is intended to ensure an adequate and normal supply of the commodity, and also ensure that production and supplies are not excessive. Growers who market in excess of their quotas pay penalties on the excess and are ineligible for government price-support loans. Quotas have been suspended for wheat, feed grains, and cotton since the 1960s. Rice quotas were abolished in 1981. Tobacco quotas ended after the 2004 crop (P.L. 108-357, Title VI). Authority for standby marketing allotments for domestically-produced sugar and crystalline fructose was mandated by the 1990 farm bill (P.L. 101-624), but eliminated by the 1996 farm bill (P.L. 104-127). The 2002 farm bill (P.L. 107-171) reintroduced marketing allotments for domestically-produced sugar under the modified sugar program. Marketing allotments are required to be in effect unless USDA projects sugar imports will be above 1.532 million short tons.

Marketing spread — See Farm to retail price spread.

Marketing year — The 12-month period, generally from the beginning of a new harvest, over which a crop is marketed. For example, for wool, mohair, and Hawaiian sugarcane, the marketing year is January 1-December 31; for honey, it is April 1-March 31; for wheat, barley, and oats, it is June 1-May 31; for flue-cured tobacco, it is July 1-June 30; for cotton, peanuts, and rice, it is August 1-July 31; for sugar beets, it is September 1-

August 31; for corn, sorghum, soybeans, mainland sugarcane, all tobacco but flue-cured, and milk, it is October 1-September 30. The crop marketing year beginning and ending dates are published by NASS in the Agricultural Prices annual summary. In contrast, the crop year is the calendar year during which the crop is harvested.

Marrakech accords — Used often to designate the Uruguay Round trade agreements, including the agreement to establish the World Trade Organization, because they were signed on April 15, 1994, in Marrakech, Morocco.

Maximum tolerated dose (MTD) — Loosely, the highest dose of a chemical that when administered to a group of test animals does not increase the death rate during a long-term study. The purpose of administering MTD is to determine whether long-term exposure to a chemical might lead to any adverse health effects in a population, when the level of exposure is not sufficient to cause premature mortality due to short-term toxic effects. The maximum dose is used, rather than a lower dose, to reduce the number of animals that need to be tested (and thus, the cost of animal testing), in order to detect an effect that occurs only rarely. This analysis is used in establishing chemical residue tolerances in foods.

mbf — Thousand board feet of timber or lumber.

MBM — See Meat and bone meal.

MBTA — Migratory Bird Treaty Act (16 U.S.C. 703 *et seq.*).

McGovern-Dole International Food for Education and Child Nutrition Program (IFEP) — A food aid program authorized in the 2002 farm bill (P.L. 107-171, Sec. 3107) which provides for the donation of U.S. agricultural commodities and associated financial and technical assistance to carry out preschool and school feeding programs in foreign countries. Maternal, infant and child nutrition programs also are authorized under this program. The program was first implemented in FY2003 with $100 million of CCC funds as stipulated in the 2002 farm bill. Beginning in FY2004, the authorizing statute provides for the program to be carried out with appropriated funding. The FY2004 agricultural appropriations act (P.L. 108-199) provided $50 million to carry out the program. IFPED began in FY2000 as a pilot project and was called Global Food for Education Initiative(GFEI). It used the donation of surplus agricultural commodities under Section 416 of the Agricultural Act of 1949 (P.L. 89-439, as amended) to support a global school feeding program. [www.fas.usda.gov/excredits/FoodAid/FFE/FFE.html].

McIntire-Stennis Act of 1962 — P.L. 87-788 makes funding available to the state agricultural experiment stations and to forestry schools and programs at the land grant colleges of agriculture for forestry research. The research covers such areas as reforestation, woodlands and related watershed management, outdoor recreation, wildlife habitats and wood utilization. Many of the research projects are performed cooperatively with scientists at the laboratories of the Forest Service. McIntire-Stennis funds are distributed by a formula that allocates $10,000 to each state, with 40% of the remainder being distributed according to a state's share of the nation's total commercial forest land, 40% according to the value of its timber cut annually, and 20% according to its state appropriation for forestry research.

MDM — Mechanically deboned meat.

Meat and bone meal (MBM) — Meat and bone meal basically is the protein which remains after dead animals or their parts are rendered – i.e., the excess moisture removed and the fats (used for tallow and other products) are separated. In recent decades, the use of

MBM was increasingly being added to animal feeds (cattle, hog, poultry, seafood, pet foods) as a source of essential amino acids and nutrients. Most scientists believe that bovine spongiform encephalopathy (BSE) can be spread by feeding ruminant-based MBM back to ruminants, and this practice is increasingly being banned worldwide, including in the United States and Canada. At issue here and in other countries is whether much more extensive limitations on the use of MBM in animal feeds are warranted for food safety and animal health reasons.

Mechanically separated meat (MSM) — A paste-like meat product obtained through a mechanical process that crushes bone. Until the publication of new regulations (69 FR 1862, January 12, 2004) following the U.S. BSE case, USDA permitted MSM that met standards for calcium, iron, and central nervous system tissue to be used as an ingredient in meat products that would be heated again in manufacturing. MSM now is not permitted for use in any human food.

Medfly — A shortened name for the Mediterranean fruit fly, a destructive pest of fruits and vegetables that is found throughout most of Central America. The Animal and Plant Health Inspection Service (APHIS) is involved in programs to keep the Medfly from spreading north and becoming established in Mexico, where it could easily enter the United States on imported winter fruits and vegetables. Eradication efforts in California, Florida and Texas have prevented infestations from becoming established. Hawaii currently is the only state with a Medfly infestation. Travelers returning to the continental United States from Hawaii or a foreign country are prohibited from bringing into the country fresh fruits that may harbor the Medfly.

Mega-reg — An exceptionally large set of regulations. In agriculture, this term usually refers to the extensive rules that the Food Safety and Inspection Service (FSIS) issued in July 1996, establishing the hazard analysis and critical control point (HACCP) approach as a complementary inspection system to the traditional organoleptic approach.

Memorandum of agreement (MOA) — An agreement between federal agencies, or divisions/units within an agency or department, or between federal and state agencies, which delineate tasks, jurisdiction, standard operating procedures or other matters that the agencies or units are duly authorized and directed to conduct. Sometimes referred to as a memorandum of understanding (MOU).

Merchant Marine Act of 1920 — P.L. 66-261 provides for the promotion and maintenance of a U.S. merchant marine. Section 27, also known as the Jones Act, deals with cabotage (i.e., coastal shipping) and requires that all goods transported by water between U.S. ports be carried in U.S.-flag ships, constructed in the United States, owned by U.S. citizens, and crewed wholly by U.S. citizens. In addition, amendments to the Jones Act, known as the Cargo Preference Act (P.L. 83-644), provide permanent legislation for the transportation of waterborne cargoes in U.S.-flag vessels.

MERCOSUR — Mercado Comun del Sur (Southern Cone Common Market).

Methane — A gas created by anaerobic decomposition of organic compounds. Natural gas is composed mostly of methane. Methane is a so-called greenhouse gas (see Greenhouse effect). Agricultural wastes, especially animal wastes, are a major source of methane releases to the atmosphere.

Methanol — A liquid alcohol, (CH_3OH, also known as methyl alcohol or wood alcohol), formed in the destructive distillation of wood or made synthetically, and used especially as an alternative fuel, a gasoline additive, a solvent, an antifreeze, or a denaturant for ethyl alcohol. As a gasoline additive it lowers the carbon monoxide emissions but increases hydrocarbon emissions. Methanol is also seen as a potential transitional fuel for fuel cells.

Methyl bromide — A fumigant used for soil treatment, to control pests in postharvest storage, for killing pests on fruits, vegetables, and grain going into export trade, for plant quarantine treatment, and for fumigation of buildings. Because methyl bromide contributes to depletion of stratospheric ozone, it is subject to phase out requirements of the 1987 Montreal Protocol on Ozone Depleting Substances and of the Clean Air Act (CAA) (42 U.S.C. 7401 *et seq.*). The Montreal Protocol requires a complete phase out in industrialized countries by the year 2005, and a future freeze in developing country use. A 1998 amendment (P.L. 105-178, Title VI) conformed the Clean Air Act phase out date with that of the Montreal Protocol. Methyl bromide regulations exempt quarantine and pre-shipment treatment of agricultural commodities. Applications for additional exemptions for critical agricultural uses, for which there are no alternative pesticides, are being evaluated by EPA. EPA will nominate uses that meet international criteria for exemptions, which may be granted by parties to the Montreal Protocol at their next meeting. In addition, a limited number of emergency exemptions will be allowed after 2005. Methyl bromide is regulated as a pesticide under the Federal Insecticide, Fungicide, and Rodenticide Act (FIFRA; 7 U.S.C. 136 *et seq.*) and as a hazardous substance under the Resource Conservation and Recovery Act (RCRA; 42 U.S.C. 6901 *et seq.*), and is subject to reporting requirements under the Emergency Planning and Community Right to Know Act (EPCRA; (42 U.S.C. 11001 *et seq.*). [www.epa.gov/spdpublc/mbr].

Methyl tertiary butyl ether — See MTBE.

Metric ton (MT) — Usually abbreviated MT, a metric ton is 2,204.62 pounds, compared to a short ton of 2,000 pounds. Generally, international agricultural trade data are cited in metric tons.

MFN — Most-favored-nation.

MGEX — Minneapolis Grain Exchange. [www.mgex.com/index.cfm].

MicroLoan Program — A program administered by the Small Business Administration (SBA). The MicroLoan Program provides very small loans to start-up, newly established, or growing small business concerns. Under this program, SBA makes funds available to non-profit community based lenders (intermediaries) that in turn, make loans to eligible borrowers in amounts up to a maximum of $35,000. The average loan size is about $10,500. Applications are submitted to the local intermediary and all credit decisions are made on the local level. The maximum term allowed for a microloan is six years. Loan terms, however, vary according to the size of the loan, the planned use of funds, the requirements of the intermediary lender, and the needs of the small business borrower. Interest rates vary, depending upon the intermediary lender and costs to the intermediary from the U.S. Treasury.

Micronutrient Fortification Programs — The 2002 farm bill (P.L. 107-171, Sec. 3013) requires the Administrator of USAID, in consultation with the Secretary of Agriculture, to establish micronutrient fortification programs under P.L. 480 food aid. Section 3013 replaces a pilot program similarly named and authorized in the 1996 farm bill (P.L. 104-127, Sec. 415). Under the programs, grains and other commodities made available to countries selected for participation will be fortified with micronutrients (e.g., iron, vitamin A, iodine, and folic acid).

Mid-term Review Reforms — A set of reforms of the EU's Common Agricultural Policy adopted by the EU Council of Ministers on June 26, 2003. In summary, the main elements of the reformed CAP include: a single farm payment for EU farmers independent (decoupled) from production (although limited coupling may be permitted in some circumstances); compliance with environmental, food safety, animal and plant health, and animal welfare standards in order to receive the single payment; increased funding for rural development and accompanying environmental or animal welfare measures; a reduction in direct payments (modulation) for larger farms to finance the expanded rural development program; an agreement to limit the CAP (including rural development) budget at the 2007 level through 2013; and some additional market price support cuts.

MIF — Milk Industry Foundation. [www.idfa.org].

Migrant and Seasonal Agricultural Workers Protection Act (MSPA) — P.L. 97-470 (January 14, 1983) repealed and replaced the Farm Labor Contractor Registration Act (P.L. 88-582). The Act provides federal labor-related standards with respect to the transportation and housing of agricultural workers, sets specific requirements concerning payment of wages and conditions of employment, and deals with the relationship between agricultural workers, their employers and farm labor contractors. The Act requires certification of farm labor contractors through the U.S. Department of Labor. (29 U.S.C. 1801 *et seq.*).

Migratory Bird Treaty Act of 1918 — This Act, as amended (16 U.S.C. 703 *et seq.*), regulates the taking of wild birds and implements the provisions of four different bilateral treaties for bird conservation (with Canada, Mexico, Japan, and Russia). Very few of its provisions affect farmers more than any other citizen, save when bird populations become pests. The Act and the associated treaties allow taking of birds to prevent serious injury "to the agricultural or other interests in any particular community." As implemented, the practice has been to use non-lethal methods where possible, especially for native species. The control of bird pests is managed by the Animal and Plant Health Inspection Service.

Milk equivalent — A measure of the quantity of fluid milk used in a processed dairy product. Measured on a milkfat basis, it takes about 21.8 pounds of farm milk to make a pound of butter, and about 9.2 pounds to make a pound of American cheese. Measured on a skim solids basis, it takes about 11.6 pounds of farm milk to make a pound of nonfat dry milk. Farm milk weighs about 8.6 pounds per gallon.

Milk Income Loss Contract (MILC) Payments — Name given by USDA to the dairy farmer counter-cyclical payments program, authorized by the 2002 farm bill (P.L. 107-171, Sec. 1502, 7 U.S.C. 7982). Under the program, dairy farmers nationwide are eligible for a federal payment whenever the minimum monthly market price for farm milk used for fluid consumption in Boston falls below $16.94/cwt. A producer potentially can receive a payment equal to 45% of the difference between the $16.94 per cwt. target price and the market price, in any month that the Boston market price falls below $16.94. A producer can receive a payment on all milk production during that month, but no payments will be made on any annual production in excess of 2.4 million pounds per dairy operation. On average this limit is reached by a milking herd of 130 cows. MILC payments apply to production beginning December 1, 2002, through September 30, 2005.

Milk marketing orders — Administered by the Agricultural Marketing Service (AMS), federal milk marketing orders were first instituted in the 1930s to promote orderly marketing conditions by, among other things, applying a uniform system of classified pricing throughout the farm milk market. Federal milk marketing orders regulate handlers that sell milk or milk products within an order region by requiring them to pay not less than an established minimum price for the Grade A milk they purchase from dairy producers, depending on how the milk is used. This classified pricing system requires handlers to pay a higher price for milk used for fluid consumption (Class I) than for milk used in manufactured dairy products such as yogurt, ice cream, cheese, butter and nonfat dry milk (Class II, Class III and Class IV products). The 1996 farm bill (P.L. 104-127) required USDA to consolidate the number of federal milk marketing orders, and to revise the method by which minimum class prices are determined. USDA implemented these changes in 2000. There now are 11 milk marketing orders, down from 31 when the law was enacted.

Milk protein concentrate (MPC) — Any type of concentrated milk that contains 40-90% milk protein. In addition to ultrafiltered milk products, the MPC classification includes concentrates made through other processes, such as blending nonfat dry milk with highly concentrated proteins, such as casein. Currently, almost all MPC used in the U.S. is imported and is not subject to a tariff rate quota. Many dairy producer groups are concerned that foreign manufacturers are using nonfat dry milk in the production of MPC, and hence circumventing existing quotas on nonfat dry milk.

Milk-feed price ratio — A measure of the value of 16% protein ration (feed) to one pound of whole milk. As with the hog-corn ratio, this relationship is an indicator of the profitability of milk production.

Milling yield — A general term that can refer to the milling of any cereal crop. Wheat milling yield refers to the percent of flour obtained from a given unit of whole wheat kernels (flour yield or flour extraction rate), averaging 70-75% in the United States. Rice milling yield refers to the amount of polished white rice obtained from unhusked rough rice. Rice milling rates for polished white rice vary by crop variety and quality, but tend to average about 72% of rough rice weight in the United States. Byproducts from rice milling include rice hulls (about 20% of rough rice weight) and bran, polish, and germ (about 8%).

Minimal nutritional value — Refers to foods that may not be sold in competition with the school lunch and breakfast programs. These are foods that USDA has determined contain little if any nutritional value. For example, sugar candy, soda pop without fruit juices, and chewing gum are considered to be foods of minimal nutritional value. Candy containing nuts or chocolate is considered to have some nutritional value.

Minimum access — In the Uruguay Round Agreement on Agriculture, countries are obliged to provide minimum levels of imports for products subject to tariffication. Access is assured by tariff-rate quotas.

Minimum tillage — The minimum soil manipulation necessary for crop production. Conservation tillage, reduced tillage, and no-till farming are related terms.

Minnesota-Wisconsin price (M-W price) — Prior to May 1995, a component of the basic formula price for farm milk formerly used in federal milk marketing orders. It represented a survey of the average price Minnesota and Wisconsin plants were paying farmers for Grade B milk to be used in processed dairy products. In 1995, the M-W price was replaced with the basic formula price as the price mover under federal milk marketing orders.

Minor crops — Crops that may be high in value but that are not widely grown. Many fruits, vegetables, and tree nuts come under this definition. The IR-4 program is one publicly funded program to help producers of minor crops with their unique problems.

Minor oilseeds — Oilseed crops other than soybeans and peanuts; usually a reference to the "other oilseeds" eligible for marketing assistance loans under the 2002 farm bill (P.L. 101-171, Sec. 1201-1205) (sunflower seed, rapeseed, canola, safflower, flaxseed, and mustard seed).

mmt — Million metric tons.

MOA — Memorandum of agreement.

Modalities — Modalities are the formulas, targets, or specific measures used to accomplish objectives in trade negotiations. An example of modalities in the current WTO agriculture negotiations would be a percentage phase-out over a specified time period of agricultural export subsidies or the use of the Swiss formula for tariff reduction and harmonization.

Model Good Samaritan Food Donation Act — See Bill Emerson Good Samaritan Act of 1996 (P.L. 104-210).

Modulation — As part of its Agenda 2000 reforms, and effective January 1, 2000, EU member-countries may reduce direct aid to producers (by a maximum of 20%) in cases where: the labor employed in a farm falls below a threshold set by national authorities; the overall prosperity of the holding is above a certain limit; and, the total payments granted under support programs exceed a limit set by national authorities. The savings that result and those from cross-compliance or econ-conditionality (observance of environmental criteria) may be used by the member countries to supplement EU funding for early retirement measures, payments for less favored areas and areas subject to environmental restrictions, agri-environmental provisions, afforestation and rural development. Modulation is an essential element of the EU Commission's Mid-term Review (MTR) proposals. Funds acquired from the reduction of payments to farms will be allocated to rural development under the MTR.

Mohair Recourse Loan Program — A program authorized by the emergency provisions of the FY1999 USDA appropriations act (P.L. 105-277) that made interest-free recourse loans of $2.00 per pound on mohair produced prior to October 1, 1998. Final date to obtain a loan was September 30, 1999. The producer-owned mohair used as loan security had to be stored in approved bonded warehouses. Loans matured not later than 1 year following disbursement. Under the 2002 farm bill (P.L. 107-171, Sec. 1201-1205), mohair was designated a "loan commodity" and made eligible for marketing assistance loans and loan deficiency payments (LDPs).

Moisture content — Refers to the portion of grain weight made up of water. Moisture content plays a key role in grain storage and processing. Excess moisture content will cause grain to spoil quickly limiting its storability. Overly dry grains will crack or split easily during milling thus lowering the value of the end product, particularly for rice where the percent of whole grains is highly valued. Managing moisture and pest control are two important problems associated with grain storage. For some grains (i.e., corn), moisture is a grading factor.

Monetary compensatory amounts (MCAs) — Border measures in the EU consisting of taxes and subsidies formerly applicable to intra-EC trade in agricultural and food products for which intervention prices were set. These border measures were made necessary by the fact that intervention prices were set in ECUs and converted into national currency terms at green rates, set at levels different from commercial market rates. This gave rise to price differentials between member nations (in market ECUs) that would influence intra-EC trade if not offset by the MCAs. The system worked by subsidizing exports (and taxing imports) from strong-currency countries, and taxing exports (subsidizing imports) from weak-currency countries. MCAs were abolished in 1993, when border controls were removed with the advent of the Single Market.

Monetization — A P.L. 480 provision (section 203) first included in the Food Security Act of 1985 (P.L. 99-198) that allows private voluntary organizations and cooperatives to sell a percentage of donated P.L. 480 commodities in the recipient country or in countries in the same region. Under section 203, private voluntary organizations or cooperatives are permitted to sell (i.e., monetize) for local currencies or dollars an amount of commodities equal to not less than 15% of the total amount of commodities distributed in any fiscal year in a country. The currency generated by these sales can then be used: to finance internal transportation, storage, or distribution of commodities; to implement development projects; or to invest and with the interest earned used to finance distribution costs or projects.

Monoculture — A pattern of crop or tree production that relies on a single plant variety.

Montreal Protocol on Ozone Depleting Substances — An international agreement, to which the U.S. is a signatory, for controlling emissions of chemicals that deplete stratospheric ozone (including methyl bromide). The Clean Air Act Amendments of 1990 (P.L. 101-549) contain provisions for implementing the Montreal Protocol, as well as explicit, separate authority for the EPA to regulate ozone depleting chemicals.

Morbidity — Rate of disease incidence; an important measure in epidemiological studies.

Morrill Act of 1862 — This first Morrill act (7 U.S.C. 301 *et seq.*) allocated federal land to each state and directed the states to sell the land and use the proceeds to establish a college dedicated to the agricultural and mechanical arts. States without federal lands within their borders received land in scrip, giving them the right to sell federal land located in other states. The act resulted in the establishment of the land grant colleges of agriculture. The purpose of the Act was not only to improve the economic and social welfare of farmers, but also to make higher education with a practical application generally available to all segments of U.S. society. The Act pertained only to the original establishment of the colleges of agriculture, and is not an authority under which the colleges currently receive federal funds.

Morrill Act of 1890 — This popularly named Agricultural College Act of 1890 (7 U.S.C. 321 *et seq.*) authorized additional direct appropriations for the land grant colleges of agriculture that had been established under the Morrill Act of 1862 (7 U.S.C. 301 *et seq.*). However, the most significant feature of this second Morrill Act was that the 1862 schools could receive the additional funds only if they admitted blacks into their programs or if they provided separate but equal agricultural higher education to black students. In the period following the Civil War, sixteen southern states established separate land grant colleges of agriculture for black students under this Act; Congress designated Tuskegee University an 1890 institution at a later date. Federal funds for research and extension at the 1890 schools are provided under subsequent laws, not the second Morrill Act.

Most-favored-nation treatment (MFN) — A commitment that a country will extend to another country any concessions (reduced tariff rates) it grants to other countries. MFN is a basic principle and obligation of the General Agreement on Tariffs and Trade (GATT, 1947). Almost all countries are accorded permanent MFN status by U.S. law. Since 1998, the term normal trade relations (NTR) has replaced most-favored-nation in all U.S. statutes. This change was included in section 5003 of the Internal Revenue Service Restructuring and Reform Act of 1998 (P.L. 105-206). However, Title IV of the Trade Act of 1974 (P.L. 93-618) established conditions on U.S. MFN/NTR tariff treatment to certain non-market economies, one of which is certain freedom-of-emigration requirements (better known as the Jackson-Vanik amendment). The Act authorizes the President to waive a country's full compliance with Jackson-Vanik under specified conditions, and this must be renewed by June 3 of each year. Once the President does so, the waiver is automatic unless Congress passes (and sustains a Presidential veto of) a disapproval resolution. MFN/NTR status for China, a non-market economy, which had been originally suspended in 1951, was restored in 1980 and was continued in effect through subsequent annual Presidential extensions. Following the brutal suppression of pro-democracy demonstrators in Tiananmen Square in 1989, however, the annual renewal of China's MFN status became a source of considerable debate in the Congress; and legislation was introduced to terminate China's MFN/NTR status or to impose additional conditions relating to improvements in China's actions on various trade and non-trade issues. Agricultural interests generally opposed attempts to block MFN /NTR renewal for China, contending that several billion dollars annually in current and future U.S. agricultural exports could be jeopardized if that country retaliated. In China's case, permanent normal trade relations (PNTR) status was accorded in P.L. 106-286. PNTR paved the way for China's accession to the WTO in December 2000; it provides U.S. exporters of agricultural products the opportunity to benefit from China's WTO agreements to reduce trade barriers and open its agricultural markets.

MOU — Memorandum of understanding (see Memorandum of agreement).

MPC — Milk protein concentrate

MPP — Market Promotion Program.

MPS — Market price support.

MRL — Maximum-residue limit (see Registration, and Pesticide).

MS(S) — Mechanically separated (S), where (S) denotes the species, e.g., beef, pork, or chicken.

MTBE — Methyl tertiary butyl ether. A byproduct of petroleum refining and natural gas production, MTBE is used to extend gasoline stocks, boost octane, and comply with reformulated gasoline (RFG) oxygen requirements. While MTBE use has reduced ozone-forming and toxic pollutant emissions from motor vehicles, its use has led to contamination of underground wells. Because of this contamination, several states have banned or will ban its use, and there is interest in a national ban, as well. Ethanol is being used to replace MTBE.

MTD — Maximum tolerated dose.

MTN — Multilateral trade negotiations.

Mulch — A natural or artificial layer of plant residue or other material on the soil surface. Mulch reduces erosion, conserves soil moisture, inhibits weed growth, and can provide the soil with organic matter as it breaks down. Mulch till prepares the soil so as to leave plant residues (or other mulching materials) on or near the surface.

Multi-Peril Crop Insurance (MPCI) — The oldest and most common form of federal crop insurance. MPCI protects against crop yield losses by allowing participating producers to insure a certain percentage of historical crop production. A single policy protects crops against all natural perils including adverse weather, fire, insects, disease, wildlife, earthquake, volcanic eruption and failure of irrigation water due to unavoidable causes. It is delivered by private companies and reinsured by the federal government.

Multifunctionality — The term used to characterize a broad range of economic, social, and environmental attributes generated by the agricultural sector, in addition to its primary role of producing food and fiber. These additional attributes, according to proponents of supporting agricultural multifunctionality, include long-term food security, rural viability, cultural heritage, land conservation, agricultural landscape, agro-biological diversity, and sanitary/phytosanitary health. In agricultural trade discussions in the WTO, the EU and Japan, among others, argue that multifunctionality justifies continued protection and subsidization of agriculture. The United States and the Cairns Group argue that support of multifunctionality should be specific, targeted, and provided in a non-trade distorting manner.

Multilateral agreement — A trade agreement involving many countries (as with the World Trade Organization) in contrast to a bilateral agreement (as with the US-Israel Free Trade agreement) involving only two countries.

Multilateral aid — See Bilateral aid.

Multilateral trade negotiations (MTN) — Initially this applied to negotiations between General Agreement on Tariffs and Trade (GATT) member nations conducted under the auspices of the GATT and aimed at reducing tariff and nontariff trade barriers. The World Trade Organization has now replaced the GATT as the administrative body. A current round of multilateral negotiations to further reform trade, including trade in agriculture, is under way in the Doha Development Agenda round.

Multiple basing points — A method of regional pricing in milk marketing orders that would allow more than one basing point, or surplus area, to be used. Surplus areas are administratively defined as areas with low Class I utilization, meaning that a relatively small percentage of the milk produced in an area is used in that area as Class I (fluid) milk. In a multiple basing point system, the order used as the basing point has the smallest Class I differential (the difference between the Class I price and the Class III price). The Class I differential for other orders is then based on transportation costs to the nearest basing point plus the minimum differential. Pricing now largely reflects the Upper Midwest as the only basing point, even though the northeast and southwest are surplus areas.

Multiple component pricing — The practice of valuing farm milk according to the value of its protein, fat, and mineral content. This practice has been adopted by many regions for federal milk marketing orders. Historically, milk was priced solely on the basis of fat content.

Multiple use — According to the Multiple Use, Sustained Yield Act of 1960 (P.L. 86-517), multiple use of the national forests means the "harmonious and coordinated management of the various resources, each with the other, without impairment of the productivity of the land, with consideration being given to the relative values of the various resources, and not necessarily the combination of uses that will give the greatest dollar return or the greatest unit output." Multiple use implies a sustained yield of outdoor recreation, range, timber, watershed, and wildlife and fish values.

Multiple Use, Sustained Yield Act of 1960 (MUSY) — P.L. 86-517 specifies that the national forests are to achieve a coordinated high level of annual or periodic output through harmonious management of outdoor recreation, range, timber, watershed, wildlife and fish, and wilderness, but without impairing the long-term productivity.

MUSY — Multiple Use and Sustained Yield Act of 1960 (P.L. 86-517; 16 U.S.C. 528 *et seq.*).

Mutagen — An agent that causes a permanent genetic change in a cell other than that which occurs during normal growth. Testing to determine mutagenicity is one component of assessing the potential chronic toxicity of pesticides and other chemicals.

Mutual self-help housing — A program to assist groups of low-income families in building their own homes. Each family is expected to contribute at least 700 hours of labor in building homes for each other. Participating families generally have low income and are unable to pay for homes built by the contract method. The homes generally are financed by Section 502 loans.

Mycotoxins — Toxic substances produced by fungi or molds on agricultural crops that may cause sickness in animals or humans that eat feed or food made from contaminated crops. There are between 300 and 400 known mycotoxins, but of most concern, based on toxicity and occurrence, are aflatoxin, vomitoxin, zearalenone, fumonisin, T-2 toxin, and T-2-like toxins (trichothecenes). The Grain Inspection Packers and Stockyards Administration (GIPSA) currently measures for aflatoxin in all exports shipments of corn, and measures for aflatoxin and vomitoxin on a voluntary basis in domestic shipments.

NAA — National Aquaculture Act of 1980 (P.L. 96-362; 16 U.S.C. 2801 *et seq.*).

NACD — National Association of Conservation Districts. [www.nacdnet.org].

NACD — National Association of Conservation Districts. [www.nacdnet.org].

NACMCF — National Advisory Committee on Microbiological Criteria for Foods. [www.fsis.usda.gov/OPHS/nacmcf/index.htm].

NACO — National Association of Counties. [www.naco.org].

NAD — National Appeals Division. [www.nad.usda.gov].

NADA — New animal drug applications.

NAEGA — North American Export Grain Association.

NAFTA — North American Free Trade Agreement. [www.nafta-sec-alena.org/english/index.htm].

NAHLN — National Animal Health Laboratory Network. [www.aphis.usda.gov/vs/highlights/section6/section6-6.html].

NAHMS — National Animal Health Monitoring System. [nahms.aphis.usda.gov].

NAHRS — National Animal Health Reporting System. [www.aphis.usda.gov/vs/ceah/ncahs/nahrs].

NAICS — North American Industry Classification System (replaces standard industrial classification (SIC). [www.census.gov/epcd/www/naics.html].

NAL — National Agricultural Library. [www.nalusda.gov].

NAP — Noninsured Assistance Payments.

NARC&DC — National Association of Resource Conservation and Development Councils. [www.rcdnet.org].

NAS — National Academy of Sciences. [www.nas.edu].

NASDA — National Association of State Departments of Agriculture. [www.nasda-hq.org].

NASS — National Agricultural Statistics Service. [www.usda.gov/nass].

NASULGC — National Association of State Universities and Land Grant Colleges. [www.nasulgc.org].

National Academy of Sciences (NAS) — An institution created by Congress in 1863 to provide science-based advice to the government. The sister organizations associated with the Academy are the National Academy of Engineers, Institute of Medicine, and the National Research Council. The Academies and the Institute are honorary societies that elect new members to their ranks each year. The bulk of the institutions' science policy and technical work is conducted by the National Research Council (NRC), created expressly for that purpose. The NRC's Board on Agriculture and Natural Resources addresses issues confronting agriculture, food, and related environmental topics. [www.nas.edu].

National Advisory Committee on Microbiological Criteria for Foods — This committee was established in 1987 to advise the Secretaries of Agriculture and Health and Human Services concerning the development of science-based, microbiological standards by which the safety of foods can be evaluated and by which plant sanitation and processing systems can be improved.

National Agricultural Library (NAL) — A national depository of scientific and popular agricultural information located at the Agricultural Research Service's research center in Beltsville, Maryland. NAL's administration was merged with ARS in 1994. [www.nalusda.gov].

National Agricultural Research, Extension, and Teaching Policy Act of 1977 — Title XIV of the 1977 farm bill (P.L. 95-113; 7 U.S.C. 3101 *et seq.*) made USDA the leading federal agency for agricultural research, extension, and teaching programs and consolidated the funding for these programs. This Act and other statutes relating to the research mission area are reauthorized every 4 to 7 years as part of omnibus legislation that provides funding authority and policy guidance for nearly all of USDA's agencies.

National Agricultural Research, Extension, Education, and Economics Advisory Board — A 30-member board established by the 1996 farm bill (P.L. 104-127) to replace three existing advisory committees. The Board, which was reauthorized through 2007 by the 2002 farm bill (P.L. 107-171, Sec. 7133), advises USDA on national priorities and policies related to agricultural research, extension, and education.

National Agricultural Statistics Service (NASS) — A USDA agency that collects and publishes statistics on the U.S. food and fiber system, with offices located in each state's department of agriculture. [www.usda.gov/nass].

National Ambient Air Quality Standards (NAAQS) — National standards set by the EPA under authority of the Clean Air Act (42 U.S.C. 7401 *et seq.*). NAAQS define the maximum allowable concentrations of specified air pollutants in outdoor (ambient) air.

NAAQS have been set for carbon monoxide, particulate matter, sulfur oxides, nitrogen dioxide, lead, and ozone. Primary NAAQS protect human health, with a margin of safety; secondary NAAQS protect human welfare, which includes effects on soils, water, crops, vegetation, materials, etc.

National Animal Health Laboratory Network (NAHLN) — A network of federal and state resources intended to enable a rapid and sufficient response to animal health emergencies. The concept of the NAHLN reconfigures animal health diagnostic services in the United States by positioning National Veterinary Services Laboratory as the lead U.S. animal health laboratory and allowing select laboratories operated by state and university officials to cooperate in foreign animal disease surveillance and related services. [www.aphis.usda.gov/vs/highlights/section6/section6-6.html].

National Animal Health Reporting System (NAHRS) — The National Animal Health Reporting System is a joint effort of the U.S. Animal Health Association (USAHA), American Association of Veterinary Laboratory Diagnosticians (AAVLD), and USDA's Animal and Plant Health Inspection Service (APHIS). NAHRS was designed to provide data from chief state animal health officials on the presence of confirmed Office International des Epizooties (OIE) LIST A and B clinical diseases in specific commercial livestock, poultry, and aquaculture species in the United States. It is intended to be one part of a comprehensive and integrated animal-health surveillance system. [www.aphis.usda.gov/vs/ceah/ncahs/nahrs].

National Animal Identification System (NAIS) — See United States Animal Identification Plan (USAIP).

National Appeals Division (NAD) — The National Appeals Division of USDA was established by the Department of Agriculture Reorganization Act of 1994 (Title II of P.L. 103-354) to consolidate and improve the hearing procedures for USDA claims and disputes. The statute and regulations provide that certain sections of the Administrative Procedure Act (APA) (P.L. 79-404, as amended), including the hearing requirements, do not apply to NAD proceedings. The NAD procedures govern informal and formal hearings covering appeals of decisions made by the rural development agencies, Natural Resources Conservation Service, Risk Management Agency, and the Farm Service Agency. The statute and regulations set forth the procedures for hearings, requirements for the presiding officers, requirements for communications between the decision-maker and persons interested in the matter, and other important issues. After a decision is made by a hearing officer, both the appellant and the agency have the right to a review by the NAD director, who then issues a final determination. The final determination of the NAD is reviewable and enforceable by the U.S. District Court in accordance with the judicial review provisions of the APA. [www.nad.usda.gov].

National Aquaculture Act of 1980 — P.L. 96-362, as amended, is intended to promote and support the development of private aquaculture and to ensure coordination among the various federal agencies that have aquaculture programs and policies. It provided for a national aquaculture policy, including a formal National Aquaculture Development Plan; established a Joint Subcommittee on Aquaculture on which officials of USDA, Commerce, the Interior, and nine other federal agencies sit; designated USDA as the lead agency for coordination; and authorized the National Aquaculture Information Center within the National Agricultural Library (accessible now through the Alternative Farming Systems Information Center at [www.nal.usda.gov/afsic/afsaqua.htm]).

National Bank for Cooperatives (CoBank) — An institution of the Farm Credit System that provides financial services to approximately 2,000 agricultural cooperatives, rural utility systems, Farm Credit System associations and other businesses serving rural America. CoBank also finances agricultural exports and provides international banking services primarily to U.S. agricultural cooperatives. [www.cobank.com].

National Cheese Exchange (NCE) — A now defunct private non-profit corporation that operated in Green Bay, Wisconsin. Every Friday morning for one-half hour, members of the NCE met to buy or sell cheddar cheese in 40-pound blocks and 500-pound barrels on the exchange. The closing prices were published and widely circulated throughout the dairy industry, and were used as the basis for buying and selling cheese throughout the food distribution system. Up until April 1997 the USDA used changes in the NCE price as a principal component in determining the basic formula price for all milk sold under federal milk marketing orders. Activity on the NCE was regulated by the Wisconsin Department of Agriculture and the Wisconsin Attorney General.

National Dairy Promotion & Research Program — See Dairy Promotion Program. [http://www.ams.usda.gov/dairy/dairyrp.htm].

National Drought Policy Commission — P.L. 105-199, the National Drought Policy Act of 1998, authorized creation of the Commission to conduct a study of current federal, state, local and tribal drought preparedness, and review laws and programs to determine if deficiencies exist in current relief policies and resources. The Commission's report to the President and Congress, Preparing for Drought in the 21st Century, May 2000, is available online. [www.fsa.usda.gov/drought/default.htm].

National Early Warning System — A program run by the Centers for Disease Control to increase federal support to state health departments to detect foodborne diseases by increasing the number of scientists available to investigate outbreaks and by enhancing laboratory-based surveillance of important foodborne pathogens.

National Environmental Policy Act of 1970 — P.L. 91-190 made a declaration of national environmental policy and established a continuing responsibility of the federal government to reach a number of substantive goals that embody nationwide improvements in environmental quality. Federal policies, regulations, and laws must be administered in accordance with NEPA. To insure this is accomplished, all federal agencies must consider the environmental consequences of their actions through the preparation of environmental impact statements (EIS). Also, the law creates the Council on Environmental Quality in the Executive Office of the President.

National Estuarine Research Reserve System — A program created in the 1972 Coastal Zone Management Act (P.L. 92-583) to have states nominate portions of estuaries representing the major coastal biogeographic regions of the country as sites for research and education activities. The 25 units in the system can be used for comparative research. The National Oceanic and Atmospheric Administration provides grants to states to acquire, develop, and operate these sites.

National Estuary Program — A program created by the 1987 amendments (P.L. 100-4) to the Clean Water Act (P.L. 92-500, as amended) to provide grants to states where governors identify nationally significant estuaries that are threatened by pollution, land development, or overuse. Governors have identified a total of 28 estuaries, and the Environmental Protection Agency awarded grants to these states to develop comprehensive management plans to restore and protect the estuaries.

National farm program acreage — Prior to 1996, the number of harvested acres of feed grains, wheat, and cotton needed nationally to meet domestic and export use and to accomplish any desired increase or decrease in carryover levels. The acreage base for an individual farm was calculated as the producer's share of the national farm program acreage. The 1996 farm bill (P.L. 104-127)and the 2002 farm bill (P.L. 101-171) have eliminated the need to calculate a national program acreage.

National Farm Animal Identification and Records (FAIR) — A pilot animal ID program established in 1999 by Holstein Association USA, Inc. of Brattleboro, Vermont; it has received funding through USDA's Animal and Plant Health Inspection Service.

National Fluid Milk Processor Promotion Program — This program was first authorized by the Fluid Milk Promotion Act of 1990 (7 U.S.C. 6401-6417). Fluid milk processors develop and finance generic advertising programs designed to maintain and expand markets and uses for fluid milk products produced in the United States. The mandatory national fluid milk program is financed by a 20¢ per hundredweight assessment on all fluid milk processed and marketed commercially in consumer-type packages (except by small processors). Fluid Milk Board annual revenue is approximately $110 million. [www.ams.usda.gov/dairy/dairyrp.htm].

National forest — Originally, forest reserves, established by Presidential proclamation mostly between 1891 and 1909. Today, the boundaries of the 155 national forests cannot be modified without congressional authorization, although many (especially eastern) national forests are combined for easier administration. The 114 administrative units, commonly referred to as national forests, are managed by the Forest Service for multiple use and sustained yield of renewable resources, as determined in forest plans.

National Forest Management Act (NFMA) of 1976 — P.L. 94-588 largely amended the Forest and Rangeland Renewable Resources Planning Act of 1974 (P.L. 93-378). The law was seen as necessary, because a lawsuit (commonly known as the Monongahela decision) had invalidated many timber practices in the national forests. NFMA substantially enacted detailed guidance for forest plans, particularly in regulating when, where, and how much timber could be harvested and in requiring public involvement in preparing and revising the plans. Also, NFMA established and expanded several Forest Service trust funds and special accounts.

National Forest System (NFS) — The 192 million acres administered by the Forest Service for multiple use and sustained yield; comprised of 155 national forests (in 114 administrative units) with nearly 188 million acres, 20 national grasslands with nearly 4 million acres, and 117 other units (e.g., purchase units, land utilization projects, research and experimental areas) with less than a million acres. While the NFS lands are concentrated in the West, the 25 million acres east of the 100th Meridian (the Great Plains) make the Forest Service the largest land manager in the East.

National grasslands — A type of unit designated by USDA and under Title II of the Bankhead-Jones Farm Tenant Act (P.L. 75-210) and permanently held by USDA as part of the National Forest System.

National Marine Fisheries Service (NMFS) — An agency within the National Oceanic and Atmospheric Administration (NOAA) at the Department of Commerce, now in transition to the new agency name of NOAA Fisheries [www.nmfs.noaa.gov]. NMFS/NOAA Fisheries administers programs and provides services and products to support the domestic and international conservation and management of living marine resources, fisheries development, trade and industry assistance activities, enforcement, and protected species and habitat conservation operations. Under the authority of the 1946 Agricultural Marketing Act, NMFS/NOAA Fisheries conducts voluntary seafood inspection on a fee-for-service basis, mainly as a marketing and quality program rather than as a food safety program [seafood.nmfs.noaa.gov].

National Milk Laboratory Certification Program — Under a Memorandum of Understanding with the National Conference on Interstate Milk Shipments, the Food and Drug Administration (FDA) conducts a national certification program for state centralized laboratories that test dairy products for contaminants and residues. FDA maintains accreditation of milk laboratories and sample collection surveillance procedures by making triennial on-site evaluations of laboratory facilities and equipment and by testing annually the performance skills of analysts. The FDA also standardizes, evaluates, and certifies state and territorial milk laboratory evaluation officers and state sampling surveillance officers.

National Natural Resources Conservation Foundation (NNRCF) — A nonprofit private organization established by the 1996 farm bill (P.L. 104-127) to promote and fund innovative solutions to conservation problems through effective partnerships. The Foundation can accept gifts and raise money. The NNRCF will conduct research, undertake educational activities, support demonstration projects, and make grants to state and local governments and nonprofit organizations. Appropriations were authorized at $1 million per year for 1997-99, but no appropriations were provided and the Foundation has not become operational. Similar foundations have been created for several other natural resource areas.

National Nutrition Monitoring and Related Research Act of 1990 — P.L. 101-445 was enacted to establish a comprehensive, coordinated program for nutrition monitoring and related research to improve the assessment of the health and nutrition of the U.S. population. The Act required: a program to achieve coordination of federal nutrition monitoring efforts within 10 years and assist states and local governments in participating in a nutrition monitoring network; an interagency board to develop and implement the program; and, an advisory council to provide scientific and technical advice and evaluate program effectiveness. The Act also required that dietary guidelines be issued every five years, and that any dietary guidance issued by the federal government for the general public be reviewed by the Secretaries of both Agriculture, and Health and Human Services.

National Organic Program (NOP) — A program authorized by the Organic Foods Production Act of 1990 (Title 21 of the 1990 farm bill (P.L. 101-624) that went into operation in October 2002 (7 CFR 205). The NOP sets national standards for organic farming practices, and accredits organizations and people who perform farm visits to certify that a farm meets the standards and may label its products USDA Organic. [www.ams.usda.gov:80/nop].

National Organic Standards Board (NOSB) — A board established by Title 21 of the 1990 farm bill (P.L. 101-624) to develop national standards for practices and substances to be used in implementing a National Organic Program (NOP). The Board is to evaluate new practices and inputs (e.g., fertilizers) as they develop and determine whether or not they are acceptable under the program. [www.ams.usda.gov/nosb].

National Partnership Office (NPO) — The NPO is responsible for implementing National Rural Development Partnership policies and activities. The NPO provides budgetary and financial technical assistance to State Rural Development Councils. [www.partnershipresourcecenter.org/resources/npo/pship_coord.php].

HPGS — National Plant Germplasm System. [www.ars-grin.gov/npgs].

National Research Council (NRC) — See National Academy of Sciences. [www.nas.edu/nrc].

National Research Initiative Competitive Grants Program (NRI) — Generally referred to as the NRI, this program makes grants to scientists at both public and private laboratories for basic and applied agricultural research in priority areas as designated in the research title of the1990 farm bill (P.L. 101-624, Title XVI), as amended. Grants are awarded competitively through a peer-review process. [attra.ncat.org/guide/n_z/nri.html].

National Resources Inventory (NRI) — A periodic survey of status and changing conditions of the soil, water, and related resources on private land conducted by USDA's Natural Resources Conservation Service (NRCS). The survey has been conducted at 5-year intervals since 1977, most recently in 2002. NRCS released initial results from the 1997 survey in late 1999, which can be found at [www.nrcs.usda.gov/technical/NRI].

National Rural Development Council (NRDC) — This is the federal component of the National Rural Development Partnership. The NRDC comprises representatives from various federal departments and national organizations whose activities or policies may affect rural areas. The NRDC provides guidance for the Partnership and works on behalf of State Rural Development Councils at the national level. Its administrative office is h o used in the U.S. Department of Agriculture. [www.rurdev.usda.gov/nrdp/national/nrdc/nrdc.html].

National Rural Development Partnership (NRDP) — A collaborative effort comprised of representatives of the federal, state, local, and tribal governments, the private sector, and the nonprofit sector to promote rural development across the nation. The principle component of the Partnership is the State Rural Development Councils. The Partnership was authorized by the 2002 farm bill (P.L. 107-171, 6021) as an amendment of the Consolidated Agriculture and Rural Development Act of 1972 (7 U.S.C. 2008m). The law authorizes appropriations of $10 million per fiscal year 2002-2006 for the Partnership. [www.rurdev.usda.gov/nrdp].

National Rural Economic Development Institute (NREDI) — NREDI helps develop the capacity of the National Rural Development Partnership and its constituent organizations (State Rural Development Councils and the National Rural Development Council) by providing economic development-related training and consulting services.

National School Lunch Act of 1946 — P.L. 79-396 authorized federal cash and commodity support for school lunch and milk programs, "...as a measure of national security..." in response to claims that many American men had been rejected for military service in World War II because of diet-related health problems. Beginning in the early 1930's federal support had been provided for school lunch programs by donations of surplus commodities, and when these dried up during the World War II, by grants provided under annual appropriations laws. The original Act established multi-year funding authority for grants to states to help finance school feeding programs. The law has been amended numerous times since its initial passage and now permanently authorizes the national school lunch program and the child and adult care food program (42 U.S.C. 1751 *et seq.*).

It was re-titled the Richard B. Russell National School Lunch Act in 1998 as a tribute to the former Louisiana Senator's prominent role in the creation of school feeding programs. Federally guaranteed subsidies are provided for every lunch served, with higher amounts generally provided for lunches served to low-income children who meet low income criteria set by the law. This Act also requires federal payments for meals and snacks served to children and elderly and disabled persons in day care facilities (the Child and Adult Care Food Program) and children in summer programs operated in low-income areas (the Summer Food Service Program). The law also requires a set value of commodity assistance for each lunch served under these programs. Other activities supported by this Act include meal supplements for children in after-school care, a homeless children nutrition program, meal service for Department of Defense overseas dependents schools, and an information clearinghouse.

National School Lunch Program — This child nutrition program provides cash and commodity assistance to public and private nonprofit elementary and secondary schools and residential child care institutions to support lunches served to all children in schools and institutions that choose to participate; snacks served in after-school programs also are federally subsidized. While all lunches and snacks are federally assisted, larger federal subsidies generally are provided for meals (or after-school snacks) served to children from lower-income families (i.e., free or reduced-price lunches and snacks, as opposed to paid lunches and snacks). Each meal or snack is subsidized at legislatively established rates that are annually indexed for food-price inflation. The program is permanently authorized under Sections 4 and 11 of the National School Lunch Act (P.L. 79-396, as amended; 42 U.S.C. 1751 *et seq.*), administered by the Food and Nutrition Service, and funded as an entitlement by annual agriculture appropriations acts. [www.fns.usda.gov/cnd/Lunch/default.htm].

National Sheep Industry Improvement Center — The Federal Agriculture Improvement and Reform Act of 1996 (P.L. 104-127) established a revolving fund of up to $50 million, to be used by a new, eventually privatized, center that aims to revive the declining U.S. sheep and goat industries through loans and loan guarantees for such activities as improving production and marketing methods, purchasing new equipment, and modernizing processing facilities. [www.rurdev.usda.gov/rbs/coops/cssheep.htm].

National Shellfish Sanitation Program (NSSP) — A program under which the federal Food and Drug Administration works cooperatively with the states, the Interstate Shellfish Sanitation Conference, and industry to assure the safety of molluscan shellfish (clams, oysters, mussels). Among other things, all such products entering interstate commerce must be handled by state-certified dealers, be properly tagged, be tracked by appropriate records, and be processed in plants that meet sanitation requirements. The FDA continually reviews state shellfish control programs for their effectiveness. [www.pcsga.org/Government_Pages/NSSP.html].

National Trade Estimate (NTE) book — A publication of the Office of the U.S. Trade Representative (USTR) that describes, by country, significant foreign barriers to U.S. exports, including agricultural exports. It is based on information provided by USTR, the U.S. Departments of Commerce and Agriculture, and other agencies and sources. [www.ustr.gov/Document_Library/Reports_Publications/Section_Index.html].

National treatment — A basic principle of GATT/WTO that prohibits discrimination between imported and domestically produced goods with respect to internal taxation or other government regulation. The principle of national treatment is formulated in Article 3 of the GATT 1947 (and incorporated by reference in GATT 1994); Article 17 of the General Agreement on Trade in Services (GATS); and in Article 3 of the Agreement on Trade-Related Aspects of Intellectual Property Rights (TRIPS). The aim of this trade rule is to prevent internal taxes or other regulations from being used as a substitute for tariff protection.

National Wetland Inventory — A Fish and Wildlife Service program started in the 1970s to inventory and map all wetlands, primarily for scientific purposes. The data and maps it produces have been used to track gains and losses of wetlands for more than two decades. It uses the Cowardin Classification System to classify wetlands.

National Wildlife Refuge System (NWRS) — Consists of all of the National Wildlife Refuges, the Waterfowl Production Areas, and certain other small tracts managed by states under cooperative agreements with the Fish and Wildlife Service. In general, these areas are managed primarily for conservation of wild plants and animals (particularly waterfowl). Other uses, such as recreation, grazing, energy development, etc., are permitted to the extent they are compatible with the conservation purpose. Some refuges have additional purposes defined in law. [refuges.fws.gov].

National Wool Act of 1954 — Title VII of Agricultural Act of 1954 (P.L. 83-690) was designated the National Wool Act and provided for a new and permanent price support program for wool and mohair to encourage increased domestic production through incentive payments. Wool and mohair commodity programs were in effect through marketing year 1995, at which time it was terminated under the explicit mandate of P.L. 103-130, Sec. 1.

Natural Resources Conservation Service (NRCS) — A USDA agency responsible for developing and carrying out national soil and water programs in cooperation with landowners, operators, and others. It was created in 1994 reorganization legislation by merging the Soil Conservation Service and many of the conservation cost-sharing programs of the Agricultural Stabilization and Conservation Service. A large majority of the approximately 12,000 employees work at the field level directly with land owners. The gross value all NRCS program activities was $2.3 billion for FY2003. [www.nrcs.usda.gov].

NAWD — National Association of WIC Directors has become the National WIC Association. [www.nwica.org].

NAWG — National Association of Wheat Growers. [www.wheatworld.org].

NBB — National Biodiesel Board. [www.biodiesel.org].

NBC — National Broiler Council (in 1998 the organization changed its name to the National Chicken Council). [www.eatchicken.com].

NCA — National crop acreage.

NCAMP — National Coalition Against the Misuse of Pesticides (in 1998 the organization changed its name to Beyond Pesticides). [www.beyondpesticides.org].

NCBA — National Cattlemen's Beef Association. [www.beef.org].

NCBA — National Cooperative Business Association. [www.ncba.coop/index.cfm].

NCC — National Chicken Council (formerly the National Broiler Council) [www.eat chicken.com].

NCC — National Cotton Council. [www.cotton.org/ncc].

NCE — National Cheese Exchange.

NCFAP — National Center for Food and Agricultural Policy. [www.ncfap.org].

NCFC — National Council of Farmer Cooperatives. [www.ncfc.org].

NCFH — National Center for Farmworker Health. [www.ncfh.org].

NCGA — National Corn Growers Association. [www.ncga.com].

NCIMS — National Conference on Interstate Milk Shipments.

NDC — National Dairy Council. [www.nationaldairycouncil.org].

NE — Northern Europe cotton price.

NEPA analysis — Under the National Environmental Policy Act of 1970 (NEPA) (P.L. 91-190) an environmental impact statement (EIS) must be conducted for any major federal action with significant impact on the environment. In some cases, an environmental assessment is performed prior to an EIS to evaluate how significant such impacts may be, and thus to determine whether an EIS is required.

NEPA — National Environmental Policy Act of 1970 (P.L. 91-190; 42 U.S.C. 4321 *et seq.*).

Net cash income — A farm's actual cash receipts and expenses in a given year, regardless of the year the goods sold were produced. In general, it serves as an indicator of the short-term financial condition of agricultural producers and their ability to pay household expenses, farm operating expenses, loan payments, and to purchase capital assets such as machinery. It consists of cash receipts from farm marketings of crop and livestock products, other cash income from such farm-related sources as machine hire, custom work and farm recreational activities, and direct government payments, less production expenses paid in cash. It excludes the non-monetary components of gross farm income and net farm income.

Net energy value (NEV) — In the context of fuel, NEV is the energy content minus the energy used to produce it. An NEV less than 1.0 means a negative energy balance. Estimates of the NEV of corn-ethanol range from as low as 0.71 (Pimental 2003) to as high as 1.67 (USDA 2004). Biodiesel NEV is 3.2 and petroleum diesel NEV is 0.83 according to 1998 USDA-DOE study. In the context of livestock production, the net energy value of feed is the energy available to the animal for maintenance and production of muscle or milk after deducting for the work of digestion.

Net farm income The return (both monetary and non-monetary) to farm operators for their labor, management and capital, after all production expenses have been paid (that is, gross farm income minus production expenses). It includes net income from farm production as well as net income attributed to the rental value of farm dwellings, the value of commodities consumed on the farm, depreciation, and inventory changes.

Net income (food stamps) — As related to the Food Stamp Program, net monthly income is an amount calculated for each food stamp household that, together with its size, effectively determines its food stamp benefit. It is calculated by reducing the household's total cash monthly income by a series of deductions. The lower a household's net income, the larger its food stamp benefit.

Net income stabilization account (NISA) — A voluntary farm income safety-net program used in Canada, under which farmers set aside money in individual interest bearing accounts that is matched by federal and provincial treasuries. Farmers can make withdrawals from the account when their income falls below their five-year average returns after costs, or when their taxable income falls below a fixed level.

Net pen culture — A type of aquaculture where fish remain captive throughout their lives in marine pens built from nets. This approach is used extensively for culturing Atlantic salmon.

Net ton-mile — The movement of a ton of freight one mile in contrast to the movement of freight and cars (a gross ton-mile).

NET — Nutrition Education and Training Program.

New Zealand Dairy Board (NZDB) — Until 2001,when it was merged with two New Zealand dairy cooperatives to become the Fonterra Cooperative Group, the New Zealand Dairy Board was the quasi-government corporation explicitly created by New Zealand statute to purchase and export all New Zealand dairy products (thus, it was a state trading enterprise). (New Zealand accounts for about 3% of world milk production, but has an export share of about 25% of the market.) It operated through a global network of marketing subsidiaries, based in the countries where New Zealand dairy products are sold. Through its purchasing activities and its system of premiums and penalties, it encouraged the production of dairy products that are in high demand and discouraged those that are in surplus. [www.fonterra.com/default.jsp].

NFI — National Fisheries Institute. [www.nfi.org].

NFMA — National Forest Management Act of 1976 (P.L. 94-588; 16 U.S.C. 1600 *et seq.*).

NFO — National Farmers Organization. [nfo.org].

NFPA — National Food Processors Association. [www.nfpa-food.org].

NFS — National Forest System.

NFSMI — National Food Service Management Institute. [www.olemiss.edu/depts/nfsmi].

NFU — National Farmers Union. [www.nfu.org].

NGA — National Governors Association. [www.nga.org].

NGA — National Grocers Association. [www.nationalgrocers.org].

NGFA — National Grain and Feed Association. [www.ngfa.org].

NGTC — National Grain Trade Council. [www.ngtc.org].

NHCP — National Handbook of Conservation Practices. [www.bse.vt.edu/swcs/nh cp_2.html].

NIAA — National Institute for Animal Agriculture. [www.animalagriculture.org].

NIS — New Independent States of the Former Soviet Union.

NISA — Net income stabilization account.

Nitrate — The nitrogen ion, NO^{3-}, is derived from nitric acid and is an important source of nitrogen in fertilizers. Most nitrogenous materials in water tend to be converted to nitrate, so all sources of combined nitrogen, particularly organic nitrogen and ammonia, should be considered as potential nitrate sources. Primary sources of organic nitrates include human sewage and livestock manure, especially from feedlots. The primary source of inorganic nitrate is fertilizer. Consequently, wells in rural areas are at greater risk of nitrate contamination. Nitrate pollution of drinking water, shallow wells being particularly vulnerable, is of concern because excessive levels of nitrate in drinking water can cause serious illness and sometimes death in infants. EPA has set a drinking water standard for nitrate at10 milligrams per liter (mg/L) under the Safe Drinking Water Act (P.L. 93-523, as amended; 42 U.S.C. 300f-300j).

Nitrogen — An element found in the air and in all plant and animal tissues. For many crops, nitrogen fertilizer is essential for economic yields. However, nitrogen can also be a pollutant when nitrogen compounds are mobilized in the environment (e.g., leach from fertilized or manured fields), are discharged from septic tanks or feedlots, volatilize to the air, or are emitted from combustion engines. As pollutants, nitrogen compounds can have adverse health effects (see Nitrate, and Air pollution) and contribute to degradation of waters (see Eutrophication).

NLEA — Nutrition Labeling and Education Act of 1990 (P.L. 101-535).

NMA — National Meat Association. [www.nmaonline.org].

NMFS — National Marine Fisheries Service. [www.nmfs.noaa.gov].

NMPF — National Milk Producers Federation. [www.nmpf.org].

No net cost — A requirement that certain commodity programs operate at no net cost to the federal government. The No-Net-Cost Tobacco Act of 1982 (P.L. 97-218) required an assessment on 1982 and subsequent tobacco crops to cover potential tobacco price support program losses. The 1985 farm bill (P.L. 99-198) required that USDA operate the sugar program for the first time at no cost; a provision repealed by the 1996 farm bill (P.L. 104-127) and reinstated by the 2002 farm bill (P.L. 107-171, Sec. 1401(a)). The 1996 changes to the peanut price support program were designed to ensure that it also operated at no cost. Subsequently, the peanut program was completely changed by the 2002 farm bill, but not in a manner to make it no-net-cost.

No net loss wetlands policy — An overall policy goal for wetland protection first adopted by the George Bush Administration (1989-1993), and endorsed and updated by the Clinton Administration (1993-2001). The George W. Bush Administration (2001-)has not stated a position. The goal is to halt the decline in the overall number of wetland acres in the country. The Clinton Administration stated a more aggressive goal of adding 100,000 acres of wetland per year by 2005 in its Clean Water Action Plan. The no net loss policy refers only to acres and does not compare the functions and values of wetlands gained and lost. Also, this goal does not address the question of whether it is acceptable to destroy some wetlands if at least the same number of acres are created or restored at another site. Currently there are about 100 million wetland acres, compared to about 200 million when the country was first settled.

No Net Cost Tobacco Act of 1982 — P.L. 97-218 required that the tobacco price support program operate at no net cost to taxpayers, other than for the administrative expenses common to all price support programs. To satisfy this mandate, sellers and buyers (including importers) of tobacco were assessed equally to build a capital account that was drawn upon to reimburse the Commodity Credit Corporation (CCC) for any losses of principal and interest resulting from nonrecourse loan operations. Other provisions of this law provided for reducing the level of support for tobacco and made various modifications to the marketing quota and acreage allotment programs. No net cost assessments ended when price support was terminated after the 2004 crop.

No observable adverse effect level (NOAEL) — From long-term toxicological studies of agricultural chemical active ingredients, levels which indicate a safe, lifetime exposure level. Used in setting pesticide residue tolerances.

No-till farming — A method of planting crops that involves no seed bed preparation other than opening the soil to place individual seeds in holes or small slits; usually no cultivation during crop production; chemical weed control is normally used. No-till may be referred to as slot tillage or zero cultivation. See, for comparison, conservation tillage and minimum tillage.

NOAA — National Oceanic and Atmospheric Administration. [www.noaa.gov].

NOAEL — No observable adverse effect level.

Non-Attainment Area — An area considered to have air quality worse than the National Ambient Air Quality Standards as defined in the Clean Air Act Amendments of 1970 (P.L. 91-604, Sec. 109). Non attainment areas must have and implement a plan to meet the standard, or risk losing some forms of federal financial assistance. An area may be a non-attainment area for one pollutant and an attainment area for others.

Non-indigenous species — See Invasive species, and Exotic species.

Non-native species — See Invasive species, and Exotic species.

Non-program crops — Sometimes the phrase implied any agricultural commodity not covered by federal commodity programs. More often it implied crops not eligible for target price deficiency payments prior to 1996 or not eligible for contract payments after 1996 (the program crops), but eligible for other forms of support. The phrase was replaced in law by the 2002 farm bill (P.L. 101-171), which adopted the phrases "covered commodities" and "loan commodities" to distinguish between programs.

Non-road emissions — Pollutants emitted by non-road engines and non-road vehicles, e.g., farm and construction equipment, gasoline-powered lawn and garden equipment, and power boats and outboard motors. Section 213 of the Clean Air Act (42 U.S.C. 7547), provides that the EPA can issue regulations to limit emissions from these sources.

Nonambulatory — See Downer (or downed animals).

Nonbasic commodities — Commodities, other than the basic commodities, for which USDA is authorized to provide price support in permanent law. This includes soybeans and other oilseeds, milk, sugar beets, and sugarcane.

Noncompetitive imports — A term used by the Economic Research Service in its reporting of agricultural trade statistics to refer to imports of commodities not widely produced in the United States. Commodities such as tea, bananas, or coffee are considered noncompetitive imports. In contrast, imported commodities that are also produced in the United States are referred to as competitive imports.

Nondegradation standard — Defined in the 2002 farm bill provisions establishing the Conservation Security Program (CSP; P.L. 107-171, Sec. 2001) to mean the level of measures required to protect and prevent degradation of 1 or more natural resources, as determined by the NRCS. Participating farmers' conservation security plans must address resources of concern and meet the appropriate nondegradation standard.

Noninsured Assistance Program (NAP) — Producers who grow a crop that is currently not eligible for crop insurance may be eligible for a direct payment under the Farm Service Agency's noninsured assistance program (NAP). NAP has permanent authority under the Federal Crop Insurance Reform Act of 1994, (P.L. 103-354, as amended). To be eligible for a NAP payment, a producer first must apply for coverage under the program by the application closing date, which varies by crop, but is generally about 30 days prior to the final planting date for an annual crop. Like catastrophic crop insurance, NAP applicants also must pay a \$100 per crop service fee at the time of application. In order to receive a NAP payment, a producer must experience at least a 50% crop loss caused by a natural disaster, or be prevented from planting more than 35% of intended crop acreage. For any losses in excess of the minimum loss threshold, a producer can receive 55% of the average market price for the covered commodity. Hence, NAP is similar to catastrophic crop insurance coverage in that it pays 55% of the market price for losses in excess of 50% of normal historic production. A producer of a noninsured crop is subject to a payment limit of $100,000 per person and is ineligible for a payment if the producer's qualifying gross revenues exceed $2 million. [www.fsa.usda.gov/pas/disaster/nap.htm].

Nonmoney income — A statistical allowance used in farm income compilations to credit farmers with income for the value of farm products used on the farm (instead of being sold for cash) and the rental value of farm dwellings. It assumes farmers otherwise live rent-free on their farm business premises.

Nonpoint source pollution — Pollutants that are not discharged or emitted from a specific point source, such as a pipe or smokestack. Nonpoint water pollutants are often carried from dispersed, diverse sources into water channels by rain-induced runoff. Runoff from streets, open pit and strip mines, and agricultural fields are prominent examples (see Agricultural pollution). Nonpoint source air pollutants (often called fugitive emissions) include small dispersed sources (e.g., fireplace smoke, and uncontained emissions, like dust blown from fields and unpaved roads).

Nonrecourse loan — Farmers or processors participating in government commodity programs may pledge certain stored commodities as collateral and obtain a loan from the Commodity Credit Corporation (CCC) at a commodity-specific, per-unit loan rate. The borrower may repay the loan, with interest, within a specified period and regain control of the commodity. Alternatively, the commodity can be forfeited to the CCC at the end of the term with no penalty. The government takes no recourse beyond accepting the commodity as full settlement of the loan. Under the 2002 farm bill (P.L. 101-171, Sec. 1001), "loan commodities" include wheat, corn, grain sorghum, barley, oats, upland cotton, extra long staple cotton, rice, soybeans, other oilseeds, wool, mohair, honey, dry peas, lentils, and small chickpeas. Other commodities eligible for nonrecourse loans include peanuts, and sugar. For those commodities eligible for marketing assistance loan benefits, producers may repay the loan at the adjusted world price (rice and upland cotton) or posted county price (wheat, corn, grain sorghum, barley, oats, soybeans, other oilseeds, wool, mohair, honey, dry peas, lentils, and small chickpeas) or a loan repayment rate (peanuts). For those producers with commodities eligible for marketing assistance loans, but who agree to not utilize the loan program, loan deficiency payments provide an equivalent benefit. Historically, CCC acquisition of forfeited loan commodities became the method used to support market prices at the loan rate. Beginning with target price deficiency payments in the mid-1970s, generally it has been a goal to avoid distorting market prices when subsidizing farm incomes.

Nonrenewable resources — Nonrenewable resources, in contrast to renewable resources, do not naturally replenish themselves within time limits that permit sustained yield (i.e., minerals and hydrocarbons, such as phosphate rock, limestone, petroleum). Nonrenewable resources may be called stock resources because of their fixed supply. Some resources, such as soil and water, can be termed either nonrenewable or renewable depending on circumstances. For example, some underground reservoirs replenish so slowly they are effectively nonrenewable, such as the Ogallala Aquifer. Soil that is eroding faster than its T value faces eventual depletion.

Nontariff barriers (NTBs) — Any restriction, charge, or policy other than a tariff, that limits access of imported goods. Examples of nontariff barriers include quantitative restrictions, mainly import quotas and embargoes; import licenses; exchange controls; state trading enterprises; bilateral agreements; and certain rules and regulations on health, safety, and sanitation. The Uruguay Round Agreement on Agriculture requires conversion of NTBs to bound tariffs and tariff-rate quotas, and that sanitary and phytosanitary measures be based on scientific principles.

NOP — National Organic Program. [www.ams.usda.gov:80/nop].

Normal crop acreage — The acreage on a farm normally devoted to a group of designated crops. When a set-aside program was in effect, a participating farm's total planted acreage of such designated crops plus set-aside acreage could not exceed the normal crop acreage. The authority for set-asides was eliminated by the 1996 farm bill (P.L. 104-127).

Normal flex acreage — A provision of the Omnibus Budget Reconciliation Act of 1990 requiring a mandatory 15% reduction in payment acreage. Under this provision, producers were ineligible to receive deficiency payments on 15% of their crop acreage base (not including any acreage removed from production under any production adjustment program). Producers, however, were allowed to plant any crop on this acreage, except fruits, vegetables, and other prohibited crops. Flex acreage was eliminated by the 1996 farm bill (P.L. 104-127).

Normal trade relations (NTR) — A phrase recently adopted in the United States in place of the more generally used phrase most-favored-nation (MFN). The mandate to replace most-favored-nation in all U.S. statutes was included in the Internal Revenue Service Restructuring and Reform Act of 1998 (P.L. 105-206, Sec. 5003). Normal trade relations and most-favored-nation treatment means equal treatment of all countries with regard to tariff concessions. Under U.S. law, all countries other than those specifically denied it, have NTR status.

Normal yield — The average historic yield established for a particular farm or area. Can also describe average yields. Normal production would be the normal crop acreage planted multiplied by the normal yield. These measures, once required by commodity programs to calculate benefits are replaced by base acres, payment acres, and payment yield under the 2002 farm bill (P.L. 101-171, Sec. 1101-1102).

North American Free Trade Agreement Implementation Act — P.L. 103-182 (December 8, 1993) approved and implemented the North American Free Trade Agreement (NAFTA). NAFTA pertains to cross-border trade between the United States, Mexico, and Canada. NAFTA substantially eliminated all nontariff barriers to agricultural trade between the United States and Mexico, generally through their conversion to tariff-rate quotas or ordinary tariffs, and maintained the provisions of the United States-Canada Free Trade Agreement on agricultural trade. With respect to Mexico, the law eliminated tariffs on a broad range of agricultural products and provided for a phase-out over up to 15 years for tariffs on other products. A special safeguard provision will apply to certain products, with a designated quantity of imports allowed at a NAFTA preferential tariff rate. NAFTA increases incentives for buying within the NAFTA region.

North American Free Trade Agreement (NAFTA) — A multilateral agreement negotiated by the United States, Canada and Mexico that sets forth agreements to lower and/or eliminate unfair trade barriers that affect the trade of goods and services between the three countries. The President signed the North American Free Trade Agreement Implementation Act (P.L. 103-182) in December 1993 and NAFTA entered into force on January 1, 1994. The agriculture portion of NAFTA effectively is three bilateral agreements; U.S./Mexico, Mexico/Canada, and U.S./Canada. The U.S.-Canada agricultural agreement in NAFTA was negotiated previously as part of the U.S.-Canada Free Trade Agreement.

North American Industry Classification System (NAICS) — Standard industrial classification codes have been replaced in the 1997 Census of Agriculture by the new North American Industry Classification System (NAICS). NAICS is a unique, all-new system for classifying business establishments. It is the first economic classification system to be constructed based on a single economic concept. Economic units that use like processes to produce goods or services are grouped together. In the case of the Census of Agriculture, the shift from SIC codes to NAICS codes involves few changes. Farms are grouped into crop or livestock production and then into the subcategory that accounts for 50% or more of the total value of sales of agricultural products (i.e., cotton, tobacco, dairy). [www.census.gov/epcd/www/naics.html].

North American Waterfowl Management Plan — An international program in cooperation with Mexico and Canada to protect, restore, enhance, and manage wetland ecosystems for migratory birds and other wildlife and fish. It was authorized by the North American Wetlands Conservation Act of 1989 (P.L. 101-233). This program is administered by the Fish and Wildlife Service, and USDA agencies participate as appropriate. By the end of 2001, the U.S. and its partners had invested $1.7 billion affecting more than 5 million acres of waterfowl habitat.

North American Wetlands Conservation Act — P.L. 101-233 (December 13, 1989) authorizes a wetlands habitat program, administered by the Fish and Wildlife Service, which provides grants to protect and manage wetland habitats for migratory birds and other wetland wildlife in the United States, Mexico, and Canada. A nine-member council meets periodically to decide which projects to fund. The program encourages private-public cost-sharing projects. It must allocate between 50% and 70% of all funds to projects in Mexico and Canada, and no more than 50% of the U.S. share for projects in these countries can come from federal funds. The Act was recently reauthorized through FY2007 in P.L. 107-308, which gradually increases the funding level to $75 million in the final year. Agricultural wetlands are not specifically identified in the law, and agricultural interests are not expressly represented on the council.

Northeast Interstate Dairy Compact — An agreement among the six New England states to support the farm price of milk used for fluid consumption at a higher level than under federally mandated minimum prices in the region. From 1997 until its expiration on September 30, 2001, the Northeast compact required processors in the region to pay dairy farmers at least $16.94/cwt. for farm milk used for fluid consumption. The law allowed membership in the compact to expand to New York, New Jersey, Pennsylvania, Delaware, Maryland and Virginia, if the prospective state was contiguous to a member state, and if the compact was approved by the state legislature of the prospective state and the U.S. Congress. Although many states expressed interest in joining the compact or forming a separate compact, no congressional approval was granted beyond the six New England states, because of strong opposition from Upper Midwest dairy farmers and dairy processors. The dairy compact has been replaced with Milk Income Loss Contract (MILC) payments under the 2002 farm bill (P.L. 107-171, Sec. 1502), which mandates direct federal payments to all participating dairy farmers nationwide when the minimum price for fluid farm milk in the Northeast falls below $16.94 per cwt.

Northern Europe (NE) cotton price — An average of the five lowest prices of several internationally-traded cottons (including cost, insurance, and freight) quoted for delivery in Northern Europe. The NE price is used by USDA in its formula for calculating the adjusted world price, used in administering marketing assistance loan and step 2 payment benefits under the cotton price support program.

Northwest Cattle Project — An agreement between the United States and Canada, first announced on October 22, 1997, that was initially intended to facilitate shipments of live cattle from the states of Montana and Washington to Canada. Such shipments had been blocked due to Canadian animal health rules. Canada revised its animal health import requirements for live cattle entering from these U.S. states among other things by recognizing areas deemed to be at low risk for certain animal diseases. The project has since been expanded to many more U.S. states, particularly after the two countries signed a joint cooperation agreement on a variety of agricultural matters in late 1998.

NOSB — National Organic Standards Board. [www.ams.usda.gov/nosb].

Noxious weeds — Undesirable plants that infest either land or water resources and cause physical and economic damage. Under the Plant Protection Act of 2000 (7 U.S.C. 7701 *et seq.*), which consolidated Animal and Plant Health Inspection Service (APHIS) authorities including almost all of the provisions of the Noxious Weeds Act, P.L. 93-629; 7 U.S.C. 2801 *et seq.*), the APHIS works to prevent noxious weeds from entering the country, and conducts cooperative control/eradication programs with the states. Local governments frequently impose taxes on landowners to carry out noxious weed control programs.

NPB — National Pork Board. [www.porkboard.org/Home/default.asp].

NPE — Nutrition Program for the Elderly.

NPHAP — National Pesticide Hazard Assessment Program.

NPPC — National Pork Producers Council. [www.nppc.org].

NPR — National Performance Review.

NPS — Non-point source.

NRA — National Renderers Association. [www.renderers.org].

NRA — National Restaurant Association. [www.restaurant.org].

NRC — National Research Council (see National Academy of Sciences). [www.nas.edu/nrc].

NRCS — Natural Resources Conservation Service. [www.nrcs.usda.gov].

NRDP — National Rural Development Partnership. [www.rurdev.usda.gov/nrdp].

NRECA — National Rural Electric Cooperative Association. [www.nreca.org].

NRI — National Research Initiative Competitive Grants Program; National Resources Inventory.

NRSP-7 — National Research Support Project 7, also know as the Minor Use Animal Drug Program, is the counterpart for animals of the IR-4 plant program. The program targets development of therapeutic drugs for minor species, such as small ruminants and aquatic species, plus support for drugs for minor use within major species. It is carried out in partnership with the Food and Drug Administration's (FDA) Center for Veterinary Medicine. [www.nrsp-7.org].

NSLA — National School Lunch Act (P.L. 79-396; 42 U.S.C. 751 *et seq.*).

NSLP — National School Lunch Program.

NSSC — National Soil Survey Center. [soils.usda.gov].

NSSP — National Shellfish Sanitation Program.

NTB's — Nontariff barriers.

NTF — National Turkey Federation. [www.turkeyfed.org].

NTP — National Toxicology Program.

NTR — Normal trade relations.

Nutraceuticals — See Farmaceuticals, and Functional foods.

Nutrient management plan — A farm conservation plan developed by the Natural Resources Conservation Service (NRCS) for a landowner that describes how nutrients will be stored, used, and disposed of to minimize environmental problems such as water pollution. In animal agriculture, managing nutrients often accomplished through manure management.

Nutrient pollution — Contamination by excessive inputs of nutrient. A primary cause of eutrophication of surface waters, in which excess nutrients, usually nitrogen or phosphorus, stimulate algal growth. Sources of nutrient pollution include runoff from fields and pastures, discharges from septic tanks and feedlots, and emissions from combustion.

Nutrition Assistance Grants — Federal programs in Puerto Rico and American Samoa that provide food assistance through block grant funds in lieu of food stamps, and to the Northern Marianas under a covenant governing U.S. relations with that jurisdiction.

Nutrition Education and Training (NET) Program — A program authorized through FY2003 under Section 19 of the Child Nutrition Act (P.L. 89-642). Funds are authorized to make grants to all states for a nutrition education program that targets school children, teachers, parents, and food service workers. Appropriators have not funded the program since FY1998. Between its inception in 1977 and 1994, the NET program had a time-limited authorization and funding (averaging $5 million per year), which was provided under annual appropriations laws. In 1994, however, the Child Nutrition Act was amended (P.L. 103-448) to make NET permanent and funding of $10 million annually was mandated for the program. Two years later, amendments to the Child Nutrition Act (P.L. 104-193) restored NET to temporary status and again made funding for it subject to appropriations. The change in the authorizing statute occurred after the FY1997 appropriations had been enacted so it was necessary to reprogram funds from Team Nutrition ($3.75 million) to continue NET activities for that year. The FY1998 appropriation for NET was $3.75 million. Subsequently, appropriators have funded nutrition education and training activities under Team Nutrition; no funding has been provided explicitly for the NET program.

Nutrition guidelines — Federal guidelines established for meals served in child nutrition meal service programs governing the types of foods and nutrient content required in order to be eligible for reimbursement.

Nutrition Labeling and Education Act (NLEA) of 1990 — P.L. 101-535 mandated for the first time that most foods regulated by the Food and Drug Administration must be sold with nutrition labeling. The Act changed the list of nutrients that must appear on nutrition labels to focus on those of primary public health concern, adding saturated fat, cholesterol, total and subgroups of carbohydrates, and dietary fiber. Nutrition information was required on packaged and bulk foods, with few exceptions. A voluntary program was established for produce and seafood. Although NLEA did not directly affect the labeling of meat and poultry, which is under the jurisdiction of USDA, comparable changes were made in the USDA labeling program for meat and poultry products in order to provide consistent nutrition labeling information on all food products for consumers.

Nutrition Program for the Elderly (NPE) — This program, authorized under Title III of the Older Americans Act (P.L. 89-73, as amended), provides (1) basic grants to operate nutrition programs for the elderly, such as meals-on-wheels and congregate dining programs, and (2) additional cash payments or commodity assistance for each meal served. The basic grants are funded under annual appropriations to the Department of Health and Human Services and distributed by formula. The additional per-meal cash or commodity assistance is funded under annual appropriations to the Food and Nutrition Service (FNS). While states may opt for either cash or commodity assistance, most choose per-meal cash payments.

NWA — National WIC Association. [www.nwica.org].

NWF — National Wildlife Federation. [www.nwf.org].

NWR — National Wildlife Refuge.

NWRS — National Wildlife Refuge System. [refuges.fws.gov].

NWS — National Weather Service. [www.nws.noaa.gov].

NYBOT — New York Board of Trade. [www.nybot.com].

OAQ — Overall allotment quantity.

OBPA — Office of Budget and Program Analysis. [www.usda.gov/agency/obpa/Home-Page/obpa.html].

Occupational Safety and Health Administration (OSHA) — The U.S. Department of Labor agency responsible for administering the Occupational Safety and Health Act of 1970 (P.L. 91-596; 29 U.S.C. 651 *et seq.*). According to OSHA, farming is the nation's most hazardous occupation. The intrinsically seasonal nature of many segments of agriculture not only causes the size of this workforce to vary temporally and often geographically via migrant work groups, but usually also has major effects on the nature and intensity of the work itself. OSHA has issued safety standards relating to agricultural operations. [www.osha-slc.gov/SLTC/agriculturaloperations/index.html].

Ocean freight differential (under P.L. 480) — The difference between the cost of P.L. 480 shipments that are required to be carried on U.S. flag vessels compared to the cost that would have been incurred had they been carried on lower cost foreign bottoms or vessels. The U.S. government pays this difference either by paying the total freight, if the sale is made under Title II of P.L. 480, or by reimbursing the recipient country or private grain company (whichever pays the shipping) if the sale is made under Title I of P.L. 480.

Ocean ranching — A type of aquaculture, used mainly in culturing Pacific salmon, wherein juvenile fish are hatched and reared, released to mature in the open ocean, and caught when they return as adults to spawn.

OECD — Organization for Economic Cooperation and Development. [www.oecd.org].

Off-farm (non-farm) income — That portion of farm household income obtained off the farm, including nonfarm wages and salaries, pensions, and interest income earned by farm families. On average for all farms in the United States, off-farm income accounts for over 90% of farm operator household income.

Offal — Generally, the viscera (organs) and trimmings of butchered animals. Usually, edible offal are referred to as variety meats or organ meats. Inedible offal and many of the other inedible animal byproducts from livestock processing are rendered into tallow, lard, grease, protein meal, bone meal, feather meal, and blood meal. The broader term, animal byproducts, includes offal as well as such other items as blood, hide, bones, horns, hooves.

Offer versus serve — Refers to the option children may be given to refuse up to two items offered as part of a federally subsidized school lunch or breakfast without the meal service operation losing the federal reimbursement for the meal. Enacted to reduce plate waste, which, some contended, was exacerbated by forcing children to take items they did not want as part of a meal in order for the meal to qualify for a federal subsidy.

Office International des Epizooties (OIE) — The intergovernmental organization created by the International Agreement of January 25, 1924, initially signed by 28 countries, and as of May 2002, counting 164 member countries. The OIE provides an international clearinghouse for the reporting of animal diseases worldwide; provides technical expertise to help with their control; collects, analyzes, and disseminates veterinary scientific information; and develops scientific-based standards for protecting against and eradicating animal diseases which are recognized by the World Trade Organization in international trade negotiations and disputes. [www.oie.int].

Office of Migrant Education (OME) — An Office of the U.S. Department of Education that works to improve teaching and learning for migratory children. Programs and projects administered by OME are designed to enable children whose families migrate to find work in agricultural, fishing, and timber industries to meet the same academic content and student performance standards that are expected of all children. [www.ed.gov/about/offices/list/oese/ome/index.html].

Office of Risk Management — See Risk Management Agency. [www.act.fcic.usda.gov].

Office of the Chief Economist (OCE) — The Office of the Chief Economist advises the Secretary of Agriculture on the economic implications of policies and programs affecting the U.S. food and fiber system and rural areas. The Chief Economist coordinates, reviews, and approves the USDA's commodity and farm sector forecasts. In addition, the Chief Economist oversees the activities of the World Agricultural Outlook Board, the Coordinator of Agricultural Labor Affairs, the Office of Risk Assessment & Cost-Benefit Analysis, the Global Change Program Office, the Director of Sustainable Development, and the Office of Energy Policy and New Uses. [www.usda.gov/oce].

Offset — Liquidating a purchase of futures contracts through the sale of an equal number of contracts of the same delivery month, or liquidating a short sale of futures through the purchase of an equal number of contracts of the same delivery month. In other words, selling if one has bought, or buying if one has sold, a futures or option contract.

Offsetting compliance — A requirement that a farmer owning multiple farms who wishes to participate in a crop program must comply with the program's provisions on all farms under the farmer's ownership in order to be eligible for program benefits. This provision did not apply to production flexibility contracts enacted under the 1996 farm bill (P.L. 104-127), or to direct and counter-cyclical program payments under the 2002 farm bill (P.L. 101-171, Sec. 1101-1108).

OGC — Office of General Counsel.

OIE — Organization of International Epizootics, the international veterinary and animal disease control organization. [www.oie.int/eng/en_index.htm].

OIG — Office of Inspector General. [www.usda.gov/oig].

Oilseeds — Vegetable oils, in contrast to mineral oils, are produced from oilseeds. In the United States, the largest oilseed crop is soybeans. "Other oilseeds" are defined for purposes of federal support to include sunflower seed, rapeseed, canola, safflower, flaxseed, mustard seed, or another oilseed if so designated by the Secretary of Agriculture. Soybeans and these other oilseeds are eligible for Direct and Counter-cyclical Program (DCP) payments, and for marketing assistance loans and loan deficiency payments(LDPs). Peanuts are another major oilseed crop for which federal support is provided under rules that match soybeans. Additional U.S. oilseed crops include castor beans, sesame, and crambe. Oilseeds are used for the production of cooking oils, protein meals for livestock, and industrial uses. In tropical regions of the world, palm nuts provide vegetable oil. The different crops produce oils with different chemical and nutritional characteristics, making some more valuable than others.

OL — Farm operating loans.

Omnibus Consolidated and Emergency Appropriations Act, FY1999 — P.L. 105-277, among its numerous provisions that include the regular annual appropriations for most USDA programs, provided $5.9 billion in emergency spending for USDA programs to shore up farm income and to compensate farmers for natural disasters. More than one-half of this amount ($3.1 billion) was in the form of direct market loss payments to grain, cotton, and dairy farmers for income assistance. Most of the balance was for disaster payments made to farmers who experienced large crop losses in either 1998 or in 3 of the 5 years between 1994 through 1998.

Omnibus Trade and Competitiveness Act of 1988 — P.L. 100-418 provided the President with negotiating authority for the General Agreement on Tariffs and Trade (GATT) Uruguay Round, U.S.-Canada Free Trade Agreement, and the North American Free Trade Agreement, and specified U.S. negotiating objectives regarding agriculture. The law revised statutory procedures for dealing with unfair trade practices and import damage to U.S. industries. It gave USDA discretionary authority to trigger marketing loans for wheat, feed grains, and soybeans, if it is determined that unfair trade practices exist.

Onion market loss assistance — Assistance provided to onion producers in Orange County, New York, that suffered losses to onion crops during one or more of the 1996 through 2000 drop years. The 2002 farm bill (P.L. 107-171, Title X, Subtitle A, Section 10106) provided $10 million of Commodity Credit Corporation (CCC) funds for this purpose.

Open position — Ownership of a fixed-price forward contract, especially a futures contract.

Option premium, futures — The amount an option buyer pays the option writer for an option contract.

Option writer, futures — A person who sells an option contract, receives the premium, and bears the obligation to buy or sell the asset at the strike price.

Optional flex acreage — Under the planting flexibility provision of the Agricultural Act of 1949, as amended by the 1990 farm bill, producers could choose to plant up to 25% of the crop acreage base in other Commodity Credit Corporation (CCC)-specified crops (except fruits and vegetables) without a reduction in crop acreage bases on the farm, but receive no deficiency payments on this acreage. The Omnibus Budget Reconciliation Act of 1990 further amended the 1949 Act to make a 15% reduction in payment acreage mandatory. The remaining 10% was optional flex acreage. Optional flex acreage was eligible for deficiency payments when planted to the program crop. The 1996 farm bill (P.L. 104-127) expanded planting flexibility to all of the base acres, and this policy was continued by the 2002 farm bill (P.L. 101-171, Sec. 1106).

Options contract — An option contract gives the buyer the right, but not the obligation, to buy (call option) or sell (put option) a futures contract at a specific price within a specified period of time, regardless of the market price of that commodity.

Options contracts, futures — A contract traded on a commodity exchange that gives the buyer the right without obligation to buy or sell a futures contract over a specified time period. The 1996 farm bill (P.L. 104-127) required USDA to conduct research through pilot programs to determine if futures and options contracts could provide producers with reasonable protection from the financial risks of fluctuations in price, yield, and income inherent in the production and marketing of agricultural commodities.

ORACBA — Office of Risk Assessment and Cost Benefit Analysis, USDA. [www.usda.gov/agency/oce/oracba].

Oral toxicity — Ability of a chemical to cause injury when ingested by mouth.

Orderly marketing — Coordination of the total supply of a commodity in order to achieve sellers' joint market objectives. This is an activity carried out by some marketing order programs.

Organic farming — An approach to farming based on biological methods that avoid the use of synthetic crop or livestock production inputs and on a broadly defined philosophy of farming that puts value on ecological harmony, resource efficiency, and non-intensive animal husbandry practices. USDA launched the National Organic Program (NOP) in October 2002. Farmers who wish to label their products as organically produced must have their operations certified by private or state certification organizations as following the standards set in the regulations (7 CFR 205).

Organic foods — Food products produced by organic farming practices and handled or processed under organic handling and manufacturing processes as defined in federal regulations under the National Organic Program (7 CFR 205).

Organic — Chemically, a compound or molecule containing carbon bound to hydrogen. Organic compounds make up all living matter. The term organic frequently is used to distinguish natural products or processes from man-made synthetic ones. Thus natural fertilizers include manures or rock phosphate, as opposed to fertilizers synthesized from chemical feedstocks. Likewise, organic farming and organic foods refer to the growing of food crops without the use of synthetic chemical pesticides or fertilizers; pests are controlled by cultivation techniques and the use of pesticides derived from natural sources (e.g., rotenone and pyrethrins, both from plants) and the use of natural fertilizers (e.g., manure and compost).

Organization for Economic Development and Cooperation (OECD) — An international organization established by the United States, Canada and certain Western European countries in 1960. The OECD studies and discusses trade and related matters. Its current 30 members include the United States, Canada, the 15 countries of the European Union, Australia, New Zealand, Japan, and Turkey. More recent members include Mexico, the Czech Republic, Hungary, Poland, and South Korea, among others. [www.oecd.org/home].

Organoleptic — Relating to the senses (taste, color, odor, feel). Traditional USDA meat and poultry inspection techniques are considered organoleptic because inspectors perform a variety of such procedures (involving visually examining, feeling, and smelling animal parts) to detect signs of disease or contamination. These inspection techniques alone are not adequate to detect invisible foodborne pathogens that now are the leading causes of food poisoning.

Organophosphates — Insecticides that contain phosphorus, carbon, and hydrogen. They are cholinesterase inhibitors; some are highly acutely toxic, but they usually are not persistent in the environment. Parathion is an example of an organophosphate.

Orthophotography — Aerial photographs that more precisely show the features of the landscape, including those that might be important for agriculture such as slope or size of gullies, because they are corrected for distortion caused by tilt, curvature, and ground relief.

OSHA — Occupational Safety and Health Administration. [www.osha.gov].

Other oilseed — The 2002 farm bill (P.L. 101-171, Sec. 1001) defines "other oilseed" as sunflower seed, rapeseed, canola, safflower, flaxseed, mustard seed, or another oilseed if so designated by the Secretary of Agriculture. Previous laws referred to these as minor oilseeds because of their comparatively small production compared to soybeans. These other oilseeds are eligible for Direct and Counter-cyclical Program payments and marketing assistance loans and loan deficiency payments (LDPs).

Outlays — Spending made to pay a federal obligation. This differs from budget authority in that it reflects money the federal government actually spends, not the amount that has been appropriated by Congress. Outlays may pay for obligations incurred in previous fiscal years or in the current year; therefore, they flow in part from unexpended balances of prior-year budget authority and in part from budget authority provided for the current year. For most categories of spending, outlays are recorded when payments are made or when cash is disbursed from the Treasury. However, outlays for direct and guaranteed loans reflect estimated subsidy costs instead of cash transactions. USDA operates many such loan programs for farmers, for rural areas, and to promote agricultural exports.

Outreach and Assistance for Socially Disadvantaged Farmers and Ranchers —Program operated by the Cooperative State Research, Education, and Extension Service (CSREES) of the USDA that provides grants to eligible organizations assist socially disadvantaged farmers and ranchers to own and operate farms and ranches and to participate in agriculture programs. Supports a wide range of outreach and assistance activities, including farm management, marketing, application, and bidding procedures to members of a socially disadvantaged group, generally defined as persons whose race, ethnicity, or gender has subjected them to prejudice. See Socially disadvantaged farmers and ranchers.

Overall allotment quantity (OAQ) — Under the marketing allotment provisions of the sugar program authorized by the 2002 farm bill (P.L. 107-171, Sec. 1403), the OAQ is the amount of domestically-produced sugar that processors of sugar cane and refiners of sugar beets can sell into the U.S. market during a fiscal year. The 2002 farm bill requires USDA to set the OAQ using the following formula: (estimated sugar consumption + reasonable carryover or ending stocks) *minus* (1,532,000 short tons + carry-in or beginning stocks). What USDA decides is a reasonable carryover stock level is closely watched by the sugar industry because of its influence on prices. Sugar production in excess of the OAQ (sometimes referred to as "blocked stocks") cannot be marketed.

Ozone (O_3) — A highly reactive molecule composed of three oxygen atoms. Environmentally, ozone is important in two completely separate contexts: 1) as a naturally occurring screen in the outer atmosphere (i.e., stratospheric ozone) that partially blocks the bombardment of earth by of harmful radiation, and 2) as a component of polluting smog formed from emissions resulting from human activities (i.e., urban smog).

In the stratosphere 7 to 10 miles above the Earth, naturally occurring ozone acts to shield the Earth from harmful radiation. In the 1970s and 1980s, it was discovered that emissions of certain chemicals catalyze destruction of stratospheric ozone, allowing more radiation to reach the Earth's surface. The U.S. is a signatory to the 1987 Montreal Protocol on Ozone Depleting Substances, which bans or limits uses of chemicals whose emissions deplete stratospheric ozone. Among the chemicals being phased out as ozone depleters are chlorofluorocarbons used in refrigeration and air conditioning and methyl bromide, a pesticide. In the lower atmosphere (troposphere), ozone is a major air pollutant that contributes to smog, adversely affects human health, and is toxic to some plants, damaging forests and crops. Tropospheric ozone forms from reactions between nitrogen oxides and volatile organic compounds in the presence of sunlight. The precursor pollutants are emitted by combustion sources such as motor vehicles and utilities, use of solvents, and petrochemical facilities. Tropospheric ozone is regulated under a National Ambient Air Quality Standard.

P.L. 480 (or Public Law 480) — P.L. 83-480 (July 10, 1954), also called Food for Peace, is the common name for food aid programs established by the Agricultural Trade Development and Assistance Act of 1954, that seeks to expand foreign markets for U.S. agricultural products, combat hunger, and encourage economic development in developing countries. Title I makes export credit available on concessional terms, for example, at low interest rates for up to 30 years. Donations for emergency food relief and non-emergency humanitarian assistance are provided under Title II. Title III authorizes a Food for Development program that provides government-to-government grant food assistance to least developed countries. The 2002 farm bill (P.L. 107-171) extends the authority to enter into new P.L. 480 agreements through 2007.

P.L. 566 — Public Law 83-566 (August 4, 1954), Watershed Protection and Flood Prevention Act of 1954. See Watershed and Flood Prevention Operations. (16 U.S.C. 1001 *et seq.*).

P&S — A shorthand reference either to the Packers and Stockyards Act of 1921,(7 U.S.C. 181 *et seq.*) or to the branch of Grain Inspection, Packers & Stockyards Administration (GIPSA) that administers the Act.

PACA — Perishable Agricultural Commodities Act (P.L. 71-325; 7 U.S.C. 499a *et seq.*).

Pacific Rim — An informal but imprecise term generally referring to countries and economies bordering the Pacific Ocean. Broadly, it may include as many as the following countries: Australia, Brunei, Cambodia, Canada, Hong Kong/Macau, Indonesia, Japan, Laos, Mexico, New Zealand, North Korea, South Korea, Malaysia, New Zealand, the Pacific Islands, the Philippines, People's Republic of China, Russia (or the Commonwealth of Independent States), Singapore, Taiwan, Thailand, United States, and Vietnam, as well as the countries of Central and South America that border the Pacific Ocean.

Packer concentration — The degree to which a few large firms dominate total sales within segments of the meat packing industry, which, some farmers and other critics contend, can cause or at least contribute to lower prices for their animals. Market control by five large packers in the early 1900s led to passage of the Packers and Stockyards Act of 1921 (P.L. 67-51; 7 U.S.C. 181 *et seq.*). Concentration declined after that, but has increased sharply in more recent years. According to USDA, for example, the four largest firms accounted for 82% of the steer and heifer slaughter in 2000, compared with 36% in 1980.

Four-firm concentration in hog slaughter was 56% in 2000, compared with 34% in 1980. Numerous government-sponsored studies and investigations have been inconclusive on the relationship in recent years between concentration and prices.

Packer ownership — see captive supply, packer concentration, and WORC petition.

Packers and Stockyards Act of 1921 — P.L. 67-51 remains, in amended form, the basic authority for USDA to regulate marketing practices in the livestock, poultry, and meat industries. The law was enacted to prevent unfair, deceptive, and monopolistic trade practices, focusing on livestock terminal and auction markets, livestock marketing agencies, dealers, meat packers, and live poultry dealers. The law also includes provisions to ensure that livestock and poultry producers are promptly paid when they sell their animals. (7 U.S.C. 181 *et seq.*).

Paid diversion — A program, repealed by the 1996 farm bill (P.L. 104-127), under which farmers were paid to voluntarily take acreage out of production. The diverted land was devoted to approved conservation practices. Unlike acreage reduction and set-aside programs, participation in a paid diversion program was not normally a condition of eligibility for other support program benefits. While the Conservation Reserve Program pays farmers to take land out of production, the objective is to achieve certain environmental objectives.

Paid lunch (or breakfast, supper, or snack) — Refers to a federally reimbursed meal (or snack) bought by a child who does not qualify for a free or reduced price meal. Also often referred to as a full-price lunch, or a Section 4 lunch. See Free lunch.

Palmer Drought Index — A measure of the severity of prolonged periods (months or years) of abnormal dryness or wetness over large regions, published periodically by the Joint Agricultural Weather Facility. The Palmer Index has long been used by the agriculture community to anticipate the potential effects of weather on crop development and yield.

Parity price — A measurement of the purchasing power of a unit of a particular commodity. Originally, parity was the price per bushel, bale, pound, or hundredweight that would be necessary for a unit of a commodity today to buy the same quantity of other goods (from a standard list) that the commodity could have purchased in the 1910-14 base period. Under permanent law, prices of some commodities would be supported at 50 to 90% of parity through direct government purchases or nonrecourse loans. In 1948, the parity price formula was revised to make parity prices dependent on the relationship of farm and nonfarm prices during the most recent 10-year period for nonbasic commodities. Basic commodities, including wheat, corn, rice, peanuts, and cotton use the higher of the historical or the new formula.

Parity ratio — The ratio of the prices received index, 1910-14 = 100, to the prices paid index on a 1910-14 = 100 base (called the parity index). The parity ratio is a measure of relative price relationships. It is not a measure of farm income, of farmers' total purchasing power, or of farmers' economic welfare. The well-being of the farm community depends upon a number of factors other than price relationships, such as changes in production efficiency and technology, quantities of farm products sold, and supplementary income (including income from off-farm jobs and federal farm programs). Over time the parity ratio has declined due to greater efficiency gains in agriculture. Compared to a parity ratio of 100 in the 1910-14 time period, the January 2003 parity ratio was 38.

Partial payments — If it is anticipated that counter-cyclical payments will be made, producers have the option to receive partial payment in advance of the final payment determination at the end of the marketing year. If the total payment is eventually calculated to be less than the advance payment, the producer is required to refund the difference. This term is analogous to the term advance deficiency payments used prior to 1996, when counter-cyclical payments were called target price deficiency payments.

Particulates — The EPA has set National Ambient Air Quality Standards for particulates. One, in effect since 1987, regulates particles smaller than 10 microns in diameter (PM_{10}). The other, promulgated in 1997, would regulate particles smaller than 2.5 microns in diameter ($PM_{2.5}$), but court challenges delayed its implementation. EPA has completed putting its monitoring network in place, and anticipates that it will be able to start designating non-attainment areas by late 2004. These are of interest to agriculture because dust from tillage and smoke from burning field residues may contribute to pollutant levels. Whether controls might be imposed on agricultural activities depends largely on how each state chooses to meet the standards.

Partners for Fish and Wildlife — A voluntary partnership program administered by the Fish and Wildlife Service to provide financial and technical assistance to private landowners who wish to protect or restore wetlands, uplands, and riparian and instream habitats. Through 2002, the program entered into nearly 29,000 land owner agreements to protect or restore about 640,000 acres of wetlands and almost 1.1million acres of uplands. This program has been widely used by rural landowners, including farmers. [www.fws.gov/partners].

Partners in Quality — One of the Agricultural Marketing Service (AMS) process verification programs, partners in quality enables fresh produce packing houses that incorporate specified, rigorous quality standards and requirements (monitored by periodic unannounced AMS audits) into their ongoing daily operations, to forgo the traditional, more costly, and less flexible end-of-the-line inspections that AMS conducts before awarding a quality grade to the company's products. [www.ams.usda.gov/fv/fpb piq.html].

Partnerships and Cooperation Program — This program, enacted in the 2002 farm bill (P.L. 107-171, Sec. 2003), funds special projects recommended by a state conservationist to meet the requirements of three specified federal environmental laws or address conservation needs in watersheds or other areas with significant environmental problems. Participants agree to increase environmental benefits through implementation of conservation programs in return for incentives payments. The statute specifies participation criteria, which include: conservation practices that affect multiple producers; sharing information and resources among producers; cumulative conservation benefits in geographic areas; and, demonstrating innovative conservation methods. The total made available for this program annually may be up to 5% of the mandatory funding for conservation programs. [www.nrcs.usda.gov/programs/cpi].

Pasture Recovery Program — A program, authorized in the 2001 agriculture appropriations act (P.L. 106-387, Sec. 806 in H.R. 5426), to assist producers in reestablishing permanent vegetative forage crops on pastureland affected by natural disasters during calender year 2000. The Farm Service Agency (FSA) was authorized to spend up to $40 million from commodity Credit Corporation (CCC) funds in counties with emergency designations by the USDA.

Pastureland — Land used primarily for the production of domesticated forage plants for livestock (in contrast to rangeland, where vegetation is naturally-occurring and is dominated by grasses and perhaps shrubs). Rotation pasture or cropland under winter cover is not included in this definition. The 1997 National Resources Inventory recorded 120 million acres of pastureland, nearly 9% of all nonfederal rural land.

Pathogen; pathogenic — Pathogens are infectious or toxin-forming microorganisms causing disease. A foodborne pathogen is a microorganism that causes illness through the ingestion of food.

Payment acres — Under 2002 farm bill (P.L. 107-171, Sec. 1101(f)) each farm's payment acres for the Direct and Counter-cyclical Program (DCP) are 85% of the farm base acres. The law specifies how producers establish each farm's base acres.

Payment amount — The size of each farm's payments under 2002 farm bill (P.L. 107-171) for the Direct and Counter-cyclical Program (DCP) are the payment rates times payment acres times payment yield. The direct payment rate for each covered commodity is specified in Section 1103. The counter-cyclical payment rate is specified in Section 1104.

Payment-in-kind (PIK) — In general, a payment made in the form of Commodity Credit Corporation (CCC)-owned commodities (or title to them) in lieu of cash. This form of payment was widely used during the 1980s for paid diversion, deficiency payments, and export subsidy payments as a means of disposing of or avoiding the acquisition of commodity inventories. PIK certificates entitled the holder to a specific quantity of commodities.

Payment limitation — The maximum annual amount of farm program benefits a person can receive by law. Persons are defined under payment limitation regulations, established by USDA, to be individuals, members of joint operations, or entities such as limited partnerships, corporations, associations, trusts, and estates that are actively engaged in farming. The three entity rule allows payments for up to three farms (two of which are subsidized at half the normal level). Also, provisions exist to treat spouses separately as persons. For covered commodities, the 2002 farm bill (P.L. 107-171, Sec. 1603) sets limits at $40,000 per person per fiscal year on fixed, decoupled direct payments, and $65,000 per person per year on counter-cyclical payments. Separately, peanuts have the same limits. The limit on marketing assistance loan gains and loan deficiency payments for loan commodities is $75,000 per person per year (this limit applies separately to wool, mohair, honey and peanuts). Farmers are not subject to any limits on the use of commodity certificates to repay marketing assistance loans. The Conservation Reserve Program has a limit of $50,000 per person per year, the Environmental Quality Incentive Program (EQIP) limits total payments to $450,000 to any participating producer, and the Conservation Security Program has annual payment limits for each of the three alternative levels of participation. Section 1604 of the Act imposes a prohibition on making commodity payments or conservation payments to individuals or entities that have 3-year average adjusted gross incomes exceeding $2.5 million (unless 75% or more of the income is from farming, ranching, or forestry).

Payment quantity — The quantity of production eligible for direct payments or separately for counter-cyclical payments under the 2002 farm bill (P.L. 101-171). Payment quantity is calculated as the farm's payment yield (per acre) multiplied by the farm's payment acres (where payment acres equal 85% of the base acres).

Payment rate — Generally, the amount paid per unit of production (i.e., $/bu., $/lb., $/cwt.) to each participating farmer for eligible production under commodity income and price support programs. Under the 2002 farm bill (P.L. 101-171), separate payment rates for each commodity are specified for direct payments, and are determined by formulas for counter-cyclical payments and loan deficiency payments.

Payment yield — The farm commodity yield of record determined by averaging the yield of a particular commodity for a specified time period for purposes of making certain commodity support payments. Under the 2002 farm bill (P.L. 101-171, Sec. 1102) the payment yield is used in calculating direct payments and counter-cyclical payments. For six years of direct payments, the payment yield is the yield applied to 1995-crop payments, which is the same as the yield calculated for 1985-crop target price payments. For counter-cyclical payments, producers have the opportunity to use an updated calculation.

Payments in lieu of taxes (PILT) — A program administered by the Bureau of Land Management of the Department of the Interior to compensate counties for the tax-exempt status of federal lands; the fixed payments per entitlement acre (on most but not all federal lands) are adjusted for low county populations and for other revenue-sharing payments (e.g., Forest Service county payments) in a complicated formula.

Payments to States — See Federal-State Marketing Improvement Program.

PBIS — Performance Based Inspection System.

PC — Prior converted wetland.

PCA — Production Credit Association.

PCC — Prior converted cropland.

PDP — Pesticide data program.

Peace clause — Term used to refer to Article 13 of the Uruguay Round Agreement on Agriculture, which protects countries using subsidies which comply with the Agreement from challenges under other WTO agreements. The peace clause was to have expired at the end of 2003. Extension of the peace clause is an issue in the Doha Development Agenda negotiations. Some countries propose extending the peace clause so as to facilitate adherence to any new agreements limiting domestic subsidies. Other countries want to see it lapse in order to strengthen the WTO Member countries' ability to take legal action against subsidies. Some developing countries propose eliminating the peace clause for industrialized countries like the United States or the European Union, but keeping it for developing countries.

Peanut poundage quota — Poundage quotas, authorized by the Agricultural Adjustment Act of 1938, were the supply control mechanism for the peanut price support program until its revision in the 2002 farm bill (P.L. 107-171, Sec. 1301-1310). The 1996 farm bill (P.L. 104-127) required that (for the 1996-2002 crops) the poundage quota be set equal to projected food demand and related uses. The national quota was allocated among states based on historical shares, and then divided among farms based on production history. Owners (via inheritance or purchase) of quota were allowed to sell peanuts produced against their quota, or sell, lease and transfer their quota to other producers. Peanuts marketed above the quota limits (called additional peanuts) had to be crushed for non-edible uses or exported. The 2002 farm bill eliminated peanut quotas and the two-tiered pricing structure and replaced this with a support program comparable to that for wheat, feedgrains, cotton, and rice, as well as buy-out funding.

Peanut program — The 2002 farm bill (P.L. 107-171, Sec. 1301-1310) replaced the longtime (65-year) support program for peanuts with a framework identical in structure to the program for the so-called covered commodities.(wheat, corn, grain sorghum, barley, oats, upland cotton, rice, soybeans, and other oilseeds). The three components of peanut program are fixed direct payments (at $36/ton), counter-cyclical payments (based on a target price of $495/ton), and marketing assistance loans or loan deficiency payments (LDPs) (based on a loan rate of $355/ton). The peanut poundage quota and the two-tiered pricing features of the old program were repealed. Only historic peanut producers are eligible for the direct and counter-cyclical program (DCP). All current production is eligible for marketing assistance loans and LDPs. Previous owners of peanut quota were compensated through a buy-out program at a rate of 55¢/lb. ($1,100/ton) over a 5-year period.

Percolation — The movement of water downward and radially through subsurface soil layers, usually continuing downward to groundwater. The rate at which soils permit percolation is a measure of the vulnerability of groundwaters to contamination, as well as a determinant in the siting of septic fields.

Performance-based inspection system (PBIS) — A computer-based system used by USDA's meat and poultry inspection agency, the Food Safety and Inspection Service (FSIS). The system organizes inspection requirements, schedules inspection activities, and maintains records of findings for meat and poultry processing operations under federal inspection. PBIS has been at issue because consumer advocates and some inspectors have contended that it is not flexible and second-guesses inspectors' more reliable experience and judgment. USDA views it as an objective tool for inspection that enhances rather than undermines inspectors' roles.

Perishable Agricultural Commodities Act (PACA) of 1930 — P.L. 71-325 (June 10, 1930), as amended, regulates the buying and selling of fresh and frozen fruits and vegetables to prevent unfair trading practices and to assure that sellers will be paid promptly. Both produce sellers and buyers must pay fees for a license in order to do business, and these license fees are the source of funding for a trust program that resolves disputes and protects sellers from non-payment when buyers become bankrupt. Amendments to the Act in 1995 (P.L. 104-48, Sec. 3) include a 3-year phase out of the annual license fees for retailers and grocery wholesaler-dealers to be replaced by one-time fee. (7 U.S.C. 499a *et seq.*).

Perishable commodities — Farm goods that prior to processing cannot be stored for a substantial period of time without excessive loss through deterioration or spoilage. Examples of perishable commodities are fresh fruits and vegetables, meat and poultry. Most of the commodities purchased by the Agricultural Marketing Service under Section 32 authority are perishable items. Perishable commodities have a more precise statutory definition under the Perishable Agricultural Commodities Act (PACA) of 1930 (7 U.S.C. 499a *et seq.*).

Permanent law — Legislation that would be in force in the absence of all temporary or short-term laws (e.g., farm bills). The Agricultural Adjustment Act of 1938, the Agricultural Act of 1949, and the Commodity Credit Corporation (CCC) Charter Act of 1948 serve as the basic laws authorizing the major commodity programs. Technically, each new farm bill (including the 2002 farm bill (P.L. 101-171)) amends the permanent law for a specified period. Many programs and activities of USDA are authorized by permanent laws that are periodically amended.

Permanent vegetative cover — Trees, or perennial grasses, legumes, or shrubs with an expected life span of at least 5 years. Permanent cover is required on cropland entered into the Conservation Reserve Program.

Permitted acreage — The acreage on which a farm program participant was permitted to grow a program crop after satisfying acreage reduction requirements. For example, when a 10% acreage reduction program was in effect for wheat, a farmer with a 100-acre wheat base could grow wheat on 90 acres, the permitted acres. Limits on production were eliminated under the 2002 farm bill (P.L. 101-171) through crop year 2007, as also was done under the 1996 farm bill (P.L. 104-127).

Persistent organic pollutants (POPs) — Generally, these pollutants are complex, synthetic chemicals, and many are chlorinated hydrocarbon pesticides. They can harm human health and wildlife, do not break down easily in the environment, and tend to accumulate as they move up the food chain. POPs may be transported in the air and water across international boundaries.

Persistent pesticides — Pesticides that do not readily break down in the environment. Becoming long-lived components of the ecosystem, these chemicals may have enduring effects at low concentrations or may bioaccumulate, posing hazards to higher predators.

Persistent Poverty Counties — The Economic Research Service (ERS) of USDA categorizes non-metropolitan counties by their dominant economic foundation and by characteristic policy type. Persistent poverty counties are those where 20% or more of the county population in each of four Census years (1960, 1970,1980, 1990) had poverty level household incomes. In 1989, there were 535 such counties concentrated largely in the Delta South, Central Appalachia, Rio Grande Valley, the Northern Great Plains, and western Alaska. The average poverty rate in these counties was approximately 29% in 1989.

Person — An entity defined by USDA as being eligible to receive federal farm program benefits, subject to annual payment limitation constraints. A person may be an individual farmer, an individual member of a joint operation, a corporation, a joint stock company, an association, a limited partnership, a trust, an estate, or a charitable organization. A husband and wife generally are considered one person for payment limitation purposes. A joint operation is not a person; neither is a cooperative association of producers that markets commodities for producers.

Personal Responsibility & Work Opportunity Reconciliation Act of 1996 — P.L. 104-193 was labeled as a major welfare reform initiative. In addition to provisions making major changes to federal cash welfare, medicaid, work, and child care development programs, this law revised the Food Stamp Program and several commodity distribution programs (notably the Emergency Food Assistance Program and the Soup Kitchen Food Bank Program).

Pest Management in Schools — A proposal that was approved by the Senate but not adopted in the finally enacted 2002 farm bill (P.L. 107-171). The proposal would have created a new School Environment Protection Act of 2002" under section 33 of the Federal Insecticide, Fungicide, and Rodenticide Act. It would have required states to develop pest management plans for schools, established guidelines for these plans and required that they be part of state cooperative agreements with the EPA.

Pest resistance management (PRM) plans — To protect the continued use of biopesticides, the EPA is requiring companies developing transgenic crops (see Genetic engineering) to submit and implement pest resistance management (PRM) plans as a requirement of product registration. If they are exposed to a toxin excessively, most insect populations can develop resistance, making pest control products less effective. With new biopesticide technologies comes the concern that pests will rapidly develop resistance to natural insecticides, because plant pesticides tend to produce the pesticidal active ingredient throughout a growing season, increasing the selection pressure upon both the target pests and any other susceptible insects feeding on the transformed crop. A resistance management plan is intended to sustain the useful life of transgenic technology and well as the utility of the toxin for organic farmers.

Pest scouting — Inspecting a field for pests, including insects, weeds, and pathogens. Pest scouting is a basic component of integrated pest management programs. It is used to determine whether pest populations are at levels that warrant control intervention and also may help to determine the most appropriate method of control.

Pest — An animal or plant that is directly or indirectly detrimental to human interests, causing harm or reducing the quality and value of a harvestable crop or other resource. Weeds, termites, rats, and mildew are examples of pests.

Pesticide Data Program (PDP) — A program initiated in 1991 by the Agricultural Marketing Service to collect pesticide residue data on selected food commodities, primarily fruits and vegetables. PDP data are used by the EPA to support its dietary risk assessment process and pesticide registration process, by the Food and Drug Administration to refine sampling for enforcement of tolerances; by the Foreign Agricultural Service, to support export of U.S. commodities in a competitive global market; by the Economic Research Service to evaluate pesticide alternatives; and by the public sector to address food safety issues. [www.ams.usda.gov/science/pdp].

Pesticide Recordkeeping Program (PRP) — Authorized by the 1990 farm bill (P.L. 101-624, Sec. 1491), the program requires that private pesticide applicators keep records of the pesticides they use in agricultural production and that the records be surveyed to provide a database on restricted-use pesticides. [www.ams.usda.gov/science/sdpr.htm].

Pesticide — A substance used to kill, control, repel, or mitigate any pest. Insecticides, fungicides, rodenticides, herbicides, and germicides are all pesticides. EPA regulates pesticides under authority of the Federal Insecticide, Fungicide, and Rodenticide Act (FIFRA; 7 U.S.C. 136 *et seq.*). In addition, a substance used as a plant regulator, defoliant, or desiccant is defined as a pesticide and regulated accordingly under FIFRA. All pesticides must be registered and carry a label approved by EPA.

PETA — People for the Ethical Treatment of Animals. [www.peta-online.org].

PFC — Production flexibility contract.

Pfiesteria piscicida — A microbe that has been linked to massive fish kills in Maryland, Delaware and North Carolina. Some scientists believe *pfiesteria* growth is fostered by nutrients such as nitrogen and phosphorus from air pollution, cities, suburbs, sewage plants and farms. Both nitrogen and phosphorus are concentrated in chicken manure, which is spread liberally on farmland as fertilizer in watersheds in those states. While there is no proven direct connection between farm runoff and *pfiesteria*, the farm community is involved in efforts to identify the causal linkages and remedy the problem.

pH — An expression of the intensity of the basic or acidic condition of a liquid or of soil; the logarithmic scale ranges from 0 to 14, where 0 is the most acid, 7 is neutral, and above 7 is alkaline. Natural waters usually have a pH between 6.5 and 8.5. Plants have differing tolerances for acidity and alkalinity.

Phase II Tobacco Settlement — As a followup, but separate from the MSA, the tobacco companies agreed in 1999 to pay 14 tobacco producing states $5.15 billion over 12 years to offset the adverse economic on farmers and rural communities of the MSA. This is called Phase II of the tobacco settlement. Adoption of the tobacco quota buyout, financed by fees on tobacco manufacturers, ended the Phase II payments after 2004.

Phosphorus — An essential nutrient for plants and animals that is commonly applied to crops as a phosphate fertilizer. Phosphorus can contribute to the eutrophication of lakes and other water bodies. Sources of excess phosphorus in streams and lakes include sewage and agricultural runoff.

Phytoremediation — The systematic use of plants to treat environmental contamination. It is being investigated as a potential low-cost technology to help meet environmental regulations. For example, it has been discovered that young seedlings of Indian mustard (*Brassica juncea*) grown in aerated water are very effective at removing toxic metals from water.

Phytosanitary — See Sanitary and phytosanitary standards (SPS).

Pigford Case — This refers to two discrimination suits brought by African-American farmers against the U.S. Department of Agriculture (Pigford v. Veneman, No. 97-1978; and Brewington v. Veneman, No. 98-1693) and their subsequent settlement agreement. The suits, filed in 1997 and 1998, claimed that the USDA since 1981 had discriminated against African-American farmers on the basis of race and had failed to investigate or properly respond to complaints of discrimination in USDA farm credit and non-credit benefit programs. In January of 1999, the parties reached a settlement agreement for the class action lawsuit and a consent decree was approved by the court in April of 1999. The decree established the category of eligible class recipients: (1)African American farmers who farmed or attempted to farm between January 1, 1981 and September 31, 1996; (2) applied to USDA for farm credit or farm program benefits and believe they were discriminated against on the basis of race; and (3) made a complaint against the USDA on or before July 1, 1997, regarding USDA's treatment of their farm credit or benefit application. The agreement established a two-track dispute resolution mechanism for those seeking relief and a system for notice, claims submission, consideration, and review that involved a facilitator, arbitrator, adjudicator, and monitor. There has been some controversy with the actual amount of claims paid by the USDA pursuant to the consent decree.

PIK — Payment-in-kind.

PILT — Payments in lieu of taxes.

PIN — Pesticide information network.

Plant breeding — The traditional practice of cross-breeding to produce varieties with particular characteristics (traits) that are carried in future generations. Traditional plant breeding works more slowly than genetic engineering and so far has not been effective at producing transgenic crops.

Plant hardiness zones — The USDA has divided North America into 11 hardiness zones based on average annual minimum temperatures. Horticulturalists and nurseries rate plants by their hardiness; the hardiness zone maps can then be used to determine the likely survivability of particular plant species and varieties according to one's local growing area.

Plant quarantine — A technique for insuring disease- and pest-free plants by isolating them during a period while performing tests for the presence of these problems. Animal and Plant Health Inspection Service (APHIS) retains this function although the agency's closely allied border inspection function was transferred to the Department of Homeland Security (DHS) by P.L. 107-296.

Plant Quarantine Act — Originally enacted in 1912 (7 U.S.C. 151 *et seq.*), this Act gave the Animal and Plant Health Inspection Service (APHIS) authority to regulate the importation and interstate movement of nursery stock and other plants that may carry pests and diseases that are harmful to agriculture. This Act has been superseded by the consolidated APHIS statute, the Plant Protection Act of 2000 (7 U.S.C. 7701 *et seq.*). This authority is particularly important to the agency's ability to prevent or limit the spread of harmful invasive species within or to a state or region of the United States.

Plant regulator — A chemical that affects the physiological behavior of plants, for example through accelerating or retarding the rate of growth or maturation of produce. Typically the definition of plant regulator excludes nutrients. Plant regulators must be registered as pesticides.

Plant Variety Protection Act of 1970 — P.L. 91-577 was enacted to create an incentive for public and private research on new commercial plant varieties by making it possible for scientists to benefit financially from developing them. The Act provides patent-like protection for new non-hybrid seed varieties. The PVPA Amendments of 1994 (P.L. 103-349 made the law consistent with the International Convention for the Protection of New Varieties of Plants (UPOV) of March 19, 1991, to which the United States is a signatory. In February 1999, UPOV formally accepted the 1994 PVPA amendments as being in conformance with the International Convention. USDA's Agricultural Marketing Service administers the law rather than the U.S. Patent and Trademark Office (7 U.S.C. 2321 *et seq.*).

Plant-pesticide — As proposed by the EPA (November 23, 1994), plant-pesticides are all substances responsible for pest resistance in plants, as well as the genes needed for production of these substances. This term was rejected in favor of plant-incorporated protectant when EPA promulgated its final rule (66 *Federal Register* 37772, July 19, 2001). EPA regulates plant-incorporated protectants introduced into plants using recombinant DNA techniques under legal requirements of FIFRA (7 U.S.C. 136 *et seq.*) and FFDCA (21 U.S.C. 321 *et seq.*). Exempt from tolerance requirements are those defense substances and genes evolved naturally or transferred to the plant by traditional plant breeding methods. Regulations for plant-incorporated protectants are found at 40 CFR 174.

Plasticulture – According to the American Society for Plasticulture, the term plasticulture is defined as the use of plastics in agriculture. This broad term would include plastic film mulches, drip irrigation tape, row covers, low tunnels, high tunnels, silage bags, hay bale wraps, and plastic trays and pots used in transplant and bedding plant production. The use of plasticulture in the production of horticultural crops (vegetables, small fruits, flowers, tree fruits, and ornamentals) helps to mitigate the sometime extreme fluctuations in weather, especially temperature, rainfall and wind. Production costs for vegetable plasticulture generally are much higher than for conventional row-cropped vegetables.

Plate Waste — See Offer versus serve.

Playa Lake — A temporary lake created in the lowest elevation of a basin in an arid area that has no surface drain into another water body, such as a perennial stream or river. Lake water is removed either by evaporation into the air or seepage into the ground. Playa lakes may be considered isolated wetlands, and may be eligible to enroll in the new wetlands component of the Conservation Reserve Program, enacted in the 2002 farm bill (P.L. 107-171, Sec. 2101).

PLD — Paid land diversion. See Paid diversion.

Plum Island Animal Disease Center (PIADC) — Located off the northeastern tip of New York's Long Island, Plum Island is the only place in the U.S. to study certain highly infectious foreign animal diseases (FAD), such as foot-and-mouth disease, that could be accidentally or deliberately introduced into the U.S. The site has a biosafety level-3 (BL-3) facility. The land and buildings of Plum Island were transferred to the Department of Homeland Security (DHS) by the Homeland Security Act of 2002 (P.L. 107-296). USDA retains is research and diagnostic mission at Plum Island, employing scientists and veterinarians from the Agricultural Research Service (ARS) and Animal and Plant Health Inspection Service (APHIS). Plum Island also contributes to DHS's biological countermeasures program in the DHS Science and Technology Directorate. [www.ars.usda.gov/plum].

PMA — Produce Marketing Association. [www.pma.com].

PMO — Pasturized Milk Ordinance.

PNTR — Permanent normal trade relations. See Normal trade relations (NTR).

Point farm — The official definition of a farm for census purposes is "any place from which $1,000 or more of agricultural products were produced and sold or normally would have been sold during the census year." If a place does not have $1,000 in sales, a point system assigns values for acres of various crops and head of various livestock species to estimate a normal level of sales. Point farms are farms with fewer than $1,000 in sales but have points worth at least $1,000. Point farms tend to be very small. Some, however, may normally have large sales, but experience low sales in a particular year due to bad weather, disease, or other factors. Both the Economic Research Service (ERS) Agricultural Resource Management Survey (ARMS) and the census of agriculture use the point system to help identify farms meeting the current definition.

Point source pollution — Pollutants that are discharged or emitted from discrete point sources, such as pipes and smokestacks. Both the Clean Water Act (P.L. 92-500; 33, U.S.C. 1251-1387) and the Clean Air Act (42 U.S.C. 7401 *et seq.*) focus control requirements on point sources and both require permits for major sources of discharges from point sources. While much agricultural pollution is nonpoint source, some agricultural activities are affected: for example, feedlots of over 1000 animal units (CAFOs) are considered point sources requiring permits under the Clean Water Act. However, irrigation return flows, although considered point sources, are expressly exempted from the permit requirement.

Point — A measure of price change equal to 1/100 of one cent in most futures contracts traded in decimal units. In grains, it is one cent; in T-bonds, it is one percent of par.

Pollution — Alteration of the environment, as through the introduction of hazardous or detrimental substances, heat, or noise whose nature, location, or quantity produces adverse health or environmental effects. Under Section 502 of the Clean Water Act, (P.L.

92-500; 33 U.S.C. 1362) for example, pollution means the man-made or man-induced alteration of the physical, biological, chemical, and radiological integrity of water.

POP — Producer option payment; persistent organic pollutant.

POPs — Persistent organic pollutants.

Pork bellies — One of the major cuts of the hog carcass that, when cured, becomes bacon. Futures contracts for pork bellies are traded in the futures market.

Posted county price (PCP) — This price is calculated for the so-called loan commodities (except for rice and cotton) for each county by the Farm Service Agency. The PCP reflects changes in prices in major terminal grain markets (of which there are 18 in the country), corrected for the cost of transporting grain from the county to the terminal. It is utilized under the marketing loan repayment provisions and loan deficiency payment (LDP) provisions of the commodity programs. Rice and cotton use an adjusted world price as the proxy for local market prices.

Postharvest — Refers to activities in the food and fiber sector that occur after agricultural products are sold from, or leave, the farm or ranch. In total, about 75% of the retail cost of the market basket of foods is added in postharvest activities.

Postmortem inspection — As used in the meat and poultry inspection program, the phrase refers to the inspection that Food Safety Inspection Service (FSIS) inspectors are required to conduct of all animal carcasses immediately after slaughter.

Potato Diversion Program (PDP) — A USDA program under which farmers are paid to divert potatoes to charitable institutions, livestock feed, ethanol production, and/or render them nonmarketable and destroyed. The most recent program was for 2000 crop fresh russet potatoes with expenditures limited to $10.25 million. There also was a 1997 program to divert fresh Irish round white and russet potatoes to charitable institutions or for use as livestock feed. The program is administered by the Agricultural Marketing Service and implemented in the field by the Farm Service Agency. The objective of PDP is to reduce supplies and raise farm prices.

Poultry Products Inspection Act of 1957 — P.L. 85-172, as amended, requires USDA's Food Safety and Inspection Service (FSIS) to inspect all domesticated birds when slaughtered and processed into products for human consumption. By regulation, FSIS has defined domesticated birds as chickens, turkeys, ducks, geese, and guinea fowl. Ratites were added in 2001. The primary goals of the law are to prevent adulterated or misbranded poultry and products from being sold as food, and to ensure that poultry and poultry products are slaughtered and processed under sanitary conditions. These requirements also apply to products produced and sold within states as well as to imports, which must be inspected under equivalent foreign standards (21 U.S.C. 451 *et seq.*).

Poundage quota — A quantitative limit on the amount of a commodity that can be marketed (also called a marketing quota) under the provisions of a permanent law. Once a common feature of price support programs, this supply control mechanism ended with the quota buyouts for peanuts in 2002 and tobacco in 2004.

Poverty guidelines — These are monthly or annual income amounts that are used to help determine eligibility for a number of federal food assistance and other programs. They are derived from the income poverty thresholds used by the Census Bureau in counting the number of poor persons. They differ by household size, are adjusted upward for Alaska and Hawaii, are published by the Department of Health and Human Services early each spring, and are updated annually for overall inflation. For example, 130% of the income poverty guidelines is the limit for food stamp and free school lunch income eligibility. Sometimes referred to as the Office of Management and Budget (OMB) poverty guidelines.

Poverty thresholds — For statistical purposes (e.g., counting the poor population), the Census Bureau uses a set of annual income levels, the poverty thresholds, slightly different than the federal poverty guidelines. As with the poverty guidelines, they represent a federal government estimate of the point below which a household of a given size has cash income insufficient to meet minimal food and other basic needs. They were developed in the 1960s, based largely on estimates of the minimal cost of food needs, to measure changes in the poor population. The thresholds form the basis for calculating the poverty guidelines, and, like them differ by household size and are adjusted annually for overall inflation. They do not include different levels for Alaska and Hawaii; they do include separate levels for small elderly households.

Powder — A synonymous term for nonfat dry milk, which is one of three storable dairy products purchased by USDA (butter, cheese, and powder) in order to support the farm price of milk, when wholesale prices for these products are below specified levels.

PPB — Parts per billion.

PPI — Prices Paid Index.

PPI — Producer Price Index.

PPIA — Poultry Products Inspection Act (P.L. 85-172; 21 U.S.C. 451 *et seq.*).

PPM — Parts per million.

PPT — Parts per trillion.

Prairie potholes — A type of wetland that is at the center of a shallow depression characteristic of glaciated areas in the Upper Midwest (North Dakota especially). Potholes are a type of isolated wetland (geographically) because they do not have surface outlets to drain into rivers. Some isolated wetlands were ruled exempt from Clean Water Act (P.L. 92-500) permitting requirements by a January 2001 Supreme Court decision (Solid Waste Agency of Northern Cook County v. U.S. Army Corps of Engineers (121S.Ct.675)(2001). Swampbuster remains the only law the federal government can apply to limit altering potholes to produce crops. Many potholes are wet during only a portion of the year, usually early spring. They provide important nesting habitat for migratory waterfowl, and were designated as a national priority area by the Secretary of Agriculture under the Conservation Reserve Program.

Precautionary Principle — An EU concept that, even if scientific evidence is insufficient or inconclusive regarding a practice's or product's potential dangers to human, environmental, animal, or plant health, it should be prohibited if there are reasonable grounds for concern. The precautionary principle has been raised by the EU in trade disagreements hindering U.S. exports there of several agricultural products that the United States deems safe.

Precision farming — Farmers use global positioning (GPS) technology involving satellites and sensors on the ground and intensive information management tools to understand variations in resource conditions within fields. They use this information to more precisely apply fertilizers and other inputs and to more accurately predict crop yields.

Preferential tariff — A tariff that benefits one or more, but not all, countries within the scope of bilateral, regional, or preferential trade agreements (e.g., the North American Free Trade Agreement, the Generalized System of Preferences, the Europe Agreements, the European Economic Area, the Cotonou Agreement). These tariff preferences have created numerous departures from the normal trade relations principal, namely that World Trade Organization (WTO) members should apply the same tariff to imports from other WTO members.

Preferential trade agreements — Agreements among a group of countries to extend special trading advantages, usually tariff rates that are lower than normal trade relations rates. The U.S.'s North American Free Trade Agreement and the EU's Cotonou Agreement that provide preferential access for exports of former EU member country colonies in Africa, the Pacific and the Caribbean (APC countries) are examples of preferential trade agreements.

Preharvest — Refers to activities on the farm or ranch that occur before crop or livestock products are sold. Preharvest food safety activities, for example, is a phrase often used to describe USDA's efforts, through research and cooperative work, to foster changes in on-farm production that can reduce public health risks in live animals before they are sent to slaughter.

Premium, crop insurance — The annual amount that is paid or agreed to be paid in return for insurance coverage. The total premium consists of two components: the producer-paid premium, or the amount required to be paid by the participating producer; and, the premium subsidy, or the portion of the premium paid by the federal government on behalf of the producer. Total premiums for each crop are set to balance premiums with expected losses considering risks associated with the historical loss experiences of producers in each region.

Preproduction expenses — Expenses incurred prior to the period when a farm activity begins producing, primarily raising orchard trees or breeding animals.

Prescribed burning — The practice of intentionally setting fires within identified areas under specified conditions to reduce fuels and produce other ecological benefits with less risk than from wildfires. Often called controlled burning, but this term implies less risk and greater control than commonly exist (as demonstrated by the Cerro Grande fire, an escaped prescribed fire that burned 235 houses in Los Alamos, NM, in May 2000).

Presidential trade negotiating authority — See Trade promotion authority.

Prevented planting acreage — Land on which a farmer intended to plant a program crop or insurable crop, but was unable to because of drought, flood, or other natural disaster. Used in the calculation of disaster payments and crop insurance indemnity payments.

Prevented planting — Under crop insurance, acreage that cannot be planted because of flood, drought, or other natural disaster is eligible for indemnification. Also, prevented planting acreage may be excluded from the time frame used for calculating support program base acres.

PRIA — Public Rangelands Improvement Act (P.L. 95-514; 43 U.S.C. 1901 *et seq.*).

Price band — A policy instrument that serves to insulate domestic producers and processors when the world price for a commodity falls below a calculated reference price (e.g., a price target comparable to a commodity support level). Protection is provided by imposing a variable import levy on the imported commodity that raises the importer's cost to the reference price. Chile, some Andean Group countries, and some Central American countries use price bands to protect specific commodity and processed food sectors.

Price elasticity of demand — The relationship between the change in the price of a commodity and the corresponding change in the quantity that is sold. If a small change in the price is accompanied by a relatively large change in the quantity sold, demand is said to be elastic (responsive to price changes). But if a large change in the price is accompanied by a small change in the quantity sold, demand is said to be inelastic. The demand for many farm products is relatively price inelastic. As a result of low price elasticity of demand, shifts in supply can have large impacts on prices. For example, the presence of surpluses results in disproportionately large price declines, and conversely shortages result in large price increases. For these reasons, agriculture often is described as an inherently unstable industry.

Price index — Current price expressed as a proportion to the same price in an earlier time period, commonly called the base period. Monthly price indexes computed by the National Agricultural Statistics Service are the index of prices received by farmers and the index of prices paid by farmers for commodities and services, interest, taxes, and farm wage rates. The ratio of these two indexes is referred to as the parity ratio.

Price support — Programs operated by USDA that are intended to raise farm prices when supply exceeds demand and prices are unacceptably low. Support usually is achieved through nonrecourse loans, payments, and purchases. Some commodities are designated in the law to receive mandatory support; others may be supported at the discretion of USDA. Over time, policy changes have shifted toward farm income support and away from commodity price support. However, the term price support frequently is used to describe commodity programs that support farm income though they may not impact market prices.

Prices paid index — An index that measures changes in the prices paid for goods and services used in crop and livestock production and family living. The production component of the index accounts for over 65% of the total, and family living expenses represented by the CPI-U account for less than 20% of the index. The remaining components are interest charges on farm real estate and non-real estate debt, taxes payable on farm real estate, and wage rates paid to hired farm labor. NASS currently publishes the index on a 1990-92 = 100 base. Used in calculating the federal grazing fee, among other purposes. The index of prices paid on a 1910-14 = 100 base is called the parity index and is used in calculating the parity ratio.

Prices received index — An index that measures changes in the prices received for crops and livestock. NASS currently publishes the index on a 1990-92 = 100 base. A ratio of the prices received index to the prices paid index on the 1990-92 base that is greater than 100% indicates that farm commodity prices have increased at a faster rate than farm input prices. When the ratio is less than 100%, farm input prices are increasing a more rapid pace than farm commodity prices. The prices received index and the prices paid index are used to calculate the parity ratio.

Prime farmland — Land that is best suited to and available for the production of food, feed, forage, fiber, and oilseed crops. It can be cropland, pastureland, rangeland, forestland, or other land. It has the soil quality, growing season, and moisture needed to produce high yields of crops each year economically, if managed according to acceptable farm practices. Prime farmland produces the highest yields with minimal expenditure of energy and economic resources and does so with the least damage to the environment. Of the 332 million acres of prime farmland, 212 million are in cropland use, according to the 1997 national resources inventory. See Unique farmland.

Prions — Abnormal proteins generally believed to cause a number of degenerative brain diseases called transmissible spongiform encephalopathies (TSEs) in livestock, including bovine spongiform encephalopathy (BSE) in cattle, and Creutzfeldt-Jakob disease in humans.

Prior appropriations — The system of water allocation under state law used primarily in the arid western United States, where water is scarce. Under this system, earlier claims have priority over later claims, and claims are associated with specific volumes of water that are diverted and put to beneficial uses. Rights to water can be lost if they are not used. The riparian rights system is primarily used in the East.

Prior converted wetland — Under the swampbuster program, these are wetlands that were converted to cropland before swampbuster was enacted on December 23, 1985, and meet wetland criteria for saturated soils or water-loving plants. Under swampbuster, there are no restrictions on either drainage maintenance or additional drainage on prior converted wetlands, which are estimated to total more than 50 million acres.

Priority areas — Starting with the Conservation Reserve Program in the Food Security Act of 1985 (P.L. 99-198), either agencies in USDA or statutory language could identify regions or portions of states where enhanced assistance would be provided through specified conservation programs because either problems were concentrated in those areas or because conservation programs could provide more benefits for the level of effort expended in those areas. The 2002 farm bill (P.L. 107-171, Sec. 2006) repeals geographic priority areas, but national priority areas are retained for the Conservation Reserve Program.

Private grazing land lease rate index — See Forage value index.

Private nonindustrial forest lands — Forest land owned by a private individual or organization that does not also own a wood processing facility.

Private voluntary organization (PVO) — A nongovernmental, nonprofit organization that provides economic and social assistance to people in need, often in foreign countries. PVOs play an important role, along with cooperatives, in distributing U.S. food aid and implementing development projects under P.L. 480 Title II.

PRN — Pesticide registration notice.

Process verification programs — The emergence of value-enhanced commodities and a niche market for non-biotechnologically derived commodities has created an increased need to differentiate products in the handling system. In response to these needs, USDA's Agriculture Marketing Service (AMS) and Grain Inspection, Packers & Stockyards Administration (GIPSA) have begun to provide and are planning to develop a variety of programs and services to facilitate the marketing of agriculture, such as the Qualified Through Verification, Partners in Quality, and Animal Protein Free Certification.

Producer agreement — To be eligible for the Direct and Counter-cyclical Program (DCP) payments under the 2002 farm bill (P.L. 101-171, Sec. 1105), farmers annually must agree to comply: with conservation and wetland protection requirements specified in other laws; with planting flexibility requirements; with effective weed control and other sound agricultural practices on cropland devoted to conserving uses; and, to maintain the cropland in agricultural uses in contrast to commercial or industrial uses. The agreement signed by farmers is Form CCC-509, Direct and Counter-cyclical Program Contract. This new annual producer agreement replaces the previous Production Flexibility Contract required by the 1996 farm bill (P.L. 104-127).

Producer Option Payment (POP) — The original name for the loan deficiency payment (LDP), a component of the marketing assistance loan program. This phrase continues to be used by some farmers.

Producer price — The average price or unit value received by farmers for a specific agricultural commodity produced within a specified 12-month period. This price is measured at the farm gate (the point that the commodity leaves the farm) and therefore does not incorporate cost of transportation, processing, or marketing.

Producer subsidy equivalent — Former OECD term for the measurement of the annual monetary value of support to a country's agricultural producers. The current term is producer support estimate (PSE).

Producer support estimate (PSE) — An indicator of the annual monetary value of gross transfers from consumers and taxpayers to agricultural producers, measured at the farm gate level, arising from policy measures that support agriculture, regardless of their nature, objectives or impacts on farm production or income. Examples include market price support, and payments based on output, area planted, animal numbers, inputs, or farm income. PSEs, which are updated and published annually by the Organization for Economic Cooperation and Development, can be expressed in monetary terms: as a ratio to the value of gross farm receipts valued at farm gate prices, including budgetary support (percentage PSE); or, as a ratio to the value of gross farm receipts valued at world market prices, without budgetary support. See Consumer support estimate (CSE), General services support estimate (GSSE), and Total support estimate (TSE).

Producer — Generally, a producer is thought of as a farm operator. However, given the sometimes complex ownership and rental arrangements of today's farms, the 2002 farm bill (P.L. 101-171, Sec. 1001) defines a producer for purposes of farm program benefits as an owner-operator, landlord, tenant, or sharecropper that shares in the risk of producing a crop and is entitled to a share of the crop produced on the farm. Under this definition, a landlord receiving cash rent is not considered a producer and is not eligible to receive subsidy program payments. However, a landlord receiving crop share as rent is a producer.

Production contract — These contracts specify who supplies the production inputs, the quality and quantity of the commodity to be produced, and the compensation for the producer. Under some livestock production contracts, the farmer is paid to provide housing and care for the animals until they are ready for market, but the contractor actually owns the animals. In 1997, according to the USDA, about 70% of the value of poultry production was under production contracts, 33% of hogs, and 14% of cattle. See Marketing contract.

Production Credit Association (PCA) — Farm Credit System institutions that have direct lending authority to make short and intermediate term loans to eligible retail customers.

Production expenses — Measures the aggregate business cost of farming. The two main components are current farm operating expenses and overhead costs. Farm production expenses are accounted for differently in calculating farm income measures. Only production expenses paid in cash are deducted from gross farm income to derive net cash farm income. Gross farm income and net farm income include both cash and noncash production expenses. Non-cash expenses include such overhead costs as charges for depreciation and other capital consumption associated with farm buildings and other structures, motor vehicles, farm machinery and equipment, and expenses associated with farm operator dwellings.

Production flexibility contract — A 7-year contract covering crop years 1996-2002, authorized by the 1996 farm bill (P.L. 104-127) between the Commodity Credit Corporation (CCC) and farmers, which makes fixed income support payments. Farmers were given production flexibility and diversification options on their contract acres not previously allowed on base acres. Each farm's total payment was the payment rate times the payment quantity for participating base acres. In exchange for annual fixed payments, the owner or operator agreed to comply with the applicable conservation plan for the farm, the wetland protection requirements currently in law, and the constraints on growing fruits and vegetables on contract acres. Land enrolled in a contract had to be maintained in an agricultural or related activity. The law stated that not more than $35.6 billion would be paid over the 7-year period, in declining annual amounts from $5.3 billion in FY1996 to $4.0 billion in FY in 2002. The annual payments were allocated among commodities similar to historical deficiency payments, with 53.6% going to feed grains, 26.3% for wheat, 11.6% for upland cotton, and 8.5% for rice. Target prices and deficiency payments, authorized in the 1973 farm bill, were eliminated. The 2002 farm bill (P.L. 101-171, Sec. 1105) replaced this 7-year contract with an annual producer agreement (contract) required for participation in the Direct and Counter-cyclical Program (DCP).

Productivity — A measure of technical efficiency, typically expressed as the added output for an additional unit of input or the average output per unit of input, i.e., labor, land, capital productivity.

Program crops — Not a legally defined phrase, but generally it refers to what are now legally defined as covered commodities. Covered commodities are crops eligible for the Direct and Counter-cyclical Program (DCP) under commodity program provisions under the 2002 farm bill (P.L. 101-171, Sec. 1101-1108). These are wheat, corn, grain sorghum, barley, oats, upland cotton, rice, soybeans, and other oilseeds. Peanuts are under the same payments scheme but are not included in the phrase "covered commodities." Other commodities mandated to receive price support in the 2002 farm bill include extra long staple cotton, wool, mohair, dry peas, lentils, small chickpeas, peanuts, sugar, and milk.

Projected yield — The number of bushels (or pounds or hundredweight) per acre that, based on current weather estimates and other factors, USDA analysts estimate farmers will harvest.

Proposition 65, California — The ballot label given California's Safe Drinking Water and Toxic Enforcement Act, which was passed by voters in a statewide referendum in 1986. The law requires businesses to post warnings when knowingly exposing the public to a substance posing a significant risk of cancer or adverse reproductive or developmental effects. The state is directed to list such chemicals, and businesses have 12 months thereafter before a label, sign, or other warning is required for consumer products and discharges to air or drinking water that pose a significant risk to the public or workers. Twenty months after listing, a company can no longer discharge the chemical into a potential source of drinking water at a level that would pose a significant risk. The state defines significant risk as one excess cancer per 100,000 people with a reasonable lifetime exposure. It is up to businesses to determine whether their use of a listed toxic chemical poses a significant risk, and to defend their determinations in court in the event of a lawsuit.

Prorate — A quantity provision in a fruit or vegetable marketing order that is intended to even out weekly shipments (or shipments for some other specified periods of time); it aims to prevent too much of the regulated commodity from entering the commercial market at the same time and thereby depressing prices.

Protectionism — The use of import or domestic policies that enable countries to protect domestic producers from competition with foreign producers.

Protein crops — Generally, crops that provide any of a large class of naturally occurring complex combinations of amino acids. Such crops, including various oilseeds and grains, are important in meeting the nutrient requirements of farm animals. EU Common Agricultural Policy designates certain protein crops as eligible for support: peas, field beans, and sweet lupins.

Protein premium (wheat) — The price differential (expressed as cents or dollars per bushel) that a high-protein wheat normally commands over wheat of the same grade specification with lower protein content. Typically, northern dark spring wheat has higher protein and brings a premium price over hard red winter wheat.

PRV — The pseudorabies virus.

PSA — Packers and Stockyards Act (P.L. 67-51; 7 U.S.C. 181 *et seq.*).

PSE — Producer support estimate.

Pseudorabies — A disease of swine that can cause severe economic losses due to reproductive problems and fatal infection of other domestic livestock. The Animal and Plant Health Inspection Service (APHIS) began a voluntary, cooperative federal-state-industry pseudorabies eradication program in 1989 with a target completion date at the end of 2000. APHIS accelerated the program in January 1999 as part of a USDA-wide effort to combat historically low market prices for hogs by reducing the size of the U.S. herd. January 2002 marked the first time the United States had no known pseudorabies infected domestic swine herds. However, subsequent outbreaks have required APHIS to set annual performance targets for achieving disease-free status instead of setting a final eradication date. The program includes herd testing and surveillance, and it pays producers a fair market value for their hogs if they decide to destroy an infected herd.

Public elevators — Grain elevators in which bulk storage of grain is provided for the public for a fee. Grain of the same grade but owned by different persons is usually mixed or commingled as opposed to storing it identity preserved. Some elevators are approved by exchanges as "regular" for delivery on futures contracts.

Public lands — As defined in the Federal Land Policy and Management Act (P.L. 94-579), public lands are any land and interest in land outside of Alaska owned by the United States and administered by the Bureau of Land Management. In common usage, public lands may refer to all federal land no matter what agency has responsibility for its management or may refer even to state- and local municipality-owned lands.

Public Rangelands Improvement Act (PRIA) of 1978 — P.L. 95-514 defines the current grazing fee formula and establishes rangeland monitoring and inventory procedures for Bureau of Land Management and Forest Service rangelands. The National Grasslands are exempt from PRIA.

Puerto Rico Block Grant — A federal nutrition assistance grant program provided in lieu of the regular food stamp program to low income persons in Puerto Rico.

Pulse crops — Peas, beans and lentils are known as pulses. They are the seeds of plants belonging to the family Leguminosae, which gets its name from the characteristic pod or legume that protects the seeds while they are forming and ripening. Pulses are valuable because they contain a higher percentage of protein than most other plant foods and low in fat. In spite of its common name, the peanut or groundnut is a legume rather than a nut. The Food and Agriculture Organization (FAO), in its world production and trade data on pulses includes: dry beans, dry peas, dry broad beans, chickpeas, lentils, cowpeas, pigeon peas, bambara beans, lupins, vetches and pulses. The 2002 farm bill (P.L. 107-171, Sec. 1201) includes dry peas, lentils, and small chickpeas as loan commodities eligible for marketing assistance loan program benefits.

Purchase of development rights (PDR) — The acquisition of property development rights through voluntary sale by the landowner to a government agency or land trust. The government agency or land trust acquiring development rights typically restricts future uses of the land to farming or open space.

Purchase requirement — Refers to the pre-1977 requirement that food stamp recipients use a portion of their own funds to buy food stamps in order to qualify for food stamp benefits.

Put option — An option contract to sell a futures contract at an agreed price and time at any time until the expiration of the option. A put option is purchased to protect against a fall in price. The buyer pays a premium to the seller of this option. The buyer has the right to sell the futures contract or enter into a short position in the futures market if the option is exercised. See Call option.

PVO — Private voluntary organization.

QA / QC — Quality assistance / quality control.

QR — Quantitative restriction.

QTV — Qualified Through Verification.

Qualified Through Verification (QTV) — One of the USDA process verification programs, QTV enables those in the processed fruit and vegetable industry to gain official certification of the wholesomeness of their products to improve marketing opportunities. Under this voluntary, fee-for-service program, AMS, using HACCP-based principles, first inspects the company's facilities to ensure they are properly designed, are consistent with the Food and Drug Administration's good manufacturing practices, have on-site microbiological testing, follow accepted sanitary operating procedures, and so forth. Ongoing monitoring, including periodic unannounced site visits, sampling, and audits by AMS are intended to ensure that the firm maintains its QTV status, which then authorizes it to use an official AMS mark, the QTV shield, on product labels and in advertisements.

Quality enhanced crops — Crop varieties that have been developed by the use of conventional plant breeding or genetic engineering techniques to satisfy specific needs by producers, industry or consumers. Examples include: vitamin-A enriched rice, high oleic-acid corn, and long fiber cotton.

Quality Samples Program (QSP) — A pilot program, started in 2000 and administered by the Foreign Agricultural Service (FAS), that utilizes Commodity Credit Corporation (CCC) funds to reimburse U.S. agricultural trade organizations for the cost of providing small samples of U.S. products to potential importers located in overseas emerging markets.

Quantitative restrictions, trade — Limitations on the quantity or value of a product that may be permitted to enter a country. They probably are the most familiar of the nontariff barriers and include quotas, embargoes, restrictive licensing, and other means of limiting imports. The Uruguay Round Agreement on Agriculture requires the conversion of quantitative restrictions to bound tariffs and tariff-rate quotas.

Quarantine — A restraint on importation of certain animals or plants from areas where pests or contagious diseases are endemic, or isolation of imported animals or plants suspected of carrying pests or diseases. States as well as the federal government may impose quarantines or may operate quarantine facilities. See Plant Quarantine Act.

Quotas, import — A quantitative limit placed on the importation of specific commodities. The protection afforded by quotas is more certain than can be obtained by imposing import duties as the effect of the latter will depend on the price elasticities of the imported commodities. Quotas, like tariffs, also can be used to favor preferred sources of foreign supply. Quotas may be specified as an absolute limit or changed from year to year in response to changes in domestic supply and demand.

R-CALF USA — R-CALF United Stockgrowers of America. [www.r-calfusa.com].

R&D — Research and development.

Range Betterment Funds — Money collected from livestock grazing on federal lands and used for rangeland improvements. The Bureau of Land Management calls these funds Range Improvement Funds and uses them solely for labor, materials, and final survey and design of projects to improve rangelands. The Forest Service calls these funds Range Betterment Funds and uses them for planning and building rangeland improvements.

Range Improvement Fund — See Range Betterment Fund.

Range Improvement Funds — Money collected from livestock grazing on federal lands and used for rangeland improvements. The Bureau of Land Management calls these funds Range Improvement Funds and uses them solely for labor, materials, and final survey and design of projects to improve rangelands. The Forest Service calls these funds Range Betterment Funds and uses them for planning and building rangeland improvements.

Rangeland — Land on which the natural potential (climax) plant cover is principally native grasses, grasslike plants, and shrubs. It includes natural grasslands, savannahs, certain shrub and grasslike lands, most deserts, tundra, alpine communities, coastal marshlands, and wet meadows. It also includes lands that are re-vegetated naturally or artificially and are managed like native vegetation. The United States has 406 million acres of rangeland on non-federal lands and about 257 million acres on federal lands. The non-federal rangelands are about 30% of all non-federal rural lands, according to the 1992 National Resources Inventory. The federal lands are managed by the BLM (approximately 167 million acres) and the Forest Service (approximately 95 million acres).

Rapeseed — Rapeseed is the third largest source of vegetable oil in the world, after soybean and palm oil. This status is almost entirely due to the plant breeding work initiated in Canada in the 1950s and 1960s that greatly reduced the levels of two anti-nutritional compounds, erucic acid in the oil and glucosinolates in the meal, creating a new, high-value oil and protein crop known as canola in Canada and the United States. The term industrial rapeseed does not have any regulatory basis but refers to any rapeseed with a high content of erucic acid in the oil. It is used to some extent in lubricants and hydraulic fluids, especially when, because it is biodegradable, there is a significant risk of oil leaking to water ways or to ground water.

RAS — Rural Adjustment Scheme.

Ratite — A family of large flightless birds that include ostriches, emus, and rheas, which U.S. farmers are beginning to domesticate and raise for food. USDA's Food Safety and Inspection Service (FSIS) began a voluntary, fee-for-service ratite inspection program in 1995 in order to help the fledgling industry improve the marketability of the meat. A provision in the FY2001 USDA appropriations act (P.L. 106-387) amended the Poultry Products Inspection Act to make federal inspection of ratite meat mandatory (and thus supported by taxpayer dollars) as of April 2001 (21 U.S.C. 451 *et seq.*).

Raw agricultural product — The Federal Food, Drug, and Cosmetic Act defines this term as "any food in its raw or natural state, including all fruits that are washed, colored, or otherwise treated in the unpeeled natural form prior to marketing." The nonregulatory definition generally means any agricultural commodity that has undergone little or no processing.

RBS, RBCS — Rural Business-Cooperative Service. [www.rurdev.usda.gov/rbs/index.html].

RC — Regional conservationist.

RC&D — Resource Conservation and Development Program.

RCA — Soil and Water Resource Conservation Act of 1977 (P.L. 95-192; 7 U.S.C. 2001 *et seq.*).

RCAP — Rural Community Advancement Program.

RCD — Resource Conservation District.

RCRA — Resource Conservation and Recovery Act (P.L. 94-580; 42 U.S.C. 6901 *et seq.*).

RD — Rural development.

RD&D — Research, development and demonstration.

RDA — Recommended dietary allowance; Rural Development Administration.

rDNA — Recombinant DNA.

Rebating — A direct or indirect benefit provided to an agricultural producer that is contingent upon the purchase of crop insurance, that is not available to producers who do not purchase insurance, and that provides an inducement to purchase insurance from a particular person or entity. Rebating is prohibited in the federal crop insurance program. However, certain premium reductions are authorized by Congress under the Federal Crop Insurance Act (Act). As a general matter, the authority to control rebating is left to USDA's Risk Management Agency (RMA). In some instances, state law is applicable and RMA coordinates with the States to discover whether violations of state law have occurred.

REC — Rural electric cooperative.

Reclamation Act of 1902 — P.L. 57-161, as amended, appropriated the receipts from the sale and disposal of public lands and resources in 17 western states to the construction of irrigation works for the reclamation of arid lands. Amendments made by the Reclamation Project Act of 1939 gave the Department of the Interior, among other things, the authority to amend repayment contracts and to extend repayment for not more than 40 years. Amendments made by the Reclamation Reform Act of 1982 (P.L. 97-293) eliminated the residency requirement provisions of reclamation law, raised the acreage limitation on lands irrigated with water supplied by the Bureau of Reclamation, and established and required full-cost rates for land receiving water above the acreage limit.

Reclamation fund — A special fund established by Congress under the Reclamation Act of 1902, as amended, for receipts from the sale of public lands and timber, proceeds from the Mineral Leasing Act, and certain other revenues. Congress appropriates money from this fund for the investigation, construction, operation, and administration of Bureau of Reclamation projects. Collections from water users for payments made on the reimbursable costs of the federal projects are also returned to the fund.

Reclamation law — The body of law beginning with the Reclamation Act of 1902 that governs investigation, construction, and operation of Bureau of Reclamation projects.

Reclamation — The process of rehabilitating disturbed lands, or converting unproductive lands to productive uses. The term is also used for the process of recycling or reusing water or reestablishing lands disturbed during mining. In the context of the Reclamation Act and reclamation law, it means putting arid lands to use through irrigation.

Recombinant DNA (rDNA) — The technique of isolating molecules of genetic material (DNA) and inserting them into the DNA of another cell (recombinant DNA). Also known as genetic engineering.

Recommended dietary allowances (RDAs) — This term has been replaced by the term dietary reference intakes (DRIs). DRIs include: (1) Estimated Average Requirements; (2) Adequate Intakes; and (3) Tolerable Upper Intake Levels; and (4) the former RDAs. The Food and Nutrition Board of the Institute of Medicine at the National Academy of Sciences (NAS) determines DRIs for essential nutrients on the basis of scientific knowledge, for the purpose of informing the public on what levels of essential nutrients meet the needs of most healthy persons.

Reconciliation instruction — A provision in a budget resolution directing one or more House or Senate committees to recommend legislation changing existing law so as to bring spending, revenues, or the debt limit into conformity with the annual budget resolution. The instructions specify which committees are affected, the dollar changes to be achieved (usually over a 1-, 5- or 10-year period), and the deadline by which the legislation is to be reported. When the House and Senate Agriculture Committees are given reconciliation instructions, they need to report legislation reducing projected spending on certain mandatory programs under their jurisdiction, which could affect spending for the farm commodity price and income support programs, federal crop insurance, and certain USDA food and nutrition, conservation, agricultural trade, and rural development programs.

Record of Decision System (RODS) — Justification for remedial action chosen for a superfund site by EPA under the Comprehensive Environmental Response, Compensation, and Liability Act (CERCLA) of 1980 (P.L. 96-510, as amended; 42 U.S.C. 9601 *et seq.*).

Recourse loan program — Initially under the 1996 farm bill (P.L. 104-127, Sec. 137(a))and continuing under the 2002 farm bill (P.L. 101-171, Sec. 1209), recourse loans were made available to producers of high moisture corn and grain sorghum, and also for upland and ELS seed cotton. The commodity must be pledged as security, but in contrast to nonrecourse loans, the borrower must repay the loan with interest within a specified period. A honey recourse loan program and a mohair recourse loan program were authorized for 1998 production only.

Reduced price lunch (or breakfast, supper, snack) — A federally reimbursable meal (or snack) served to a child who applies for and qualifies because the family's income is between 130% and 185% of the federal income poverty level. Schools may not charge more than 40¢ for reduced price lunches, nor more than 30¢ for reduced price breakfasts.

REE — Research, education and economics (a mission area within USDA).

Reentry interval — The period of time immediately following the application of a pesticide during which unprotected workers should not enter a field.

Reference (border) price — The import cif or export fob price of a commodity, used by the Organization for Economic Cooperation and Development (OECD) in the calculation of Producer and Consumer Subsidy Equivalents (PSE/CSE). An implicit reference (border) price may be calculated as the producer price in the foreign country less the unit market price support (MPS) and may differ slightly from the explicit reference (border) price. Sometimes termed border price.

Referendum — In agriculture, referendum generally refers to a vote by farmers on whether to approve or disapprove a farm program, such as mandatory production or marketing controls, assessments for generic commodity promotion, or marketing orders.

Reformulated Gasoline (RFG) — Reformulated gasoline is required by the Clean Air Act Amendments of 1990 (P.L. 101-549) in areas with severe or extreme ozone problems. Among other requirements, RFG must contain 2% oxygen by weight. This oxygen standard is typically met using MTBE or ethanol.

Regional railroad — A non-Class 1, line-haul freight railroad that operates at least 350 miles of road and/or has operating revenues of at least $40 million annually.

Regionalization — Countries generally have prohibited imports of a foreign agricultural product if it has been associated with an unwanted pest or disease in the exporting country. Until recently, importing countries would not permit any of that product from the exporting country, even if it came from a region that did not have the disease or pest. Regionalization is a fundamental principle in the Sanitary and Phytosanitary provisions of recent trade accords. It provides for the acceptance of such imports if the exporting country can demonstrate that they are from a disease-free or a pest-free area of the country.

Registrant — Any manufacturer or formulator of a pesticide who obtains registration for an active ingredient or product.

Registration — The EPA approval of a new pesticide for specific uses before it can be sold or distributed in the United States. Under the Federal Insecticide, Fungicide, and Rodenticide Act (7 U.S.C. 136 *et seq.*), the EPA is responsible for registering (pre-market licensing) pesticides based on scientific evidence that the approved use will result in no unreasonable adverse effects on human health or the environment when applied according to approved label directions. The label on a registered pesticide specifies the legal obligations of the user in terms of uses, timing, and rates of application.

Regulatory Flexibility Act (RFA) — P.L. 96-354 (September 19,1964) requires federal agencies to consider the special needs and concerns of small business entities whenever they engage in rulemaking subject to notice and comment requirements of the Administrative Procedure Act or other laws. In most cases, when an agency publishes a proposed rule in the Federal Register, it must prepare and publish a regulatory flexibility analysis describing the impact on small businesses.

Remote sensing — The act of detecting objects when the sensor is not in direct contact. This commonly refers to using aerial photographs to observe conditions on the Earth's surface. In agriculture, this technology can be used to determine what plants are being grown and their condition.

Renderer (rendering) — An enterprise (the process) that "cooks" animal processing waste, restaurant grease, and dead animals into ingredients for a wide range of industrial and consumer goods, such as animal feed, soaps, candles, pharmaceuticals, and personal care products. U.S. regulatory actions to strengthen safeguards against bovine spongiform encephalopathy (BSE or mad cow disease) portend significant changes in renderers' business practices, and the value of their products. Also, if inedible animal byproducts have fewer market outlets, the overall economic value of the animal to the producer can decline, and questions arise about how to safely dispose of the materials.

Renewable fuels — Broadly, renewable fuels are made from replenishing feedstocks (such as biomass, sunlight, wind, water, and waste products) in contrast to exhaustible (nonrenewable) feedstocks such as petroleum and coal. Renewable fuels are a subset of alternative fuels. The Renewable Fuels Association is the trade association that promotes ethanol (www.ethanolrfa.org). Unlimited

Renewable Fuel Standard (RFS) — A policy proposal whereby motor fuels in the United States would be required to contain a certain percentage or volume of renewable fuels.

Renewable resources — Natural resources, sometimes called flow resources, that replenish themselves within time limits that permit sustained use, in contrast to nonrenewable resources. Congress recognized the renewable nature of resources managed by the Forest Service and Bureau of Land Management (including timber, livestock forage, recreation, water, and wildlife and fish) by requiring both agencies to follow the principles of multiple use and sustained yield.

Repayment contract — A contract between a water user's organization and the Bureau of Reclamation by which the organization obligates itself to repay a share of the reimbursable construction and other costs (not all construction costs are reimbursed by water users) of a Bureau project in installments determined in accordance with reclamation law in exchange for delivery of water.

Reregistration — The reevaluation and reapproval by the EPA of pesticides originally registered prior to 1984. The purpose of reregistration is to evaluate and approve the older formulations against modern scientific and regulatory standards. The process is scheduled to be completed on the older active ingredients by the end of 2006

Reserve pool — A quantity provision in a fruit or vegetable marketing order that requires that some marketable supplies be withheld from the fresh market to be used instead in a so-called secondary market such as for frozen or processed forms, for a nonfood use, or stored for sale in a future marketing year.

Residual supplier — A country that supplies the world market only after importers have met their initial needs from preferred suppliers. A residual supplier is not initially competitive because of higher prices or lower quality. The United States used to be considered a residual supplier of grains because its commodity support programs kept U.S. grain prices higher than those of competing exporters. Also, long term storage of grains under loan from the Commodity Credit Corporation (CCC) or owned by CCC caused quality to deteriorate.

Residue Management (Crop) — Leaving any portion of a plant or crop in the field after harvest to protect or improve the soil and provide wildlife habitat. Conservation tillage is a common form of residue management.

Resistance, pesticide — The evolutionary capacity of pests exposed to a pesticide to develop resistance to that pesticide. Some genetically engineered crops now being marketed are accompanied by pest resistance management plans to prevent or retard the development of resistance.

Resource Conservation and Development Program (RC&D) — The RC&D program, initiated in 1962 under authority of Food and Agriculture Act of 1962 (P.L. 87-703), assists multi-county areas in enhancing conservation, water quality, wildlife habitat, recreation and rural development. Work in each area is coordinated by a council. At present, 368 areas have been approved encompassing more than 2,500 counties. [www.nrcs.usda.gov/programs/rcd].

Resource Conservation and Recovery Act (RCRA) — This law (P.L. 94-580, as amended; 42 U.S.C. 6901*et seq.*) authorizes the EPA to regulate solid and hazardous wastes. The Act defines solid and hazardous waste, authorizes EPA to set standards for facilities that generate or manage hazardous waste, and establishes a permit program for hazardous waste treatment, storage, and disposal facilities. RCRA, enacted in 1976, made such comprehensive amendments to the Solid Waste Disposal Act (P.L. 89-272) that it became the name of reference. To date, production agriculture has not fallen under RCRA regulations or is explicitly exempted (i.e., solid or dissolved materials in irrigation water return flows).

Resource Management Plan (RMP) — A BLM planning document, prepared in accordance with Section 202 of the Federal Land Policy and Management Act (P.L. 94-579). The plan contains guidelines for making resource management decisions for specific areas managed by BLM, known as resource areas.

Resource Management System — Defined in the 2002 farm bill (P.L. 107-171, Sec. 2001) as a system of conservation practices and management specified in the field office technical guide of the Natural Resources Conservation Service (NRCS), that is designed to prevent resource degradation and permit sustained use of natural resources.

Resource of concern — Natural resources to be conserved under Conservation Security Program (CSP) contracts (P.L. 107-171, Sec. 2001).

Restitutions (or refunds) — The EU's word for export subsidies that are given to traders to cover the difference between the internal EU price of a commodity and its world market price. See also Export Enhancement Program (EEP). Export subsidies are subject to value and volume limits under the Uruguay Round Agreement on Agriculture.

Restoration Cost-sharing Agreements — An enrollment option of less than 30 years under the Wetland Reserve Program. It is the shortest term enrollment option for this program, and pays participants the lowest amount per acre.

Restricted-use pesticides — A pesticide may be classified for restricted use if it requires special handling because of its toxicity. These may be applied only by trained, certified applicators or those under their direct supervision.

Retaliation — The suspension of concessions under a trade agreement, or the imposition of other barriers to trade, by a government in response to the violation of a trade agreement or the imposition of other unfair trade barriers by another government. Section 301 of the Trade Act of 1974 (P.L. 93-618), as amended by the Omnibus Trade and Competitiveness Act of 1988 (P.L. 100-418) and the Uruguay Round Agreements Act (P.L. 103-465), provides the legal authority for the United States to impose retaliatory measures in response to trade agreement violations or other discriminatory foreign trade practices that burden or restrict U.S. commerce.

Return flow — Surface and subsurface water that leaves the field following application of irrigation water. While irrigation return flows are a point source, they are expressly exempted from permit requirements under the Clean Water Act (P.L. 92-500, as amended).

Revenue Assurance (RA) — A form of revenue insurance that protects a grower of an insurable crop whenever low prices, low yields, or a combination of both causes revenue to fall below a guaranteed level selected by the producer. It differs from other revenue insurance programs in that it allows a farmer to use the posted county price, rather than a national price, in determining a target level of revenue.

Revenue insurance — A generic term for any crop insurance program that provides coverage to producers against lost revenues (or incomes) caused by low prices, low yields, or a combination of low prices and low yields. An indemnity is paid to a producer when any combination of yield and price results in revenue that is less than a pre-specified revenue guarantee. The 1996 farm bill (P.L. 104-127) mandated a USDA pilot program for crop years 1997-2000 under which producers of feed grains, wheat, soybeans, and other crops in specified areas could elect to receive insurance against loss of revenue. Two initial pilot programs have since expanded to national scope. The two types are the Crop Revenue Coverage (CRC) program and the Income Protection (IP) program. The two programs are similar, except that the CRC allows farmers to insure their revenue at either planting-time prices or harvest prices, whichever is higher. The Agriculture Risk Protection Act of 2000 (P.L. 106-224) requires USDA to subsidize revenue insurance programs at the same rate as the traditional crop insurance program.

Revenue ton-mile — The movement of a ton of freight one mile for revenue.

Revised universal soil loss equation (RUSLE) — Revisions to the universal soil loss equation implemented in the mid 1990s to more accurately predict soil erosion caused by water. It includes the same factors as the earlier formula; climate, soils, topographic conditions, and the degree to which the use and management of the soil reduces erosion. But it takes advantage of new knowledge about these relationships and the capabilities of computer technology. The comparison between predicted erosion and T-value is important in making and carrying out conservation plans and achieving conservation compliance.

Revolving loan program — A program that uses the repayments of existing loans to make additional loans to qualified borrowers. Initial capitalization of a revolving loan fund may be derived from federal appropriations or contributions from other sources.

RFA — Regulatory Flexibility Act (P.L. 96-354; 5 U.S.C. 601 *et seq.*).

RFA — Renewable Fuels Association [www.ethanolfra.org].

RFF — Resources for the Future. [www.rff.org].

RFID — Radio frequency identification. See Animal identification (ID) and traceback.

RHS — Rural Housing Service. [www.rurdev.usda.gov/rhs/index.html].

RIA — Regulatory impact analysis/assessment.

Rice — Rice is one of the so-called covered commodities and loan commodities eligible for support under provisions of the 2002 farm bill (P.L. 101-171, Title I). In addition, rice exports routinely receive government assistance via U.S. food-aid programs such as P.L. 480 or from export credit guarantee programs. Rice an edible, starchy cereal grain grown throughout the world. In the United States rice is referred to by length of grain – indica rice is long-grained while japonica refers to the shorter grains. About 3 million acres are planted annually to rice in the United States where six states account for nearly 100% of U.S. production – Arkansas (48%), Louisiana (13%), Mississippi (8%), Texas (6%), and Missouri (5%) produce mostly long-grain rice; California (19%) produces the bulk of U.S. medium- and short-grain rice (from 2003 production data). All U.S. program provisions are on a rough rice basis; whereas USDA's PSD database records international supply and use data on a milled basis. Rough or paddy rice includes the hull, the bran (the coarse outer covering of a rice kernel), and the white endosperm. Most rice is marketed in some processed form. Depending on the extent of milling, several different products can be produced from rough rice – hulls, bran, whole-kernel brown rice, whole-kernel milled rice, and broken-kernel milled rice. Some rough rice is parboiled prior to being milled. The first stage of milling removes the outer hull leaving an intermediate product called brown rice. The final stage of milling removes the bran, leaving milled white rice. Some kernels are broken during the milling process. Milling yield refers to the amount of polished white rice obtained from rough rice. Rice milling rates for polished white rice vary by crop variety and quality, but tend to average about 72% of rough rice weight in the United States. Byproducts from rice milling include rice hulls (about 20% of rough rice weight) and bran, polish, and germ (about 8%). Moisture content plays a key role in grain storage and processing. Excess moisture content will cause rice kernels to spoil quickly limiting its storability. Overly dry kernels will crack or split easily during milling thus lowering the value of the end product; this is critical for rice where the percent of whole grains is highly valued.

Richard B. Russell National School Lunch Act — See National School Lunch Act (P.L. 79-396, as amended; 42 U.S.C. 1751 *et seq.*).

Right-to-farm laws — Right to farm laws (sometimes called nuisance laws) deny nuisance suits against farmers who use accepted and standard farming practices, even if these practices harm or bother adjacent property owners or the general public. Agricultural nuisances may include noise, odors, visual clutter and dangerous structures. Every state has some form of a right-to-farm law.

Rill erosion — One of three types of water erosion. In this erosion process, numerous small channels, typically a few inches deep, are formed. It occurs mainly on recently cultivated soils or on recent cuts and fills.

Riparian buffer — A strip of vegetation along the bank of a body of water which slows the rate of flow of runoff from adjoining uplands, causing sediment and other materials to fall out onto the land before the runoff enters and pollutes the body of water.

Riparian rights — The system of water allocation used in most eastern states. Under a purely riparian system, water may be used only by riparian land and it is recognized that all users will share shortages periodically. Few states have a purely riparian system; most states require a permit for at least large-scale uses. In contrast with the prior appropriations system used in the arid West, the right to use water is not acquired by actually putting the water to use, and access to it cannot be lost by lack of use.

Riparian — Pertaining to or situated on or along the bank of a stream or other body of water. Often referred to in the context of cattle grazing and protection of streams for fish and wildlife habitat, and water quality purposes. Riparian rights refers to the entitlement of a land owner to certain uses of water on or bordering the property, including the right to prevent diversion or misuse of upstream waters (generally a matter of state law).

Risk assessment — The qualitative and quantitative evaluation of risks posed to health or the environment that arise from an activity, chemical use, or technology. The process includes describing potential adverse effects, evaluating the magnitude of each risk (e.g., the toxicity of a chemical), estimating potential exposure to the chemical or other hazard, estimating the range of likely effects given the likely exposures, and describing uncertainties.

Risk management — The process of deciding whether and how to manage risks. Public risk management requires consideration of legal, economic, and behavioral factors, as well as environmental and human health effects of each management alternative. Management may involve regulatory and non-regulatory responses. For example, characterizing the risk to farm workers of entering a field after application of a particular pesticide is risk assessment; promulgating reentry standards is risk management. The federal government has played an active role over the years in helping farmers manage risk. Two major risks faced by agricultural producers are production risks and price risks, and the USDA has assisted with federal crop insurance and commodity programs. The Risk Management Agency is now helping farmers utilize other risk management tools.

Risk Management Agency — An independent office within USDA that is responsible for: the supervision of the Federal Crop Insurance Corporation; the administration and oversight of the federal crop insurance program and any pilot or other programs involving revenue insurance; the use of the futures contracts to manage farm risk and support income. [www.act.fcic.usda.gov].

Risk-benefit analysis — Comparison of the short- and long-term risks to the overall societal benefits of an activity, chemical use, or technology. When risks and benefits are expressed in monetary terms, this is effectively cost-benefit analysis. Both the Federal Insecticide, Fungicide, and Rodenticide Act (7 U.S.C. 136 *et seq.*) and the Toxic Substances Control Act (15 U.S.C. 2601 *et seq.*) direct the EPA to base regulatory decisions on a risk-benefit or cost-benefit basis.

RMA — Risk Management Agency. [www.act.fcic.usda.gov].

RMP — Resource Management Plan.

Rodenticide — A pesticide used to destroy, control, or deter mice and rats or other rodent pests from damaging food, crops, etc.

RODS — Record of Decision System (EPA).

Rotational grazing — Pasturing system that allows short periods of heavy use (intensive grazing), followed by a recovery period. It allows the forage to be used more fully and effectively.

Rotterdam Convention on the Prior Informed Consent Procedure for Certain Hazardous Chemicals and Pesticides in International Trade (PIC Convention) — The United States participated in negotiating this treaty under the auspices of the UN. It requires exporting countries to provide prior notification to potential importing nations of substances that are banned or severely restricted in the exporting country. Although the United States signed the treaty in 1998, prior to U.S. ratification, the Senate must give its advice and consent, and Congress must enact enabling (also known as "implementing") legislation. Legislation is needed to resolve inconsistencies between provisions of the treaty and existing U.S. laws, specifically the Toxic Substances Control Act (TSCA)and the Federal Insecticide, Fungicide, and Rodenticide Act (FIFRA). [www.pic.int].

Roundup Ready soybeans (RR soybeans) — Genetically engineered glyphosate-tolerant soybeans. Roundup is the trade name for glyphosate, a broad-spectrum herbicide. Farmers planting RR soybeans pay more for the seed, and must agree to use certain herbicide application practices as part of a pesticide resistance management plan. In addition, under patent law farmers may not save a portion of their harvest of RR soybeans to use for planting the next year's crop. The benefits of herbicide-tolerant crops include potential savings in both chemical and labor costs.

RPA Assessment / Program — As required by the Forest and Rangeland Renewable Resources Planning Act of 1974, (P.L. 93-378; 16 U.S.C. 1600 *et seq.*) the Forest Service makes a periodic (every 10 years) national assessment of renewable resource supplies, demands, and trends, to identify potential problems and opportunities. In response to the problems and opportunities identified in the Assessment, the Forest Service prepares a periodic (every 5 years) national strategic program plan.

RPA — Forest and Rangeland Renewable Resources Planning Act of 1974 (P.L. 93-378; 16 U.S.C. 1600 *et seq.*).

RPAR — Rebuttable presumption against registration.

RRA — Reclamation Reform Act of 1982 (P.L. 97-293; 43 U.S.C. 390aa *et seq.*).

RSTAC — Railroad-Shipper Transportation Advisory Council.

RTB — Rural Telephone Bank.

RUAP — Rural Utilities Assistance Program.

Rules of Origin — Rules that determine the country of origin of an imported product. Section 304 of the Tariff Act of 1930 as amended (19 U.S.C.1304) requires most imports, including many food items, to bear labels informing the ultimate purchaser of their country of origin. Meats, produce, and several other raw agricultural products generally were exempt. The 2002 farm bill (P.L. 107-171, Sec. 10816), however, contains a requirement that many retail establishments provide, starting on September 30, 2004, country-of-origin information on fresh fruits and vegetables, red meats, seafood, and peanuts. However, the consolidated FY2004 appropriation (P.L. 108-199) signed January 23, 2004, delays this requirement for two years except for seafood.

Ruminant — An animal with a stomach that has four compartments and a more complex digestive system than other mammals, which permit it to digest grass and forages. Ruminants include cattle, sheep, goats, deer, bison, elk, and camels. Swine, dogs, and humans are examples of nonruminants.

Runoff, direct — Water that flows over the ground surface or through the ground directly into streams, rivers, and lakes. Runoff is the cause of sheet, rill, and gully erosion and a source of nonpoint pollution.

RUP — Restricted use pesticide.

Rural — According to the U.S. Bureau of the Census, rural areas comprise open country and settlements with fewer than 2,500 residents. Territory outside of urbanized areas is designated rural and can have population densities as high as 999 per square mile or as low as 1 person per square mile. Rural areas consist of all territory outside of Census Bureau-defined *urbanized areas* and *urban clusters*. Urbanized areas have an urban nucleus of 50,000 or more people. They may or may not contain individual cities with 50,000 or more. In general, they must have a core with a population density generally exceeding 1,000 persons per square mile and may contain adjoining territory with at least 500 persons per square mile. The same computerized procedures and population density criteria are used to identify urban clusters of at least 2,500 but less than 50,000 persons. This delineation of built-up territory and small towns and cities was new for the 2000 census. *Metro and non-metro* areas are defined by OMB's Metropolitan Statistical Areas and Micropolitan Statistical Areas and are collectively referred to as Core Based Statistical Areas (CBSAs). Metro areas consist of (1) central counties with one or more urbanized areas and (2) outlying counties that are economically tied to the core counties as measured by worker commuting data. Outlying counties are included if 25% of workers living there commute to the core counties, or if 25% of the employment in the county consists of workers coming from the central counties. Non-metro counties are outside the boundaries of metro areas and are further subdivided into *micropolitan areas* centered on urban clusters of 10,000-50,000 residents, and all remaining non-core counties. Various programs administered by USDA's Office of Rural Development may define rural by various population thresholds. The 2002 farm bill (P.L. 107-171, Sec. 6020) defined rural and rural area as any area other than (1) a city or town that has a population of greater than 50,000 inhabitants, and (2) the urbanized areas contiguous and adjacent to such a city or town. The *rural-urban continuum codes*, *urban influence codes*, and *rural county typology codes* developed by USDA's Economic Research Service (ERS) allow researchers to break out the standard metropolitan and nonmetropolitan areas into smaller residential groups.

Rural Adjustment Scheme (RAS) — An Australian program that seeks to promote an efficient and competitive rural sector by providing assistance and services to help farmers in Australia adjust to technical, economic, and institutional changes. Assistance is provided mainly by concessional loans and interest subsidies on commercial debt. Other payments are for retraining for nonfarm jobs and for withdrawal of key agricultural assets (such as environmentally sensitive land). The RAS was terminated and replaced by the Farm Business Improvement Program (FARMBIS). FARMBIS provides financial assistance to primary producers and land managers to undertake business and natural resource management training and education activities.

Rural Business Investment Program — A new program established by the 2002 farm bill (P.L. 107-171, Sec. 6029). Program guarantees the funds raised by companies that make equity investment in rural businesses, with an emphasis on smaller businesses. Program is modeled after the SBA Small Business Investment Program. (7 U.S.C. 2009cc).

Rural Business-Cooperative Service (RBS) — One of three USDA agencies charged with implementing rural development policies and programs. RBS program objectives are to expand and retain stable employment in rural areas by providing loans, guarantees, technical assistance, and grants to rural businesses and cooperatives. See Rural Housing Service (RHS) and Rural Utilities Service (RUS). [www.rurdev.usda.gov/rbs].

Rural Community Advancement Program (RCAP) — A program established by the 1996 farm bill (P.L. 104-127, Sec. 761) under which USDA is authorized to provide state rural development block grants, direct and guaranteed loans, and other assistance to meet rural development needs across the country. Program funding is allocated to three accounts: (1) community facilities, (2) rural utilities, and (3) rural business and cooperative development. See Rural Development Trust Fund. (7 U.S.C. 2009).

Rural Development Administration (RDA) — A USDA agency established by the 1990 farm bill (P.L. 101-624, Sec. 2302), amending the Consolidated Farm and Rural Development Act of 1972 (7 U.S.C. 1921 *et seq.*), to administer FmHA community and business programs and other USDA rural development programs. These programs help fund the establishment of new businesses and industries and the construction of water and waste disposal systems and other infrastructure in rural communities. RDA was superseded by the Office of Rural Development following the 1994 reorganization of USDA authorized by P.L. 103-354. [www.rurdev.usda.gov].

Rural Development Trust Fund — Authorized under the 1996 farm bill (P.L. 104-127) the trust fund is used to distribute Rural Community Assistance Program funds. Funds are allocated among states based on such factors as rural population, income, and unemployment.

Rural Electric Cooperatives — There are 874 electric distribution cooperatives and 60 generation and transmission cooperatives in the United States that provide electric service to some 30 million people in 46 States. Reflecting their rural location, these cooperatives account for 7.4% of the kilowatt hours sold, but they maintain nearly half of the nation's electric distribution lines. Rural electric cooperatives have access to insured and guaranteed loans from the Rural Utilities Service. Insured loans primarily finance the construction of facilities for the distribution of electric power in rural areas. Guaranteed loans primarily finance generation and bulk transmission facilities for power supply borrowers.

Rural Electrification Administration (REA) — The predecessor USDA agency to the Rural Utilities Service (RUS) charged with administering certain telephone and rural electric cooperative loan programs. The reorganization of REA into RUS was authorized by the Federal Crop Insurance Reform and Department of Agriculture Reorganization Act of 1994 (P.L.103-354).

Rural home loans — Section 502 single-family loans are available through the Rural Housing Service (RHS) to low- and moderate-income rural households to purchase and repair single-family homes.

Rural home repair loans and grants — Section 504 loans and grants are available through the Rural Housing Service (RHS) to lower income rural homeowners for making repairs.

Rural housing preservation grants — Section 533 grants are available through the Rural Housing Service (RHS) for repairing and rehabilitating rural housing for low- and very low-income families.

Rural Housing Service (RHS) — The USDA agency responsible for administering the housing programs including direct loans, loan guarantees, rental assistance payments, and grants for low income families residing in rural areas. USDA's Rural Housing Service (RHS) administers various housing programs intended to aid in the development of rural America. Rural housing programs are divided into three categories: community facilities, single family housing, and multi-family housing. The agency also administers the community facility loan and grant programs under the Rural Community Assistance Program (RCAP). These programs were formerly operated by the Rural Development Administration and the Farmers Home Administration. [www.rurdev.usda.gov/rhs].

Rural multifamily housing loans — Section 515 loans are available through the Rural Housing Service (RHS) for construction of rental housing for low- and moderate-income families in rural areas, and for congregate housing for the elderly and handicapped.

Rural rental assistance — Section 521 rental assistance available through the Rural Housing Service (RHS) reduces out-of-pocket cash that very-low-income and low-income families pay for rent, including utilities.

Rural Strategic Investment Program — A new program established by the 2002 farm bill (P.L. 107-171, Sec. 6030) to fund regional investment boards. The Strategic Investment Program provides grants totaling up to $100 million for rural regional planning and plan implementation. Eligible communities are non-metropolitan counties with a population of 50,000 or less. Certain exceptions to the population threshold are allowed if the community is immediately adjacent to an eligible area. (7 U.S.C. 2009dd).

Rural Utilities Service (RUS) — An agency within USDA charged with administering activities that support rural telecommunications, distance learning and telemedicine, electrical, and water and waste disposal activities. RUS assistance can be provided in the form of direct and guaranteed loans, and grants. [www.usda.gov/rus].

Rural-Urban Commuting Areas (RUCAs) — Rural-Urban Commuting Areas (RUCA) represent a relatively new categorization of rural areas based on research at the Economic Research Service (ERS) of the USDA and the University of Washington. RUCAs are a Census tract-based classification scheme that use the standard Bureau of Census urban area definitions in combination with commuting information to characterize all of the Nation's Census tracts regarding their rural and urban status. The RUCAs are based on measures of urbanization, population density, and daily commuting. Census tracts are used to establish RUCAs because they are the smallest geographic building block for which reliable commuting data are available.

RUS — Rural Utilities Service. [www.usda.gov/rus/home/home.htm].

RUSLE — Revised universal soil loss equation.

S/U — Stocks-to-use ratio.

S.T.O.P. — Safe Tables Our Priority. [www.stop-usa.org].

S-K — Saltonstall-Kennedy Grant Program.

S&D — Special and differential treatment.

S&PF — State & Private Forestry.

SAA — Shared appreciation agreement.

SAB — Science Advisory Board of EPA.

SAC — Suspended and canceled pesticides.

SAES — State Agricultural Experiment Stations.

SAF — Society of American Foresters. [www.safnet.org].

Safe Drinking Water Act (SDWA) — P.L. 93-523, as amended, is the key federal law for protecting public water systems from harmful contaminants (42 U.S.C. 300f-300j). First enacted in 1974, the Act, as amended, is administered by the EPA through regulatory programs that establish standards and treatment requirements for drinking water contaminants, control underground injection of wastes that might contaminate water supplies, and protect ground and surface water sources. Regulated public water systems under the Act are those that have at least 15 service connections or regularly serve 25 or more individuals. The 1996 amendments (P.L. 104-182) broadened the definition of public water system to include systems that deliver water through pipes or "other constructed conveyances," which includes agricultural irrigation systems that convey water that is used for residential purposes (unless alternative water is provided for drinking and cooking; or unless water for drinking, cooking, and bathing is treated). The 1996 amendments also require states to identify, to the extent practicable, origins of contaminants in areas providing source waters for public water systems to determine the susceptibility of systems to contamination; such areas could include farmland. (42 U.S.C. 300f-300j).

Safe Meat and Poultry Inspection Panel — An advisory panel to review and evaluate inspection policies and proposed changes that the 1996 farm bill (P.L. 104-127, Sec. 918) permanently authorized by amendment to the federal meat and poultry inspection statutes. Provisions in annual USDA appropriations laws since 1996 have prohibited the department from actually establishing the advisory panel.

Safeguards, import — A trade policy tool to remedy actual or threatened injury by imports to a domestic industry. Its purpose is to allow a producing sector to adjust to changed market conditions before facing competition again without such protection. For agricultural products subject to tariffication, the Uruguay Round Agreement on Agriculture (Part I, Article 5) establishes a special agricultural safeguard that allows countries to impose an additional duty when sudden import surges (volumes) exceed, or import prices fall below, a trigger level. Free trade agreements also can include agricultural safeguards for specific products.

Sales for local currencies — A provision of P.L. 480 that permits a portion of commercial sales (under Title I) to be repaid in the country's local currency.

Salmonella — A bacterium that is the leading cause of foodborne illness in humans. It is commonly found in varying amounts in raw meats, poultry, milk, and eggs, and it can contaminate other foods if they have been improperly handled. Under 1996 rules published by USDA to control pathogens in meat and poultry, all plants that slaughter food animals and that produce raw ground meat products must test for the presence of salmonella and report the results to federal inspectors. From 1998 (the effective date of the 1996 rules) through 2001, Food Safety and Inspection Service (FSIS) personnel could withdraw inspection from a processing plant (effectively closing it down) if the plant repeatedly violated the *Salmonella* performance standard. In December 2001, a federal court upheld a lower court's ruling that the meat and poultry inspection statutes do not authorize FSIS to use the standard as a basis of enforcement action. Plants must still conduct the tests, but their results are only one of several factors that FSIS inspectors monitor to determine compliance with the 1996 sanitation and pathogen reduction regulations.

Salvage rider — A provision in the Emergency Supplemental Appropriations Act of 1995 (P.L. 104-19, Sec. 2001, July 27, 1995) to expand salvage timber sales from July 27, 1995, through December 31, 1996, by exempting them from public challenges under environmental laws. This was controversial because it reinstated numerous timber sales in Washington and Oregon that had been stopped to protect endangered and threatened species habitat.

Salvage sales — Timber sales from national forests primarily to remove dead, infested, damaged, or down trees and associated trees for stand improvement. They are controversial partly because there are no standards for the number or proportion of trees that must be dead, infested, damaged, or down and partly because the Forest Service may retain the revenues to prepare and administer future salvage sales.

Sample grade — In commodities, usually the lowest quality of a commodity, too low to be acceptable for delivery in satisfaction of futures contracts.

Sanctions — See Embargo.

Sanitary and phytosanitary (SPS) measures and agreements — Measures to protect humans, animals, and plants from diseases, pests, or contaminants. The Agreement on the Application of Sanitary and Phytosanitary Measures is one of the final documents approved at the conclusion of the Uruguay Round of the Multilateral Trade Negotiations. It applies to all sanitary (relating to animals) and phytosanitary (relating to plants) (SPS) measures that may have a direct or indirect impact on international trade. The SPS agreement includes a series of understandings (trade disciplines) on how SPS measures will be established and used by countries when they establish, revise, or apply their domestic laws and regulations. Countries agree to base their SPS standards on science, and as guidance for their actions, the agreement encourages countries to use standards set by international standard setting organizations. The SPS agreement seeks to ensure that SPS measures will not arbitrarily or unjustifiably discriminate against trade of certain other members nor be used to disguise trade restrictions. In this SPS agreement, countries maintain the sovereign right to provide the level of health protection they deem appropriate, but agree that this right will not be misused for protectionist purposes nor result in unnecessary trade barriers. A rule of equivalency rather than equality applies to the use of SPS measures.

Sanitation standard operating procedures (SSOPs) — Refers to the cleaning procedures that meat and poultry plants use, both before, during, and after production, to prevent contamination of products. Site-specific SSOPs were required to be implemented in January 1997 by all slaughter and processing plants, under the comprehensive pathogen reduction regulations issued by USDA in July 1996.

SARE — Sustainable Agriculture Research and Extension Program. [www.sare.org/htdocs /sare/about.html].

SBP — School Breakfast Program.

SC — State Conservationist.

SC; STC — State conservationist.

Scale economies — See Economies of size.

Scaled sales — A timber sale where purchasers pay the stumpage price for the amount of timber actually removed; Forest Service personnel or an independent third-party measures the volume of timber actually removed.

SCD / SWDC — Soil Conservation District; or Soil and Water Conservation District.

SCGP — Supplier credit guarantee program.

School Breakfast Program — Provides federally subsidized breakfasts to children in schools and child care facilities. Permanently authorized by the Child Nutrition Act of 1966 (P.L. 89-642, as amended; 42 U.S.C. 1771 *et seq.*). Federal funding is provided in the form of cash reimbursements for each breakfast served, varied in amount by the family income of the participating child. All children in participating schools and residential institutions are eligible for a federally subsidized meal, regardless of family income. However, free meals must be offered to children from families with incomes below 130% of the federal poverty income level, and reduced price meals to those with family incomes between 130% and 185% of the poverty level. The program is administered by the Food and Nutrition Service and funded by annual agricultural appropriations. [www.fns.usda.gov/cnd/breakfast].

School Lunch Program — See National School Lunch Program.

School meals initiative for healthy children — An initiative undertaken by the Food and Nutrition Service to revise and update the nutrition standards of school meals and bring them into compliance with the Dietary Guidelines for Americans, and to assure that school children have access to healthful nutritious meals that taste good. Includes efforts to improve the quality of commodities provided to schools and provide training and technical assistance.

Science Advisory Board — A group of independent scientists selected by the Administrator of the EPA to advise on the scientific and technical aspects of environmental problems and issues and who, at the request of the Administrator, review the scientific aspects of any reports or other written products prepared by the agency. Congress established the Board when it enacted the Environmental Research, Development, and Demonstration Authorization Act of 1978 (P.L. 95-477). [www.epa.gov/sab].

Scientific Advisory Panel — The Scientific Advisory Panel (SAP) was created on November 28, 1975, pursuant to Section 25(d) of the Federal Insecticide, Fungicide, and Rodenticide Act (7 U.S.C. 136w), to provide scientific advice on pesticides and pesticide related issues as to the impact on health and the environment of certain regulatory actions. The Food Quality Protection Act (P.L. 104-170) established a Science Review Board consisting of 60 scientists who are available to the Scientific Advisory Panel on an *ad hoc* basis to assist in reviews conducted by the Panel. The role of the SAP has been expanded to that of a peer review body for current scientific issues that may influence the direction of EPA's regulatory decisions. The Panel is composed of seven members who are selected on the basis of their professional qualifications to assess the impact of pesticides on health and the environment. Members are appointed by the EPA Administrator from a list of 12 nominees submitted by the National Institutes of Health and the National Academy of Sciences. [www.epa.gov/scipoly/sap/about.htm].

Scrapie — A fatal, degenerative neurological disease of sheep and goats. The similarity of scrapie to bovine spongiform encephalopathy (BSE) (mad cow) disease in cattle, with the possibility of subsequent transmission to humans, has caused the Food and Drug Administration (FDA) to propose regulations to prohibit using sheep and goat by-products as a component in cattle feeds. The Animal and Plant Health Inspection Service (APHIS) also conducts a voluntary scrapie flock certification program to certify scrapie-free herds.

SDWA — Safe Drinking Water Act (42 U.S.C. 300f-300j).

SE — Salmonella enteritidis.

SEARCH Grants — A program established by the 2002 farm bill (P.L. 107-171, Sec. 6301-04) to assist very small communities (under 3,000 in population) in preparing feasibility and environmental studies required to meet water and waste environmental standards. (7 U.S.C. 2009ee).

SEC — Office of the Secretary; Securities and Exchange Commission.

Second-tier Mexican sugar — The over-quota sugar exported by Mexico to the United States, subject to a North American Free Trade Agreement (NAFTA) tariff that declines 1.5¢/lb. for raw sugar, and 1.6¢/lb. for refined sugar, each year until it enters free, effective January 1, 2008. In the interim, this sugar becomes price competitive in the U. S. market whenever the applicable tariff, when added to the world sugar price and the cost of transporting it to U.S. Gulf ports (about 1.5¢/lb.), is below the loan forfeiture level created by the U.S. sugar program. Over-quota sugar entering from countries other than

Mexico is subject to a much higher tariff, which is not declining, as set under WTO rules. This prohibitive tariff, when added to the world sugar price, makes world sugar uncompetitive in price, and serves to keep it from entering the U.S. market.

Section 11 special assistance — This refers to the section of National School Lunch Act (P.L. 79-396, as amended) that provides additional cash reimbursements (or so-called special assistance payments) to schools for free lunches and reduced price lunches to children from lower income families.

Section 15 lands — These are public lands that lie outside a grazing district administered by the Bureau of Land Management under Section 15 of the Taylor Grazing Act of 1934. The BLM authorizes livestock grazing on these lands by issuing leases to private parties.

Section 201 — A section of the Trade Act of 1974 (P.L. 93-618) that permits the President to grant temporary import relief, by raising import duties or imposing nontariff barriers on goods entering the United States that injure or threaten to injure domestic industries producing like goods. This provision is the analog of GATT Article XIX, which allows GATT contracting parties to provide relief from injurious competition when temporary protection will enable the domestic industry to make adjustments to meet the competition.

Section 22 — A provision of permanent agricultural law (Agricultural Adjustment Act Amendment of 1935, P.L. 74-320) that allows the President to impose import fees or import quotas to prevent imports from non-WTO member countries from undermining the price support and supply control objectives of domestic farm programs. Legislation implementing The North American Free Trade Agreement (NAFTA) and the Uruguay Round Agreement on Agriculture exempts NAFTA partners and WTO member countries from Section 22 quotas and fees. Under both trade agreements, the United States converted then-in-effect Section 22 restrictions into tariff-rate quotas. This effectively eliminates Section 22 as a tool to shield domestic price support operations.

Section 3 lands — Public lands within a grazing district administered by the Bureau of Land Management under Section 3 of the Taylor Grazing Act of 1934. BLM authorizes livestock grazing on these lands by issuing permits. Section 3 lands make up the vast majority of BLM-administered lands.

Section 301 — A section of the Trade Act of 1974 (P.L. 93-618) that authorizes the President to take all appropriate action, including retaliation, to obtain the removal of any act, policy, or practice of a foreign government that violates an international trade agreement or is unjustified, unreasonable, or discriminatory, and that burdens or restricts U.S. commerce. This authority provides a mechanism to respond to foreign trade barriers in the hope that removal of such barriers will boost U.S. exports. Section 301 cases can be self-initiated by the United States Trade Representative (USTR) or as the result of a petition filed by a firm or industry group. If USTR initiates a Section 301 investigation, it must seek to negotiate a settlement with foreign country in the form of compensation or elimination of the trade barrier. For cases involving trade agreements, the USTR is required to request formal dispute proceedings as provided by the trade agreements.

Section 32 — Section 32 of Agricultural Adjustment Act Amendment of 1935 (P.L. 74-320) was enacted to widen market outlets for surplus agricultural commodities as one means of strengthening farm prices. Section 32 programs are financed by a permanent appropriation equal to 30% of the import duties collected on all items entering the United States under the customs laws, plus any unused balances up to $500 million. Most of this sizeable appropriation (about $6 billion annually in recent years) is transferred by appropriators to help pay for child nutrition programs. However, a portion of Section 32 also is set aside to provide USDA with a source of discretionary funds (a contingency reserve) of several hundred million dollars annually, which it uses for emergency removals of surplus agricultural commodities, disaster relief, or other unanticipated needs that might arise during a fiscal year.

Section 4 general or basic assistance — This refers to the section of National School Lunch Act (P.L. 79-396, as amended) requiring the federal government to subsidize all lunches served through the school lunch program, regardless of the income of the participant. Sometimes referred to as paid or full-price lunches because children buying meals pay most of the cost, as opposed to those getting free lunches or substantially reduced price lunches.

Section 404 — A provision of the Clean Water Act (P.L. 92-500) establishing a program regulating the discharge of dredge or fill material into the nation's waters. Permits for individual dredge or fill activities are issued by the Army Corps of Engineers, subject to guidelines prepared by the EPA. Primarily because this program can impact economic development by restricting the filling of wetlands, it has been controversial. It applies to agricultural, as well as non-agricultural lands. However, normal farming operations, silviculture, and ranching activities (such as plowing, cultivating, and minor drainage, and the construction and maintenance of farm and stock ponds, irrigation and drainage ditches, and farm and forest roads) are exempted by law from the permit requirements of this program. In addition, Section 404 authorizes general permits for certain activities, including several agricultural ones, so that landowners need not apply for individual permits. For example, there are general permits for cranberry bogs and for rice culture.

Section 416 — A section of the Agricultural Act of 1949 (P.L. 89-439)that provides for the disposition of agricultural commodities held by the Commodity Credit Corporation (CCC) to prevent waste. Disposal is usually carried out by donation of commodities to charitable groups and foreign governments.

Section 502 loans — A rural housing loan program, administered by the Rural Housing Service (RHS), authorized under Section 502 of the Housing Act of 1949. Borrowers may obtain loans for purchasing or repairing new or existing single-family housing. Loans are made directly by RHS (7 CFR 3550) or by private lenders with a USDA guarantee (7 CFR 1980). Borrowers with income of 80% or less of the area median may be eligible for 33-year direct loans and may receive interest credit to bring the interest rate to as low as 1%. In a given fiscal year, at least 40% of the units financed under this section must be made available only to very low-income families or individuals (below 60% of the area median) with terms up to 38 years. Borrowers must have the means to repay the loans, but be unable to secure reasonable credit terms elsewhere. Borrowers with income of up to 115% of the area median may be eligible for 30-year guaranteed loans from private lenders. Priority is given to first-time home buyers, and the RHS may require that borrowers complete a home ownership counseling program.

Section 504 loans and grants — A USDA rural housing repair program authorized under Section 504 of the Housing Act of 1949. Under current regulations, rural homeowners with incomes of 50% or less of the area median may qualify for the Rural Housing Service (RHS) direct loans to repair their homes. Loans are limited to $20,000 and have a 20-year term at a 1% interest rate. Owners of age 62 or more may qualify for grants of up to $7,500 to pay for needed repairs that remove a health or safety hazard. To qualify for the grants, the elderly must be unable to obtain affordable credit elsewhere. Depending on the cost of the repairs and the income of the homeowner, the owner may be eligible for a grant for the full cost of the repairs or for some combination of a loan and a grant to covers repair costs. The combination loan and grant may total no more than $20,000.

Section 514 loans — A domestic farm labor housing program authorized under Section 514 of the Housing Act of 1949 and the only nationwide program to provide housing for farm laborers. The Rural Housing Service (RHS) makes loans to farm owners, associations of farm owners, Indian tribes, or nonprofit organizations to provide modest living quarters, basic household furnishings, and related facilities. Loans also may be used to repair existing housing for farm labor use. The loans are repayable in 33 years and bear an interest rate of 1%. Applicants, who own farms or who represent farm owners, must show that the farming operations have a demonstrated need for farm labor housing and the applicants must agree to own and operate the property on a nonprofit basis. Except for state and local public agencies or political subdivisions, the applicants must be unable to provide the housing from their own resources and unable to obtain the credit from other sources on terms and conditions that they could reasonably be expected to fulfill and still provide farmworker housing at rental rates that would be affordable to the workers. The RHS may make exceptions to the "credit elsewhere" test when: (1) there is a need in the area for housing for migrant farm workers and the applicant will provide such housing; and, (2) there is no state or local body or no nonprofit organization that, within a reasonable period of time, is willing and able to provide the housing.

Section 515 loans — A USDA rural housing program authorized under Section 515 of the Housing Act of 1949 (42 U.S.C. 1485). The Rural Housing Service (RHS) is authorized to make loans to provide rental housing for low- and moderate-income families in rural areas. Section 515 loans also may be used for congregate housing for the elderly and handicapped. Loans under Section 515 are made to individuals, corporations, associations, trusts, partnerships, or public agencies. The loans are made at 1% interest for a term of 50 years. Except for public agencies, all borrowers must demonstrate that financial assistance from other sources will not enable the borrower to provide the housing at terms that are affordable to the target population.

Section 516 grants — A USDA farm labor housing program authorized by Section 516 of the Housing Act of 1949 (42 U.S.C. 1441 *et seq.*). Qualified nonprofit organizations, Indian tribes, or public bodies obtain grants for the development cost of farm labor housing. Grants may be used simultaneously with Section 514 loans if the housing, for which there is a "pressing need," will not be built without assistance from the Rural Housing Service (RHS). Grants may be made for up to 90% of the development cost of the housing. In a given fiscal year, up to 10% of the Section 516 funds shall be for domestic and migrant farm worker housing. Applicants must contribute at least 10% of the total development costs from their own resources or from other sources including Section 514 loans. Funds may be used to buy, build, or improve housing and related facilities for farm workers, and to purchase and improve the land where the housing will be located, including installation of streets, water supply and waste disposal systems, parking areas, and driveways as well as for the purchase and installation of appliances such as ranges, refrigerators, and clothes washers and dryers. Related facilities may include the maintenance workshop, recreation center, small infirmary, laundry room, day care center, and office and living quarters for the resident manager.

Section 521 rental assistance — Rental assistance authorized under Section 521 of the Housing Act of 1949 (42 U.S.C. 1441 *et seq.*). Owners of housing financed under Section 515 or Section 514 may receive rental assistance payments from the Rural Housing Service (RHS). The assistance payments enable eligible tenants to make monthly rent payments that do not exceed the greater of: (1) 30% of monthly adjusted family income; (2) 10 percent of monthly income; or, (3) the portion of the family's welfare payment that is designated for housing costs. The rental assistance payments, which are made directly to the borrowers, make up the difference between the tenants' payments and the RHS-approved rent for the units. Borrowers must agree to operate the property on a limited profit or nonprofit basis. The term of the rental assistance agreement is 20 years for new construction projects and 5 years for existing projects. Agreements may be renewed for up to 5 years. An eligible borrower who does not participate in the program may be petitioned to participate by 20 percent or more of the tenants eligible for rental assistance.

Section 523 loans — A mutual self-help rural housing program authorized under Section 523 of the Housing Act of 1949 (42 U.S.C. 1441 *et seq.*) and administered by the Rural Housing Service (RHS). Nonprofit organizations may obtain 2-year loans to purchase and develop land that is to be subdivided into building sites for housing. The interest rate is 3% for these loans. Applicants must demonstrate a need for the proposed building sites in the locality. Sponsors also may obtain technical assistance (TA) grants to pay for all or part of the cost of developing, administering, and coordinating programs of technical and supervisory assistance to the families who are building their own homes. Each family is expected to contribute at least 700 hours of labor in building homes for each other. Applicants must demonstrate that: (1) there is a need for self-help housing in the area; (2) the applicant has or can hire qualified people to carry out its responsibilities under the program; and, (3) funds for the proposed TA project are not available from other sources. The program is generally limited to very low- and low-income families. Moderate income families may be eligible to participate provided they are unable to pay for a home built by the contract method. TA funds may not be used to hire construction workers or to buy real estate or building materials. Private or public nonprofit corporations, however, may be eligible for 2-year site loans under Section 523. The loans may be used to purchase and develop land in rural areas. The land is subdivided into building sites and sold on a nonprofit basis to low and moderate income families. Generally, a land loan must result in at least 10 home sites. The sites need not be contiguous. Sites financed through Section 523 may be sold only to families who are building homes by the mutual self-help method. The homes are financed through the Section 502 program.

Section 524 loans — Land acquisition and development loans authorized under Section 524 of the Housing Act of 1949 (42 U.S.C. 1441 *et seq.*). Nonprofit organizations and Indian tribes may obtain loans from the Rural Housing Service (RHS) to purchase and develop land that is to be subdivided into building sites for housing low- and moderate-income families. The loans are made for a 2-year period. Sites financed through Section 524 have no restrictions on the methods by which the homes are financed or constructed. The interest rate on Section 524 site loans is the Treasury cost of funds.

Section 533 grants — A USDA rural housing rehabilitation program authorized under Section 533 of the Housing Act of 1949 (42 U.S.C. 1441 *et seq.*). The Rural Housing

Service (RHS) is authorized to make grants to capable organizations for: (1) rehabilitating single family housing in rural areas that is owned by low- and very low-income families; (2) rehabilitating rural rental properties; and, (3) rehabilitating rural cooperative housing structured to provide affordable housing to low- and very low-income occupants. The homes must be located in rural areas and be in need of housing preservation assistance. Assisted families must meet the income restrictions (income of 80% or less of the median income for the area) and must have occupied the property for at least one year prior to receiving assistance. Occupants of leased homes may be eligible for assistance if (1) the unexpired portion of the lease extends for 5 years or more, and (2) the lease permits the occupant to make modifications to the structure and precludes the owner from increasing the rent because of the modifications.

Secure Rural Schools and Community Self-Determination Act of 2000 — P.L. 106-393 amended the Forest Service's county payments program for FY2001-FY2006 to allow states or counties to choose to receive the average of the three highest payments for FY1986-FY1999 in lieu of the regular 25% payment, but requiring that15%-20% of those payments be used by the counties for specified purposes, in accordance with recommendations of resource advisory committees for projects on federal lands, or returned to the Treasury. The Forest Service county payments should not be confused with Bureau of Land Management "payments in lieu of taxes."

Sediments — Particulate material, including soil, sand, and minerals, transported and deposited by water or wind (see Erosion). Waterborne sediments cloud the water diminishing sunlight available to aquatic plants; sediments deposited in reservoirs, rivers, and harbors destroy fish and wildlife habitat and may fill shipping channels. Farming, mining, and building and construction activities that expose soil to wind and rain are major sources of sediments.

Seed bank — A facility used for the preservation and dissemination of seed, particularly varieties that are not in commercial use and that may be threatened with extinction. The USDA's seed bank is the U.S. National Plant Germplasm System (NPGS). [www.ars-grin.gov/npgs].

Self-Help land development loans — Section 523 loans are for acquiring and developing land that will be used for mutual self-help housing.

Septic system — An on-site system designed to treat and dispose of domestic sewage from a residence or business not connected to a sewer line. A typical septic system consists of a tank that receives waste and holds it while bacteria decompose solids, and a system of tile lines or a pit for disposal of the liquid effluent. The sludge that remains in the tank after decomposition of the solids must be pumped out periodically. By using the assimilative capacity of the land, a properly operating septic system has minimal impact, but an improperly functioning system can be a source of nitrogen pollution and of groundwater contamination.

Set-aside program — A program (not used since the late 1970s) under which farmers were required to set aside a certain percentage of their total planted acreage and devote this land to approved conservation uses (such as grasses, legumes, and small grain which is not allowed to mature) in order to be eligible for nonrecourse loans and deficiency payments. Set-aside acreage was based on the number of acres a farmer actually planted in the program year as opposed to being based on prior crop years. The authority for set-aside was eliminated by the 1996 farm bill (P.L. 104-127).

SFIREG — State FIFRA Issues Research and Evaluation Group.

SFSP — Summer Food Service Program.

Sharecropper — A tenant farmer who receives a share of the crops, livestock, or livestock products from the landowner, who in turn may extend credit to and supervises the tenant. The tenant generally supplies only labor.

Shared Appreciation Agreement (SAA) — An agreement between the USDA and farmer borrowers, instituted when a borrower is severely delinquent on making payments on Farm Service Agency (FSA) real estate loans. Such an agreement allows for the forgiveness of a portion or all of the indebtedness, in return for the borrower sharing with USDA at the end of the term a portion of any appreciation in the farmer-owned real estate that served as collateral for the loan. SAAs were instituted as part of the Agricultural Credit Act of 1987 (P. L. 100-233) for the purpose of avoiding foreclosure due to the drop in land prices at that time.

Sheep Promotion, Research, and Information Act of 1994 — P.L. 103-407 enabled sheep producers and feeders and importers of sheep and sheep products to develop, finance, and carry out a nationally coordinated program for sheep and sheep product promotion, research, and information. This law was enacted a year after passage of legislation to phase out the wool and mohair commodity programs (new support programs for wool and mohair were included in the 2002 farm bill (P.L. 107-171)). The USDA was authorized to issue a sheep and wool promotion, research, education, and information order subject to approval referenda among producers, feeders, and importers. In a 1996 referendum, the proposed check-off program was defeated. About 53% of nearly 12,000 ballots opposed the order. This group represented 67% of the production that voted.

Sheet erosion — One of three types of water erosion, where the removal of a thin, relatively uniform layer of soil from the land surface is caused by runoff. See Rill erosion, and Gully erosion.

Shelterbelt — A plant barrier of trees, shrubs, or other approved perennial vegetation designed to reduce wind erosion. Also called a windbreak.

Sherman Anti-Trust Act — The 1890 law is considered the foundation of federal anti-monopoly policy (15 U.S.C. 1 *et seq.*). Passed partly as an outgrowth of congressional investigations into alleged price collusion among large meat packers, the law generally prohibited restraint of trade and monopolistic practices in all industries, including agribusiness. The Capper-Volstead Act (7 U.S.C. 291, 292) later exempted agricultural cooperatives from certain provisions of the Sherman Act and the subsequent Clayton Act.(15 U.S.C. 12 *et seq.*).

Shipping holiday — A fruit and vegetable marketing order feature that prohibits the commercial shipping of the regulated commodity during periods following certain holidays when demand is historically low, such as the several days after Thanksgiving and Christmas.

SHOP — See Small Hog Operation Payment.

Short line railroad — Any freight railroad that cannot be classified as a Class 1 or regional railroad. A railroad company that can originate and terminate freight traffic on its line and is generally less than 100 miles in length.

Short selling — Selling a futures contract with the idea of delivering on it or offsetting it at a later date.

Short Term Export Credit Guarantee Program (GSM-102) — One of the Commodity Credit Corporation (CCC) export credit guarantee programs. See GSM-102.

Short ton — Two thousand pounds. By contrast, a long ton equals 2,240 pounds. A metric ton (MT) equals 2,204.62 pounds.

Short — (1) The selling side of an open futures contract; (2) a trader whose net position in the futures market shows an excess of open sales over open purchases. See Long.

SIC — Standard industrial classification (see North American Industry Classification System). [www.census.gov/epcd/www/naics.html].

Side letter — See Sugar side letter.

Sign-up period — A USDA-prescribed time period, usually lasting several months, when farmers can enroll in a commodity program or other farm program.

Silage — Silage is the feedstuff produced by the fermentation of a crop, forage, or agricultural byproduct of generally greater than 50% moisture content. Ensiling is the name given to the process, and the container (if used) is called a silo. Producing annuals for silage is similar to producing them for commercial seed or grain except that crop inputs are directed to optimizing total dry matter yield. Loan deficiency payments (but not marketing assistance loans) are available for silage made from loan commodities under provisions of the 2002 farm bill (P.L. 107-171, Sec. 1205(a)(2)) (7 U.S.C. 7935).

Silt — Sedimentary materials composed of fine or intermediate-sized mineral particles. Erosion carries silt from cropland to streams and rivers, which can accumulate behind dams and diminish there effectiveness.

Silvicultural activities — Silviculture is the branch of forestry that deals with the development and care of forests. Silvicultural activities (such as tree planting, thinning, prescribed burning, and timber cutting) modify forest vegetation to achieve desired goals (e.g., timber production, good forest health, wildlife habitat).

Single-line service — Service by a single railroad between two locations.

SIP — Stewardship Incentives Program.

Site loans — Section 523 loans and Section 524 loans are for acquiring and developing land for low- and moderate-income rural residents.

Slotting fees — Fees paid by manufacturers to purchase shelf space in retail food stores. Such fees are a controversial issue in the food sector. Critics regard slotting fees as unearned store discounts (or even kickbacks to stores) that give a competitive edge to larger manufacturers who can afford them, while depriving consumers of variety, new product innovations, and possibly more competitive retail pricing. Supporters of the fees contend that they enable stores to make room for the thousands of new product introductions annually. Otherwise, grocers would have to shoulder the risk of stocking items that might not sell.

Small farm — Although there is no official, widely accepted definition of a small farm, the Small Farms Commission described it, for purposes of its 1997 study, as one with less than $250,000 in gross receipts annually on which day-to-day labor and management are provided by the farmer and/or the farm family that owns the production, or owns or leases the productive assets. In 2000, such farms accounted for about 90% of the more than 2.1 million U.S. farms, but only about 40% of U.S. farm production. The concentration of production on fewer and larger operations, is a longstanding concern among some segments of the agricultural community, while others view these changes as inevitable, and even necessary to maintain the efficiency and competitiveness of the sector. Farm typology analysis by ERS divides the small family farm category into five groups: limited-resource farms; retirement farms; residential/lifestyle farms; farming occupation/lower-sales, and farming occupation/high-sales. See Farm typology.

Small Farms, National Commission on — A 30-member panel appointed by the Secretary of Agriculture in 1997 to examine the status and needs of small farms in the United States. The Commission presented its findings, which included nearly 150 specific recommendations for action, to the Secretary in a January 1998 reported entitled A Time to Act.

Small Hog Operation Payment (SHOP) — A 1999 program, funded through USDA's Section 32 account and administered by the Department's Farm Service Agency, that provided approximately $178 million in direct payments to hog producers to compensate them for low hog prices in late 1998. Under the ad hoc program, producers were eligible for up to $10 per hog on the first 500 hogs marketed during the last 6 months of 1998. Operations marketing 2,500 or more hogs during that period, those with 1998 gross income greater than $2.5 million, and those with certain marketing contracts were not eligible. USDA had first announced the program, using its existing administrative authority, in January 1999, with payments set at $5 per head and total spending to be $50 million. Both the Administration and Congress (via an emergency farm relief package that was part of the FY2000 USDA appropriation, P.L. 106-78) later sanctioned the expanded payments.

Small Watershed Program — A program created under the Watershed Protection and Flood Prevention Act (P.L. 83-566), and 1 of 3 programs that are combined into the Watershed and Flood Prevention Operations Program. The Small Watershed Program is available in watersheds that are smaller than 250,000 acres. Currently, there are 515 active projects in this program. [www.ga.nrcs.usda.gov/programs/wffp.html].

Small Watershed Rehabilitation Program — The Grain Standards and Warehouse Improvement Act of 2000 (P.L. 106-472, Sec 313) authorized cost-sharing to rehabilitate aging structural measures that are part of water resources projects (including structures built under the watershed and flood prevention operations program area of the Natural Resources Conservation Service). The 2002 farm bill (P.L. 107-171, Sec. 2505) authorizes mandatory funding from the Commodity Credit Corporation (CCC) of $45 million in FY2003, rising by $5 million each year through FY2007, and appropriations of $45 million in FY2003, increasing by $10 million each year through FY2007.

Smart Growth — A catch-all term for public policies that selectively use financial and other public incentives to influence the density and pattern of new development. The goal is to encourage development in specified areas (typically where infrastructure, such as roads, schools and public utilities, are already in place) and discourage (not prohibit) it in other areas.

Smith-Lever Act of 1914 — P.L. 63-95 authorized and provided initial funding for states to establish an educational outreach arm to extend the results of research programs at the land grant colleges of agriculture and their affiliated state agricultural experiment stations to all citizens who might benefit from them (7 U.S.C. 341 *et seq.*). In 1962 Congress amended the act to establish a formula for distributing federal funds to states for agricultural extension programs. The formula provides for each state to receive what it received in 1962 as a base. Funds appropriated in excess of the 1962 level are allocated as follows: 4% of funds go to the federal component of the Cooperative Extension

System (now part of the Cooperative State Research, Education, and Extension Service (CSREES)); of the remainder, 20% is allocated to each state equally; and 80% is allocated on the basis of a state's share of the U.S. rural and farm populations. On average, Smith-Lever formula funds account for about 30% of a state's total funding for extension programs.

Smoot-Hawley — See Tariff Act of 1930 (P.L. 71-361).

SMP — Special Milk Program.

Snapback — The reimposition of an earlier and usually higher tariff. Various indicators (price movements, quantity movements) have been used as triggers for snapbacks. Rules governing the use of snapbacks are usually spelled out in trade agreements.

SNE — Society for Nutrition Education. [www.sne.org].

SOAP — Sunflowerseed Oil Assistance Program.

Socially disadvantaged farmers and ranchers — For the purposes of USDA outreach and assistance, these are defined under the 1990 farm bill (P.L. 101-624) as members of a socially disadvantaged group, which is defined to mean those whose identity in a group has subjected them to racial or ethnic prejudice without regard to their individual identity. For the purposes of loan eligibility, Section 355(e) of the Consolidated Farm and Rural Development Act, as amended by P.L. 102-554, Sec. 21(b), adds gender to the definition of a socially disadvantaged group. (7 U.S.C. 2003).

SOCMA — Synthetic Organic Chemicals Manufacturing Association. [www.socma.com].

Sodbuster — A program created by Title 12 of the Food Security Act of 1985 (P.L. 99-198) designed to discourage the plowing up of erosion-prone grasslands for use as cropland. If such highly erodible land is used for crop production without proper conservation measures as laid out in a conservation plan, a producer may lose eligibility to participate in farm programs.

Soil and Water Conservation Act (RCA) — A 1977 law (P.L. 95-192) that requires USDA to periodically prepare a national plan for soil and water conservation on private lands based on an inventory and appraisal of existing resource conditions and trends. Natural Resources Conservation Service (NRCS) is the lead agency in this effort, and completed appraisals in the early 1980s and late 1980s; a third appraisal was initiated in the early 1990s, but not completed. Only one national plan was adopted by USDA, in 1982. Many of the activities envisioned when the RCA was enacted are being carried out, but it has not resulted in a single omnibus plan addressing conservation needs and priorities on private lands.

Soil and Water Conservation District (SWDC) — See Conservation District.

Soil Bank Act — Title I of the Agricultural Act of 1956 (P.L. 84-540), designated the Soil Bank Act, created the Acreage Reserve Program to retire land producing basic commodities under an annual agreement from 1956 through 1959, and the Conservation Reserve Program, to retire agricultural land under contracts of 3, 5, or 10 years. The Soil Bank Act was repealed by the Food and Agriculture Act of 1965 (P.L. 89-321, Sec. 601). Nevertheless, this early conservation reserve served as the model for the current Conservation Reserve Program (CRP), which was enacted in the Food Security Act of 1985 (P.L. 99-198, Sec. 1231-1236).

Soil Bank Program — A federal program (authorized by the Soil Bank Act, P.L. 84-540, Title I) of the late 1950s and early 1960s that paid farmers to retire land from production for 10 years. The predecessor to today's Conservation Reserve Program. The maximum enrollment was 28.7 million acres in 1960. Some elements in the CRP, such as a limit on CRP acres per county, were a response to the Soil Bank experience.

Soil conditioner — An organic material like humus or compost that helps soil absorb water, build a bacterial community, and take up mineral nutrients.

Soil conservation district — A legal subdivision of state government, with a locally elected governing body, responsible for developing and carrying out a program of soil and water conservation within a geographic boundary, usually coinciding with county lines. The nearly 3,000 districts in the United States have varying names (soil conservation districts, soil and water conservation districts, natural resources districts, resource districts, resource conservation districts).

Soil Conservation and Domestic Allotment Act of 1936 — P.L. 74-46 was designed to support farm income by making soil-conservation and soil-building payments to participating farmers. This design replaced the taxes on processors in the support program authorized by the Agricultural Adjustment Act of 1933, which the courts judged to be unconstitutional. The 1936 Act supported farm income and reduced surpluses by paying farmers for shifting from crops in excess supply (soil depleting crops) to soil building crops like legumes and grasses. This law, as amended, continues to serve as the enabling authority for a number of activities and programs carried out by the Natural Resources Conservation Service.

Soil Conservation Service (SCS) — Replaced by a new USDA agency, the Natural Resources Conservation Service after USDA reorganization in 1994. Responsibilities include carrying out technical assistance programs in cooperation with soil conservation districts to improve and conserve soil and water resources, and operating related programs such as the national soil survey and the natural resources inventory.

Soil loss tolerance (T value) — For a specific soil, the maximum average annual soil loss expressed as tons per acre per year that will permit current production levels to be maintained economically and indefinitely. T values range from 2 to 5 tons per acre per year. According to the 1997 national resources inventory, about 77 million acres of cropland is eroding at 2T or greater, and more than 27 million acres are eroding at more than 8T.

Soil moisture zone — Depth of soil from which plant roots extract water.

Soil quality / health — Soil quality includes consideration of measures related to both productivity for crops and environmental factors.

Soil series — A group of soils having horizons (or layers) similar in characteristics and arrangement in the soil profile, except for the texture of the surface portion. They are given proper names from place names within the areas where they occur. Thus, Norfolk, Miami, and Houston are names of some well-known soil series.

Soil solarization — Fumigating and warming soil by covering it with black plastic. This is an alternative pest control technique being investigated as an alternative to the use of methyl bromide (a chemical fumigant used in Florida tomato production and for other speciality crops, which will be phased out of use because of its ozone depleting effects).

Soil sterilant — A chemical that temporarily or permanently prevents the growth of all plants and animals, depending on the chemical. Soil sterilants must be registered as pesticides.

Soil Survey Program — A program of the Natural Resources Conservation Service to inventory soil resources as a basis for determining land capabilities and conservation treatments that are needed, provide soil information to the public (primarily through maps), and provide technical support to those who use soils information. More than 90% of the private lands have been mapped. In FY2001, maps were prepared or updated on 21.9 million acres, including 25 counties or areas for the first time.

Soil — About 70,000 kinds of soil are recognized in the nationwide system of classification. Each has a unique set of characteristics and a potential for use. These characteristics are important in designing a conservation plan to protect the soil from erosion if it is being cultivated. The Natural Resources Conservation Service is responsible for mapping the United States by soil type, through the Soil Survey Program.

Sole source bid — This refers to the required bidding process for infant formula sold through the WIC program. It offers an infant formula manufacturer the option of bidding to be the only provider of infant formula contained in WIC food packages in the state. Contracts must be awarded to the company that offers the lowest price (or largest discount) on its infant formula.

Solid waste — Non-liquid, non-soluble materials ranging from municipal garbage to industrial wastes that contain complex and sometimes hazardous substances. Solid wastes also include sewage sludge, agricultural refuse, demolition wastes, and mining residues. Technically, solid waste also refers to liquids and gases in containers. The disposal of solid waste is regulated by the EPA under the Resource Conservation and Recovery Act (P.L. 94-580, as amended).

Solid Waste Disposal Act — See Resource Conservation and Recovery Act. (42 U.S.C. 6901 *et seq.*).

SOPs — Standard operating procedures.

Sound science — A phrase generally used in a political context to rhetorically challenge the validity of the other side's arguments. The phrase has arisen in agricultural trade disputes when a country imposes a sanitary or phytosanitary (SPS) measure that an exporting group claims is an unfair trade barrier because it is not based on sound science. Often, policymakers or adjudicators are asked to make decisions based, at least in part, on which side's scientific arguments appear to be the most convincing. However, the phrase sound science is not included in public laws or international treaties. Under the Uruguay Round's Sanitary or phytosanitary (SPS) agreement, countries' SPS measures must be based on scientific principles, and may not be maintained without scientific evidence. However, the agreement does not define scientific. Under general principles of international law, the interpretation of the term is left to good faith and ordinary dictionary definitions. By requiring only that measures be based on scientific principles, as broadly understood and accepted by the scientific community, the agreement, therefore, does not expect those who are adjudicating trade disputes to choose which science is the sound science.

Soup Kitchen-Food Bank Program — Originally authorized under the Hunger Prevention Act of 1988 (P.L. 100-435) to buy commodities for soup kitchens and food banks not participating in the Emergency Food Assistance Program (EFAP). This program was consolidated with EFAP by an amendment to the Emergency Food Assistance Act of 1983 (P.L. 98-92) that was enacted as part of the 1996 welfare reform law (Personal Responsibility and Work Opportunity Reconciliation Act of 1996 (P.L. 104-193)). Program authority was extended through FY2007 by the 2002 farm bill (P.L. 107-171, Sec. 4126).

Southern Dairy Compact — An agreement among southern states, modeled on the Northeast Interstate Dairy Compact, that would allow member states to jointly establish a minimum farm price for fluid milk that is above the federally mandated minimum price level in the region. Many southern state legislatures approved state membership in the compact. However, the required congressional authority to form a compact was never granted.

Soybean Rust — Asian soybean rust (ASR) is a harmful fungal disease that can infect and reproduce on over 90 plant species, most notably soybeans. Wind-borne spores can spread rapidly and the disease can reduce soybean yields by 10-80% in infected areas. The pathogen has been most destructive in Asia, Africa, Australia, and recently (2002) South America. ASR is not known to occur in the continental United States; however, its airborne arrival by upper atmosphere wind currents is expected within the next few years. Once established, eradication is unlikely because the pathogen can infect and reproduce on so many plant species. As a result, the most effective treatment is thought to be the development and use of resistant varieties. However, no commercial soybean cultivar is resistant to or tolerant of ASR. In the short term, the only effective responses are costly fungicides and the use of early maturing soybean cultivars. As of February 2004, only two fungicides are presently registered with the Environmental Protection Agency (EPA) for control of rust on soybeans. However, independent efficacy data are still needed to confirm their effectiveness against ASR. Meanwhile some states are trying to obtain EPA Section 18 registrations (granting temporary emergency use rights) for fungicides presently in use overseas against ASR. USDA is coordinating a plan to deal with ASR that encompasses various USDA agencies, state land-grant universities, and industry participants.

Soybeans — Soybeans are considered a covered commodity and federal support is available under authority of the 2002 farm bill (P.L. 107-171, Title I) in the form of marketing assistance loan benefits, direct payments, and counter-cyclical payments. Soybeans are the largest oilseed crop in the world and the United States is the world's leading producer. Farmers harvested about 72 million acres of soybeans (slightly more than the 71 million acre corn harvest) in 2003. The six leading states account for 67% of national production (according to 2003 crop data): Illinois (15%), Iowa (14%), Minnesota (9%), Indiana (8%), Nebraska (7%), Ohio (7%), Missouri (6%). Most soybeans are crushed into oil that is refined for edible consumption and meal that is used for high protein poultry and hog feed. Comparatively small portions of the crop go to other food and industrial uses. About 35% of the crop is exported as beans. Familiar soy (or soya) foods include salad oil and margarine, tofu, soy sauce, and soymilk. Typically, in the United States, soybeans are measured and priced in bushels (1 bushel = 60 pounds).

Special Agricultural Safeguard (SSG) — A provision in the Uruguay Round Agreement on Agriculture. The SSG allows Member countries to impose additional tariffs on agricultural products if their import volume exceeds defined trigger levels, or if prices fall below specified trigger levels. Its purpose is to prevent disruption of domestic markets due to import surges or abnormally low import prices. The SSG applies only to products that are (1) subject to tarrification and (2) in cases where the country has designated a product as eligible for the SSG in its Schedule of Commitments. It can apply only to imports that exceed tariff-rate quota volumes. The SSG is an alternative to the general safeguard provision of the GATT and is easier to invoke because it does not require a test of injury or threat of injury. In the Doha Development Agenda (DDA) negotiations, the United States, the Cairns Group, and many developing countries have proposed elimination of the SSG.

Special and differential treatment (S&D) — A set of GATT provisions (GATT 1947, Article XVIII) that exempts developing countries from the same strict trade rules and disciplines of more industrialized countries. In the Uruguay Round Agreement on Agriculture, for example, developing countries are given longer time periods to phase in export subsidy and tariff reductions than the more industrialized countries. The least developed countries are exempt from any reduction commitments.

Special Milk Program — Offers federal reimbursements for each half-pint of milk served to a child in a participating outlet, which generally is any school or facility caring for children that does not participate in other federally subsidized meal programs. There is an exception from this limitation for kindergarten children in split session programs. The program is permanently authorized under the Child Nutrition Act of 1966 (P.L. 89-642, as amended; 42 U.S.C. 1771 *et seq.*). Schools may offer free milk to children meeting free lunch income requirements, if they choose, and this milk is reimbursed at full cost. Otherwise, children buy so-called paid milk, which is subsidized at a legislatively set rate for each half-pint served. This program is administered by the Food and Nutrition Service, and funded by annual agricultural appropriations.

Special review — Formerly known as Rebuttable Presumption Against Registration (RPAR), this is a regulatory process through which existing pesticides suspected of posing unreasonable risks to human health, non-target organisms, or the environment are referred for review by the EPA. Such review requires an intensive risk/benefit analysis with opportunity for public comment. If risk is found to outweigh social and economic benefits, regulatory actions (ranging from label revisions and use-restriction to cancellation or suspended registration) can be initiated.

Special Supplemental Nutrition Program for Women, Infants and Children (WIC) — WIC provides federal grant funds to state health agencies and recognized Indian tribal organizations to operate programs for low-income pregnant and postpartum mothers, infants, and children (under age 5) who are at risk due to inadequate income and diet. The program offers monthly food packages consisting of specifically authorized supplemental foods containing nutrients often found lacking in the diets of WIC participants, and is operated at the local level by public health and other agencies with a link to health care providers. WIC foods include infant formula and cereals, milk, cheese, eggs, breakfast cereals, fruit and vegetable juices, dry beans and peas, and peanut butter. Participants must meet both a low-income and nutritional risk test to be eligible. Foods are provided either directly by a local WIC clinic, or more commonly, in the form of WIC vouchers issued to participants that list specific types and amounts of foods that may be redeemed by the participant at authorized grocery stores. The program is authorized (usually for 3 to 5 years) under Section 17 of the Child Nutrition Act of 1966 (P.L. 89-642, as amended) and is funded as a discretionary program by annual agriculture appropriations acts.

Specialty crops — A crop category that is not precisely defined. The USDA's Economic Research Service classifies specialty crops as all farm commodities other than livestock and field crops (wheat, corn, rice, cotton). This definition would include all fruits and vegetables and horticultural crops. However, that same agency sometimes defines specialty crops as those not generally considered to be mainstream. Examples include minor vegetables such as okra, chile peppers, pumpkins; tropical vegetables; and tropical fruits such as carambola (star fruit).

Specialty produce — Within the produce industry this term refers to fruits and vegetables that reflect ethnic influences; that have international origins; that have limited market share; and that have limited appeal and recognition. Specialty produce is often referred to as exotic or new.

Species — Species represent the lowest and most important of the primary groupings used in classifying plants, animals, and microorganisms. While no single definition applies to all organisms, biologists rely principally on (1) morphological and genetic similarities and (2), for sexually reproducing organisms, the capability of interbreeding with one another but not other groups. If different species do interbreed, the offspring, if any, are often sterile. Biologists give species unique, binomial names: a generic name that includes closely related species, and a species-specific name. The horse, for example, is *Equus caballus;* the donkey or ass is *Equus asinus*. (Their offspring, the mule, is sterile.) As populations of organisms vary geographically and change over time (becoming extinct, or splitting or evolving into new species), species classifications are neither absolute nor immutable; where some biologists see variations within a species (and may designate subspecies), others may see separate species. About 1.5 to 2 million species have been named, but scientists estimate the total number of species could be 5 to 100 million, many of them probably undiscovered microorganisms. The Endangered Species Act (ESA) protects species designated as endangered or threatened with extinction; these protections prohibit taking endangered species and can include restrictions on habitat alterations, such as logging or water pollution (16 U.S.C. 1531 *et seq*.). Because of the way species is defined in the ESA, policy debates have arisen over whether certain groups of organisms qualify for listing (e.g., northern goshawks and the Alexander Archipelago wolf).

Specific rate duty — A tariff levied on imports, defined in terms of a specific amount per unit, such as cents per kilogram. By contrast, an ad valorem duty is a charge levied on imports defined in terms of a fixed percentage of value.

Specified risk materials (SRMs) — A term that came into being as part of USDA's regulatory response to the first confirmed U.S. BSE case in December 2003, signifying parts of cattle carcasses that cannot be inspected and passed for human food because scientists have determined that BSE-causing prions concentrate there. Under the new regulations (69 FR 1862, January 12, 2004), SRMs are: the brain, skull, eyes, trigeminal ganglia, spinal cord, vertebral column (with some exclusions), dorsal root ganglia (DRG) of cattle 30 months of age and older, and the tonsils and distal ileum of the small intestine of all cattle.

Speculator — In commodity trading, an individual who does not hedge, but who trades in futures contracts with the objective of achieving profits through the successful anticipation of price movements. Speculators serve the useful economic function of sharing the financial risks of price instability.

Spot commodity — The actual commodity as distinguished from a futures contract. Sometimes used to refer to cash commodities available for immediate delivery.

Spot market — A public or open marketplace (such as open exchanges or auction houses) where products (including agricultural products such as livestock, grain, cotton, etc.) are bought and sold. The Minneapolis Grain Exchange and the now defunct National Cheese Exchange are examples. Spot also refers to a maturing delivery month of a futures contract.

Spot price — The price at which a physical commodity for immediate delivery is selling at a given time and place. See Cash price.

Spread (or straddle) — The purchase of a futures contract of one delivery month against the sale of another futures delivery month of the same commodity; the purchase of one delivery month of one commodity against the sale of that same delivery month of a different commodity; or the purchase of one commodity in one market against the sale of the commodity in another market, to take advantage of a profit from a change in price relationships. The term spread is also used to refer to the difference between the price of a futures month and the price of another month of the same commodity. A spread can also apply to options contracts.

Sprout damage — Undesirable germination of wheat kernels that occurs on mature, unharvested wheat when wet field conditions persist just prior to and during the harvest. Mature wheat that has been cut and left laying in the field prior to threshing is particularly vulnerable to sprout damage. Early cold weather in Canada often forces wheat producers there to cut and windrow their crop to allow for drying. Wet conditions can then cause widespread sprout damage. If has occurred, there is a dramatic increase of the enzyme alpha-amylase. The falling numbers test is a measure of the presence of this enzyme. A high falling number indicates that the wheat is sound and satisfactory for most baking processes. A low falling number indicates that harmful sprouting has occurred and is suggestive of reduced baking quality. In bread, too much alpha-amylase activity will cause wet sticky bread crumb with large voids in the loaf and too little causes dry crumble bread crumb and high loaf density.

SPS — Sanitary and phytosanitary.

SRM — Society for Range Management. [www.rangelands.org/ScriptContent/Index.cfm]. Also, see Specified risk materials.

SSOPs — Sanitation standard operating procedures.

SSSA — Soil Science Society of America. [www.soils.org].

Stabilization funds — In Canada, stabilization funds are the financing mechanism for commodity-specific or multi-commodity support programs. Revenues may come from a combination of producer, federal, and in some cases provincial premiums. If a stabilization fund runs a deficit, the federal Ministry of Finance may lend it money at market interest rates.

Stabilization payments — Budgetary payments made to compensate Canadian farmers for falling farm prices and/or incomes. Stabilization programs include insurance or safety nets or underwriting schemes intended to compensate farmers for decreases in price, income or cash flow due to disturbances to yields (from drought, for example) or instability of input and commodity markets.

Staged tariff reductions — See Baskets.

Standard Reinsurance Agreement (SRA) — USDA allows private insurance companies that participate in the federal crop insurance program to transfer a portion of their risk to the federal government. The standard reinsurance agreement establishes the terms and conditions under which the federal government will provide subsidies and reinsurance on eligible crop insurance contracts sold or reinsured by the insurance company named on the agreement.

Standards of identity for food — Mandatory, federally-set requirements that determine what a food product must contain in order to be marketed under a certain name in interstate commerce. Mandatory standards (which differ from voluntary grades and standards applied to agricultural commodities) protect the consumer by ensuring that a label accurately reflects what's inside (for example, that mayonnaise is not an imitation spread, or that ice cream is not a similar, but different, frozen dessert).

Staple — A commodity that is widely and regularly produced and consumed (i.e., wheat, rice, potatoes). A term used to designated the length of fiber in cotton and wool.

StarLink — StarLink™ is a corn variety that has been genetically modified to contain an insecticidal protein derived from a naturally occurring bacterium (*Bacillus thuringiensis*, or Bt). The Environmental Protection Agency approved this gene-spliced variety of yellow corn in 1998 for use only as animal feed and set a zero-tolerance level for its use in human food based on the fact that this particular Bt protein does not break down easily in the human digestive system, is heat resistant, and could prove allergenic. Nevertheless, StarLink corn was detected in taco shells in mid-September 2000. The occurrence became the first U.S. test case of contamination of the food supply by a genetically-modified organism, raising questions about the effectiveness of U.S. biotechnology regulatory policies.

State Administrative Expenses (SAE) — Authorized under Section 7 of the Child Nutrition Act (P.L. 89-642, as amended), this program provides funding to help states meet the administrative costs of operating school lunch, breakfast, and child care food programs and related commodity donation programs authorized under the Child Nutrition Act and the National School Lunch Act (P.L. 79-396).

State agricultural experiment stations (SAES) — The Hatch Act of 1887 authorized the establishment of an agricultural experiment station, to be affiliated with the land grant college of agriculture, in each state (7 U.S.C. 361a *et seq.*). Research done at these stations underpins the curriculum of the colleges, as well as the programs of the Cooperative Extension System.

State and Private Forestry — A branch of the Forest Service providing technical and financial assistance to states and to private landowners for forest management, for fire protection, and for forest health.

State Fire Assistance — A State & Private Forestry program under which the Forest Service provides financial and technical assistance to states for grants and agreements to improve fire protection of non-federal lands.

State inspection programs — Often refers to the state-run meat and poultry inspection programs to which USDA contributes 50% of the cost. State programs (fewer than half the states have them) must be certified by USDA to be at least equal to federal inspection requirements. However, products from state-inspected plants (most of them are relatively smaller operations) cannot be sold outside of the state. Small plants and many state officials have endorsed bills in Congress that would permit state-inspected products to be sold into interstate and foreign commerce, but large meat and poultry companies (which must comply with federal inspection regulations) generally oppose such a change.

State rural development block grant — Under provisions of the Rural Community Advancement Program (RCAP), each state may receive, for direct administration, up to

10% of the funds allocated to the state. These funds may be used to establish a state administered block grant. The first 5% of the state block grant allocated does not require the community to make a matching fund contribution. A state may receive the additional 5% if it provides $2 in matching funds for every $1 in RCAP funds it would receive.

State rural development councils — A collaborative partnership comprised of representatives of the federal, state, local, and tribal governments, the private sector, and the nonprofit sector. Councils are created by a memorandum of understanding between USDA and the state Governor. The councils' purpose is to promote rural development within the state.

State technical committee — Advisory groups to state conservationists (coordinators of all Natural Resources Conservation Service activities within a state) created in the 1990 farm bill (P.L. 101-624) and amended in the 1996 farm bill (P.L. 104-127). These groups can include representatives from agencies, agriculture, agribusiness, and non-profits, as well as individuals with a demonstrated expertise. Responsibilities assigned by the 1996 farm bill include establishing procedures for evaluating petitions on new conservation practices and identifying priority areas for the Environmental Quality Incentive Program (EQIP) and Wetland Reserve Program (WRP).

State trading enterprises (STEs) — STEs are enterprises authorized to engage in trade (exporting and/or importing) that are owned, sanctioned, or otherwise supported by government. STEs are legitimate trading entities and are subject to GATT rules. Examples include the Canadian Wheat Board, the Australian Wheat Board, and the Fonterra Cooperative Group in New Zealand. Some U.S. agricultural producers think, however, that STEs through their exercise of monopoly power and government support may distort trade in their respective commodities.

STB — Surface Transportation Board. [www.stb.dot.gov].

STC — State technical committee; state conservationist.

STE — State trading enterprise.

Steagall Amendment of 1941 — P.L. 77-144 required price support for many nonbasic commodities at 85% of parity or higher. In 1942, the minimum rate was increased to 90% of parity and was required to be continued for 2 years after the end of World War II. The Steagall commodities included hogs, eggs, chickens (with certain exceptions), turkeys, milk, butterfat, certain dry peas, certain dry edible beans, soybeans, flaxseed and peanuts for oil, American-Egyptian (ELS) cotton, potatoes, and sweet potatoes.

Step 2 payments — One of the three cotton competitiveness provisions intended to keep U.S. cotton competitive in domestic and export markets (initially authorized by the 1990 farm bill (P.L. 101-624). Under the Step 2 provision, USDA is required to issue marketing certificates (or cash payments in lieu of certificates) to domestic users of upland cotton for documented purchases, and to exporters of upland cotton for documented sales, when certain U.S. cotton pricing benchmarks are exceeded. The payments provide a subsidy to U.S. cotton users and exporters so that U.S. rather than foreign cotton will be utilized, even when U.S. cotton is higher-priced. The 1996 farm bill (P.L. 104-127) and the 2002 farm bill (P.L. 101-171, Sec. 1208) continued the Step 2 provisions.

Stewardship, end-results contracting — an approach to national forest management wherein timber sales can be expanded to require purchasers to perform additional silvicultural activities. Such goods-for-services contracts have been authorized by

Congress on a limited basis in several of the annual Department of the Interior and Related Agencies Appropriations Acts. (The Forest Service has been funded in these Acts since the 1950s.).

Stewardship Incentives Program (SIP) — A program administered by the Forest Service through the Farm Service Agency that provided up to 75% cost sharing for silvicultural activities implementing approved renewable resource plans. The program was terminated in the 2002 farm bill (P.L. 107-171), and replaced with the Forest Land Enhancement Program (FLEP).

Stewart B. McKinney Homeless Assistance Act — Enacted in response to concerns about increasing hunger, homelessness, and unemployment, this Act (P.L. 100-77, July 22, 1987) authorized federal funding to support housing, food assistance, and job training.

Stocker cattle — Calves or older animals maintained, often on pasture or rangeland, to increase weight and maturity before being placed in a feedlot.

Stockholm Convention Persistent Organic Pollutants — The 2001 Stockholm Convention on Persistent Organic Pollutants is an international agreement negotiated under the auspices of the United Nations "to protect human health and the environment from persistent organic pollutants(POPs) worldwide. Although the United States has signed the treaty, prior to U.S. ratification, the Senate must give its advice and consent, and Congress must enact enabling (also known as implementing) legislation. Legislation is needed to resolve inconsistencies between provisions of the Stockholm Convention, the Rotterdam Convention on the Prior Informed Consent Procedure for Certain Hazardous Chemicals and Pesticides in International Trade (PIC Convention), and the Aarhus Protocol on Persistent Organic Pollutants, and existing U.S. laws (specifically the Toxic Substances Control Act (TSCA)and the Federal Insecticide, Fungicide, and Rodenticide Act (FIFRA)). [www.pops.int].

Stocking rate — The number of specific kinds and classes of livestock grazing or using a unit of land for a specified time. Not the same as carrying capacity.

Stocks, grain — Commercial grain stocks include domestic grain in storage in public and private elevators at important markets and grain afloat in vessels or barges in lake and seaboard ports. Commercial stocks plus government-owned stocks constitute total stocks. Information on grain stock levels is half of the supply demand equation that determines price levels.

Stocks-to-use ratio (S/U) — A convenient measure of supply and demand interrelationships of commodities. This ratio indicates the level of carryover stock for any given commodity as a percentage of the total use of the commodity.

Strategic grain reserve — National grain stocks held in reserve intentionally by government programs for the purpose of meeting future domestic and international needs. See Food Security Commodity Reserve, Food Security Wheat Reserve, Farmer-Owned Grain Reserve, Bill Emerson Humanitarian Trust.

Strike price (exercise or contract price) — The price, specified in the option contract, at which the underlying futures contract or commodity will move from seller to buyer.

Strip tillage — Planting and tillage operations that are limited to a strip not to exceed one-third of the distance between rows. The area between is left untilled with a protective cover of crop residue on the surface for erosion control.

Stripcropping — Growing crops in a systematic arrangement of strips or bands, usually parallel to the land's contour, that serve as barriers to wind and water erosion.

Structural funds — Funds intended to facilitate structural adjustment of some sectors and/or some specific regions in the EU. They include the European Regional Development Fund (ERF), the European Social Fund (ESF), the Guidance Section of European Agricultural Guidance and Guarantee Fund (EAGGF) and the Financial Instrument for Fisheries Guidance (FIFG). Assistance is concentrated on six priority objectives that are implemented through programs proposed by the appropriate authorities of the member states (National Initiative programs). The EC also co-finances other programs on subjects proposed by the EC (Community Initiatives). Rural areas mainly benefit from the EAGGF Guidance Section.

Structural practices — Conservation practices that involve either construction of a facility, such as a structure to handle animal waste, or modification of the land surface, such as building contours, to prevent degradation of soil, water, or related resources.

Structure of agriculture — The characteristics of the farm sector; most often used in describing changes in the number, size distribution, production traits, and business composition (i.e., type of ownership) of farms and agribusiness firms. Structural change is an agricultural policy issue because of concerns over its economic and social impacts on farmers, rural communities, consumers, and others.

Stubble-mulching — Leaving the stubble or crop residue essentially in place on the land as a surface cover during a fallow period. Stubble-mulching can prevent erosion from wind or water and conserve soil moisture.

Stumpage price (or stumpage rate) — The agreed-upon price, usually in dollars per thousand board feet (mbf), between a private timber purchaser and the federal agency for the right to cut and remove trees and/or logs from the federal lands.

Stunning — The practice of rendering livestock unconscious before slaughter, in keeping with the provisions of the Humane Methods of Slaughter Act (7 U.S.C. 1901), which the Food Safety and Inspection Service (FSIS) is mandated to uphold under the Federal Meat Inspection Act (21 U.S.C. 603 (b)). No similar provision exists in the Poultry Products Inspection Act (21 U.S.C. 451 *et seq.*). After confirmation of the first U.S. BSE case, FSIS issued regulations (69 FR 1887, January 12, 2004) prohibiting the use of the most widely used stunning device (air-injection captive bolt stun gun) because the compressed air (in contrast to the blank cartridge-driven or non-penetrating captive bolt) has been shown to force pieces of brain and other central nervous system (CNS) tissue into the bloodstream. Cattle blood is processed primarily for use as a protein supplement in animal feeds and milk replacer for calves, and could potentially transmit BSE if it contained specified risk materials (SRMs include brain and CNS tissue).

SUA — Sweetener Users Association.

Subscription farming — See Community supported agriculture.

Subsidy — A direct or indirect benefit granted by a government for the production or distribution (including export) of a good or to supplement other services. Generally, subsidies are thought to be production and trade distorting, resulting in an inefficient use of resources. Arguably, subsidies may be justified on grounds that they adjust for nonmarket considerations that are as important as market values. The World Trade Organization Agreement on Subsidies and Countervailing Measures categorized subsidies as prohibited (red light), actionable (amber light), and nonactionable (green light). Actionable subsidies are subject to countervailing action. The term subsidy also is used to refer to federal reimbursements for meals served through child and elderly nutrition programs.

Subsistence farm — A low-income farm where the operator is producing primarily for the family's needs rather than for sale.

Substantial equivalence — A concept, developed by OECD in 1991, that maintains that a novel food, for example, one that derives from genetic modification or engineering, should be considered the same as and as safe as a conventional food if it demonstrates the same characteristics and composition as the conventional food. Substantial equivalence is important from a regulatory point of view. If a novel food is substantially equivalent to its conventional counterpart, then it could be covered by the same regulatory framework as a conventional food.

Sucrose ethanol — Ethanol derived from sugarcane or sugar beets. Most ethanol manufactured in the United States uses corn as feedstock. Brazil following the oil crisis of the 1970s adopted a policy to produce ethanol from sugarcane as an alternative energy source to cover a larger portion of its fuel needs. More than half of Brazil's sugarcane crop is processed into ethanol each year.

Sugar program — The federal commodity support program that maintains a minimum price for sugar, authorized by the 2002 farm bill (P.L. 107-171, Sec. 1401-1403) to cover the 2002-2007 crops of sugar beets and sugarcane. Designed to protect the incomes of growers of sugarcane and sugar beets, and firms that process each crop into sugar, the program supports domestic sugar prices by: (1) making available nonrecourse loans to processors (not less than 18¢/lb. for raw cane sugar, or 22.9¢/lb. for refined beet sugar); (2) restricting sugar imports using an tariff rate quota, and (3) limiting the amount of sugar that processors can sell domestically (under marketing allotments) when imports are below 1.532 million short tons. Import restrictions are intended to meet U.S. commitments under the North American Free Trade Agreement (NAFTA) and Uruguay Round Agreement on Agriculture. Processor and refiner marketing allotments are set by USDA according to statutory requirements. Marketing allotments and new payment-in-kind authority are designed to help the USDA meet the no-cost-requirement to the federal government by avoiding the forfeiture of sugar put under loan. Other parts of the new program include a storage loan program for sugar processors, and reduced (by 1%) the USDA interest rate charged on sugar loans.

Sugar re-export programs — Administered by USDA, program regulations allow cane refiners and food manufacturers, subject to certain conditions, to import sugar exempt from tariff-rate quota provisions that apply to sugar imported for consumption in the U.S. market. Cane refiners process the imported raw sugar into refined sugar for re-export, or for transfer to food manufacturers for use in sugar-containing products for export. These programs were designed in the early 1980s to utilize excess cane refining capacity and to make U.S. refined sugar and sugar-containing products more competitive on the world market by allowing participating companies to have access to lower world-priced sugar.

Sugar side letter — A sugar side letter was added to the North American Free Trade Agreement (NAFTA) in last minute negotiations between the Clinton Administration and the Mexican Government before the Congress approved the North American Free Trade Agreement Implementation Act (P.L. 103-182). It altered NAFTA's initial sugar provisions by adding one additional factor to the formula to be used to determine how much sugar Mexico could export to the United States through 2008. Mexican access to the U.S. market was initially set to be equal to the amount of its net sugar surplus (sugar production *minus* sugar consumption), subject to a maximum of 25,000 metric tons over the 1995-2001 period and a maximum of 250,000 metric tons in 2001-2008. The side letter changed this definition to add Mexican consumption of high-fructose corn syrup (HFCS) to the "net production surplus" definition. This change effectively lowers the amount of sugar that Mexico can sell to the U.S. market.

Summer fallow — Cropland usually in semi-arid regions that is purposely kept out of production during a cropping season mainly to conserve moisture for the next season; sometimes called fallow cropland. It may be tilled or sprayed to control weeds and conserve moisture in the soil. The 1997 Census of Agriculture reported that 20.9 million acres, almost 5% of the 431 million acres of all cropland, was fallow that year.

Summer Food Service Program (SFSP) — The SFSP operates in low-income areas during the summer months to provide meals and snacks to children in summer day camps. Federal support consists of guaranteed cash and commodity assistance and support for administration and operating expenses. It is authorized under the National School Lunch Act (P.L. 79-396, as amended), administered by the Food and Nutrition Service, and funded by annual agriculture appropriations.

Sunflower Oil Assistance Program (SOAP) — Along with the Cottonseed Oil Assistance Program (COAP), SOAP was one of two programs that awarded bonuses to exporters of U.S. vegetable oil to assist with exports to targeted markets. SOAP was authorized beginning in FY1988 with funds made available under Section 32 of the Agricultural Adjustment Act Amendment of 1935 (P.L. 74-320) The provision in the Disaster Assistance Act of 1988 (P.L. 100-387), which had authorized SOAP, expired at the end of FY1995 and was not extended in the 1996 farm bill (P.L. 104-127). However, the USDA appropriations act for FY1996 (P.L. 104-37) provided authority to operate the program in FY1996. Export subsidies for sunflower oil can be financed under the Export Enhancement Program (EEP).

Super 301 — Section 310 of the 1974 Trade Act (P.L. 93-618, as amended) is commonly referred to as Super 301. As enacted, Super 301 required the USTR for 1989 and 1990 to issue a report on its trade priorities and to identify priority foreign countries that practiced unfair trade and priority practices that had the greatest effect on restricting U.S. exports. The USTR then would initiate a Section 301 investigation against the priority countries to obtain elimination of the practices that impeded U.S. exports, in the expectation that doing so would substantially expand U.S. exports. The original Super 301 provisions expired in 1990, however, President Clinton issued an executive order (EO 12901) reactivating Super 301 for two years (1994 and 1995). The Super 301 process was again extended through 1997 by EO 12973 (September 1995), but was not in operation in 1998. On March 31, 1999, Super 301 again was re-instated and revised by EO 13116. It required the USTR by April 30 to issue its Super 301 report on priority foreign trade practices and to initiate section 301 cases against such practices if agreement is not reached after 90 days. Neither the USTR's April 1999 or April 2000 Super 301 report identified any priority foreign trade practices under Super 301, but USTR did announce that it would initiate Section 301 cases against trade practices in several countries. In its April 2001 Super 301 report, USTR did not make any designations under Super 301, but did announce that consultations (the first stage in WTO dispute settlement) had been requested with Mexico on measures affecting live swine imports, with Belgium on rice import restrictions, and with the European Union on import surcharges on corn gluten feed. In a January 2002 letter report to the Senate Finance Committee on activities under Section 301, the USTR did not identify any priority foreign trade practices under Super 301, although it did report on other activities undertaken under Section 301-310 of the Trade Act of 1974.

Super Sodbuster — An amendment to sodbuster passed in the 1990 farm bill (P.L. 101-624) and repealed in the 1996 farm bill (P.L. 104-127) that made producers ineligible for specified farm program benefits on all their land if they cultivated highly erodible land that was idle.

Superfund — The principal federal program for cleaning up the nation's worst hazardous waste sites, it was created by the Comprehensive Environmental Response, Compensation, and Liability Act (CERCLA) of 1980 (P.L. 96-510, as amended; 42 U.S.C. 9601 *et seq.*). The normal application of fertilizer is explicitly excluded from liability under CERCLA.

Supima — An abbreviation for superior Pima and the name of the non-profit promotional organization of the American Pima cotton growers. [www.supimacotton.org/index.cfm].

Supplement — For child nutrition programs, this refers to federally reimbursed snacks that are served to children in participating facilities. Also refers to the addition of nutrients to the diet by the use of vitamin and mineral supplements.

Supplier Credit Guarantee Program (SCGP) — An Export Credit Guarantee Program that covers credit terms up to 180 days. The Commodity Credit Corporation (CCC) guarantees a portion (up to 65%) of payments due from foreign buyers under short-term financing that the exporters have extended directly to the buyers. These direct credits must be secured by dollar-denominated promissory notes signed by the importers. If an importer fails to make any payment as agreed, the exporter or assignee (e.g., U.S. financial institution) must submit a notice of default. The CCC pays any filed claim for loss, if found to be in good order. The SCGP targets specific U.S. agricultural products, with an emphasis on high-value products and market growth potential.

Supplier credits — A feature of the export credit guarantee programs that permits the Commodity Credit Corporation (CCC) to issue guarantees for repayment of credit that is made available by a U.S. exporter to a foreign buyer. The credits have a term of up to 180 days (or, if appropriations are made available, up to 360 days).

Supply control programs — Any of several government programs to influence the supply of farm products on the market. Some, such as acreage allotments and marketing quotas, are considered mandatory, in that farmers who produce or market in excess of assigned levels can be legally penalized. Others, such as cropland set-asides, acreage reductions and diversions, and farmer-held grain reserves, are considered voluntary, in that farmers are usually encouraged to participate through financial incentives. Supply control programs are not being used in conjunction with commodity support programs. The policy argument against supply controls is that they forfeit global markets to competing foreign producers.

Supply-chains — A tightly organized production, processing , and marketing system formed by agribusiness firms. The emergence of supply-chains is related to the longstanding trend toward fewer, larger, and more specialized commercial farms and ranches (horizontal integration). A distinguishing characteristic of supply-chains is their reliance on contractual agreements, licenses, joint ventures, integrated ownership, and other business arrangements with different segment of the agro-food system. Broiler production is the exemplar of this trend. Pork production is rapidly becoming vertically coordinated. Contracting for special quality characteristics also is growing in the cash grains market. (See also Value-chains).

Support price — A legislated minimum price for a particular commodity, maintained by USDA through a variety of mechanisms, such as nonrecourse loans and purchase programs. With the exceptions of milk and sugar, commodity programs no longer support market prices received by farmers. However, the term support price may be used inaccurately as a synonym for the marketing assistance loan rate of a loan commodity.

Surface runoff — Precipitation, snow melt, or irrigation water in excess of what can infiltrate the soil surface and be stored in small surface depressions; a major cause of erosion and transporter of nonpoint source pollutants.

Surface Transportation Board (STB) — The federal body regulating railroads created by Congress when it eliminated the Interstate Commerce Commission. Agricultural interests closely follow STB deliberations proceedings on railroad mergers, service issues, and related matters because of their potential impact on grain and other commodity transportation costs. [www.stb.dot.gov].

Surplus — The amount by which available supplies are greater than the quantity that will bring producers an adequate income. A surplus may be due to production outrunning demand, a decline in consumption, or a general decline in consumer income or buying power. Historically, commodity programs have been designed to deal with problems of surplus, and the Secretary of Agriculture has had discretion to determine whether an item is in surplus and should be removed from market channels to shore up prices. Approaches have included cropland diversion to reduce production, long-term storage of excess supplies, and purchase and donation of surplus items for foreign or domestic food program use. Sometimes the phrase surplus stocks refers to inventories owned and stored by the Commodity Credit Corporation (CCC).

Suspension — A process under the Federal Insecticide, Fungicide, and Rodenticide Act (7 U.S.C. 136d) by which the EPA can suspend the use of a pesticide in order to prevent an imminent hazard resulting from its continued use. An emergency suspension takes effect immediately. Under an ordinary suspension a registrant can request a hearing before the suspension goes into effect. Such a hearing process might take several months.

Sustainable agriculture — A systematic approach to farming intended to reduce agricultural pollution, enhance natural resource and financial sustainability, and improve efficiency. Overall, alternative agriculture emphasizes management practices that take advantage of natural processes (such as nutrient cycles, nitrogen fixation, and pest-predator relationships), improve the match between cropping patterns and agronomic practices on the one hand and the productive potential and physical characteristics of the land on the other. Commercial fertilizer and pesticides are used selectively to ensure production efficiency and conservation of soil, water, energy, and biological resources. Examples of sustainable agricultural practices include use of crop rotation, animal and green manures, soil and water conserving tillage systems, such as no-till planting methods, integrated pest management, and use of genetically improved crops and animals. Sustainable farming is not considered a midpoint on a line between conventional and organic farming. Producers who want to label their commodities as organic must be certified as complying with the list of acceptable and non-acceptable inputs and practices under the

National Organic Program (NOP). Genetically modified crops and livestock, and most commercial agricultural chemicals, are not permitted under the NOP.

Sustainable Agriculture Research and Education (SARE) Program — The Food Security Act of 1985 (P.L. 99-198) authorized a competitive grants program, now called the Sustainable Agriculture Research and Education (SARE) Program that supports farmer-scientist teams performing on-farm experiments in less chemical-intensive methods of pest control and soil fertility and other sustainable practices. The program also trains Cooperative Extension personnel to work with farmers to encourage adoption of sustainable practices. See Sustainable agriculture. [www.sare.org].

Sustainable forest management — The use of silvicultural activities to assure that the desired goals of forest management (outputs, uses, desired conditions, etc.) can continue to be achieved in perpetuity.

Sustainable Forestry Initiative (SFI) — A program of the American Forest and Paper Association to assure that the organization's members prepare plans and monitor results to realize sustainable forest management of the members' lands, and to assist others in reaching sustainable forest management on other private lands.

Sustained yield — According to the Multiple Use, Sustained Yield Act of 1960 (P.L. 86-517), sustained yield from the national forests means the "achievement and maintenance in perpetuity of a high level of annual or regular periodic output of the various renewable resources of the national forests without impairment of the productivity of the land."

Sustained yield — An output of renewable resources that does not impair the productivity of the resource; it implies a balance between harvesting and incremental growth or replenishment.

Swampbuster — A provision of the Food Security Act of 1985 (P.L. 99-198) that discourages the conversion of wetlands to cropland use. Producers converting a wetland area to cropland lose eligibility for several federal farm program benefits. Benefits are lost from when water levels are lowered to facilitate agricultural production until they have been restored. Several types of wetlands and wetlands in specified situations are exempt. Exceptions include conversions that began before December 23, 1985, conversions of wetlands that had been created artificially, crop production on wetlands that became dry through drought, and conversions that USDA has determined have minimal effect on wetland values. Swampbuster provisions were amended in the 1996 farm bill (P.L. 104-127) to provide greater flexibility for producers and landowners. The 2002 farm bill (P.L. 107-171, Sec. 2002) made only a single change prohibiting third party providers from making swampbuster determinations.

SWCD — Soil and Water Conservation District; or Soil Conservation District.

SWCS — Soil and Water Conservation Society. [www.swcs.org].

SWDA — Solid Waste Disposal Act. See Resource Conservation and Recovery Act (42 U.S.C. 6901 *et seq.*).

Sweet potato whitefly — An insect pest of cotton, fruit and vegetable, and greenhouse crops. The Animal and Plant Health Inspection Service (APHIS) works cooperatively with producers to implement pest management strategies based on integrated pest management and biological control techniques.

Swine fever — see Hog cholera.

Swiss formula — An approach to tariff reduction (a modality) used in the Tokyo Round of Multilateral Trade Negotiations to reduce and harmonize tariffs for industrial products.

Harmonization of tariffs results in a situation where all tariffs are lowered to similar levels. Use of the Swiss formula in the Doha round of agricultural trade negotiations as a means to achieve tariff reductions has been proposed by the United States and the Cairns Group of agricultural exporting countries. An alternative, the Uruguay Round formula of across-the-board percentage tariff reductions, has been proposed by the European Union. The Swiss formula results in higher tariffs being reduced more than lower tariffs. Application of the Swiss formula can be represented as follows: Final tariff = (initial tariff x *a*)/(initial tariff + *a*). The United States has proposed in the Doha agricultural negotiations that a coefficient (*a* in the above formula) of 25 be used to effect tariff reductions. Thus, to illustrate: If the initial tariff is 50%, then the final tariff = (50 x 25)/(50 + 25) = 16.7%; if the initial tariff is 10%, then the final tariff = (10 x 25)/(10 + 25) = 7.1%. In this illustration, a 50% tariff is reduced by 67%, while a10% tariff is reduced by 29%. If the Uruguay Round formula (reduction, over six years, of average tariffs by 36% and of individual tariff lines by a minimum of 15%) were used, then a 50% tariff on an individual product would be lowered to 42.5% and a 10% tariff would be reduced to 8.5%. If 50% and 10% were average tariff levels, then they would be reduced to 32% and 6.4% respectively. In this case, tariff reduction but not harmonization is achieved.

Switching arrangement — A carrier transports the railcars of a competing carrier at origin or destination (for a fee).

T value — See Soil loss tolerance.

TA — Technical assistance.

Taking endangered or threatened species — Taking, in layman's terms, means killing or removing a plant or animal of a species listed under the Endangered Species Act (ESA; 16 U.S.C. 1531 *et seq.*), or seriously damaging its chances of reproduction. Except under specified circumstances, taking is forbidden under the ESA. The definition of taking is one of the current issues in the ESA debate. Incidental taking of a listed species is a taking which occurs in the course of some other legal activity, whether carried out by a federal or a nonfederal entity.

Takings (of property) The Fifth Amendment to the Constitution and comparable provisions in many state constitutions bar the taking of private property by government unless just compensation is paid to the property owner. Initially, the Supreme Court recognized only government seizures and physical invasions of private property as takings. Thus, recurring flooding of a farm as the result of a government dam would probably be a taking. In 1922, the Court expanded the concept of takings to include government actions that merely restrict the economic use of private property, if that restriction is severe enough. However, the line between restrictions that take and those that do not has proved elusive; the Court repeatedly stresses that the determination is an *ad hoc*, case-by-case one. The agricultural community perceives a threat of takings from federal efforts to preserve wetlands and endangered species, though actual court decisions finding takings of farmland are few. The takings issue has also worked against farmers by virtue of a recent case striking down a "right to farm" law that curtailed the right of owners of residential properties adjacent to farms to sue for nuisance based on farm operations. That law was held to effect a taking of a negative easement of the adjacent properties.

Talmadge-Aiken plants — The approximately 250 meat and poultry plants in 10 states where USDA has contracted with state agency inspectors to conduct federal inspection activities. They are now formally known as Federal-State Cooperative Inspection Plants. Even though state employees conduct the inspections in these plants, they are considered to be under the federal rather than state inspection.

Target price — Arbitrary price levels used as benchmarks for making deficiency payments to producers of selected commodities when season average market prices are lower than these target levels. Target prices and deficiency payments were first adopted as commodity program income support policy in the Agriculture and Consumer Protection Act of 1973 (P.L. 93-86). After being eliminated by the 1996 farm bill (P.L. 104-127), target prices were restored by the 2002 farm bill (P.L. 101-171, Sec. 1104). Target price deficiency payments are renamed counter-cyclical payments by this law. Target prices apply to only the covered commodities of wheat, corn, grain sorghum, barley, oats, upland cotton, rice, soybeans, and other oilseeds. Prior to 1996, farmers received deficiency payments based on the difference between the target price and the higher of the national market price during a specified time period, or the nonrecourse loan rate. Under the 2002 farm bill, the counter-cyclical payment rate is the difference between the target price and the season average market price, but only down to the loan rate plus the direct payment rate.

Targeted Export Assistance Program (TEA) — A program authorized by the Food Security Act of 1985 (P.L. 99-198) to assist U.S. producer groups in promoting exports of products adversely affected by foreign governments' unfair trade practices. TEA is the predecessor of the Market Promotion Program (MPP), which was succeeded by the Market Access Program (MAP) in 1996. [www.fas.usda.gov/mos/programs/map prog.html].

Targeting — A policy concept under which government farm program benefits would be directed toward specified groups of producers. One example of targeting might be to focus farm payments on small to medium-sized family owned and operated farms. Some policy makers have argued for tighter per person payment limits as a mechanism to target smaller farms.

Tariff Act of 1930 — P.L. 71-361, also known as the Smoot-Hawley Act, raised U.S. import tariffs to their highest levels in history, prompting U.S. trading partners to adopt their own retaliatory trade barriers and exacerbating the Great Depression. Ensuing laws have virtually eliminated the Act's most onerous provisions, yet it remains as permanent authority and a vehicle for trade legislation.

Tariff schedule — A list or schedule of duties imposed in the conduct of international trade. The Harmonized Tariff Schedules of the United States (HTSUS) lists the items on which the United States levies a duty (or tariff) and/or imposes an import or tariff-rate quota. A specific duty, and/or quota amount, is assigned to each item on the schedule.

Tariff — A tariff is a list or schedule of taxes, while a duty is the tax imposed on a specific item. However, the terms duty and tariff have come to be used interchangeably. In international trade, these taxes must be paid to a government on selected imported or sometimes exported goods. The Harmonized Tariff Schedules of the United States (HTSUS) lists the items on which the United States levies duties. Tariffs may be protective of domestic producers (keeping domestic prices higher than world prices) or serve as revenue generators for the government. Tariffs are considered transparent trade barriers in contrast to several nontariff barriers. The Uruguay Round Agreement on Agriculture requires conversion of nontariff barriers to bound tariffs. See Ad valorem duty, Preferential tariff, Specific rate duty, Tariff-rate quota, Applied tariff.

Tariff-rate quota — A trade policy tool used to protect a domestically-produced commodity or product from competitive imports. A tariff-rate quota (TRQ) combines two policy instruments that nations historically have used to restrict such imports: quotas and tariffs. In a TRQ, the quota component works together with a specified tariff level to provide the desired degree of import protection. Imports entering during a specific time period under the quota portion of a TRQ are usually subject to a lower, or sometimes a zero, tariff rate. Imports above the quota's quantitative threshold face a much higher (usually prohibitive) tariff. Currently, TRQs apply to U.S. imports of certain dairy products, beef, cotton, green olives, peanuts, peanut butter, sugar, certain sugar-containing products, and tobacco.

Tariffication — The conversion of nontariff barriers to tariffs or tariff-rate quotas.

Tax Increment Financing (TIF) — Tax increment financing is an economic development tool that cities, smaller rural areas, and some regions have found promising. TIC relies on a reallocation of property tax revenues from a taxing authority's general fund to a new jurisdiction named a TIF district. TIF allows a city or community to designate an area for improvement and then earmark any future growth in property tax revenues to pay for the initial and ongoing economic development expenditures. Once an area becomes eligible for TIF, the initial assessed property valuation is held constant for a fixed number of years, e.g., 20 or more. The TIF authority uses its powers of land assembly and sale, site clearance, relocation, utility installation and street repair to improve the district and offer subsidized financing to make it more attractive to businesses and developers. As new businesses and other investments are attracted to the area, the assessed value of property and its taxes are expected to rise. The difference between the base value and new assessed value is the "tax increment." Instead of channeling the increment to the local government's general fund and to other taxing bodies with jurisdiction over the area, these funds are channeled to the TIF authority and used to finance any debt the authority accumulated when making improvements. Any increase in the assessed property values of the TIF district over the next years will go into a separate fund and will pay for TIF activities, while taxes on the base value of the properties will remain the same and will continue to be paid to the local taxing bodies.

Taylor Grazing Act (TGA) of 1934 — P.L. 73-482 provides for the regulation of grazing on the public lands (excluding Alaska) to improve rangeland conditions and regulate their use. The law initially permitted 80 million acres of previously unreserved public lands of the United States to be placed into grazing districts to be administered by the Department of the Interior. As amended, the law now sets no limit on the amount of lands in grazing districts. There are currently approximately 162 million acres inside grazing allotments.

TCDD — Dioxin.

TCK smut — A disease of soft white and hard red winter wheats caused by a fungus (*Tilletia controversa* Kuhn) that stunts plant growth and leaves smut balls in the grain head. When milled, the balls emit a fishy odor that lowers the quality of the flour. TCK smut exists in the western and northwestern states, but is not considered a major problem. The disease took on policy significance because China applied a zero tolerance on the presence of TCK spores, resulting in a ban from 1974 to 1999 on shipments from the Pacific

Northwest. Until the summer of 1996, China accepted shipments of U.S. wheat from the Gulf Coast, and negotiated price discounts with the shippers to cover the cost of decontamination if traces of TCK were found. Then in June 1996, China rejected all cargoes of U.S. wheat with traces of TCK. The November 1999 U.S.-China Agricultural Cooperation Agreement removes the ban and allows imports of U.S. wheat and other grains that meet a specific TCK tolerance level, thus improving the competitiveness of U.S. wheat with Canadian and Australian exports.

TEA — Targeted Export Assistance Program.

Team Nutrition — The USDA implementation vehicle for its School Meals Initiative for Healthy Children, begun in 1995. Among other things, it provides grants to state agencies that apply for and meet the criteria for delivering new and innovative programs that provide training to school food personnel, nutrition education to children and their parents, and support healthy eating and physical activities involving school, child care, and community groups. The program is authorized and funded under annual agriculture appropriations laws. Team Nutrition has provided $38.6 million in food service training grants to state agencies over 10 years from FY1995 through FY2004. [www.fns.usda.gov/tn].

Technical assistance provider — See Third party provider.

Technical assistance — Section 11 of the Commodity Credit Corporation (CCC) Charter Act (P.L. 80-806) was amended by the 1996 farm bill (P.L. 104-127, Sec. 162)to limit funding for technical assistance, causing some difficulties for Natural Resources Conservation Service (NRCS) to both fully support mandatory conservation programs and carry out other technical assistance responsibilities. The 2002 farm bill (P.L. 107-171, Sec. 2701) seeks to lift these limitations generally, and provisions with each conservation program seek to provide funding for any technical assistance needed to implement that program.

Technical Assistance for Specialty Crops — The 2002 farm bill (P.L. 107-171, Sec. 3205) requires use of $2 million annually in Commodity Credit Corporation (CCC) resources for resolving sanitary and phytosanitary and related barriers to exports of U.S. specialty crops. CCC resources are to go for public and private projects and for technical assistance.

Technical barriers to trade (TBTs) — A category of nontariff barriers to trade, TBTs are the widely divergent measures that countries use to regulate markets, protect their consumers, or preserve their natural resources (among other objectives), but they also can be used (or perceived by foreign countries) to discriminate against imports in order to protect domestic industries. TBTs with the greatest impact on agriculture are the various sanitary and phytosanitary (SPS) measures designed to protect humans, animals, and plants, from diseases, pests, and other contaminants. Examples of TBTs, other than SPS measures, are rules for product weight, size, or packaging; ingredient or identity standards; mandatory labeling; shelf-life restrictions; and import testing and certification procedures. The broad phrase technical barriers to trade is frequently applied to all of these types of measures, even where they might be legitimate and consistent with bilateral or multilateral trading rules. The Agreement on Technical Barriers to Trade gives rules for the use of such barriers. However, trade experts widely view TBTs as having great potential for being misused by importing countries as nontransparent (disguised or unclear) obstacles to trade. (See Transparency.).

Technical Service Providers — Another name for Third Party Providers (Vendors).

TEFAP — Temporary Emergency Food Assistance Program.

TEGMA — Transportation, Elevator and Grain Merchants Association.

Temporary Emergency Food Assistance Program (TEFAP) — A program that evolved out of surplus commodity donation efforts begun by the USDA in late 1981 to dispose of surplus foods (especially cheese) held by the Commodity Credit Corporation (CCC). This program was explicitly authorized by the Congress in 1983 when funding was provided to assist states with the costs involved in storing and distributing the commodities. The program originally was entitled the Temporary Emergency Food Assistance Program when authorized under the Temporary Emergency Food Assistance Act of 1983 (P.L. 98-8). The word temporary was dropped from statute and program title in the 1990 farm bill (P.L. 101-624). The acronym EFAP now is used for this program (7 U.S.C. 7501 *et seq.*).

Temporary Emergency Food Assistance Act of 1983 (TEFAA) — P.L. 98-8 was a supplemental appropriations act for FY1983 that among other things explicitly authorized a discretionary commodity donation effort begun in 1981 by the USDA. The initial effort was limited to disposal of excess commodities held by the Commodity Credit Corporation (CCC) by donating them to states. This law also authorized funding to help states and local emergency feeding organizations with the storage and distribution costs of handling the commodities. This is the origin of the current Emergency Food Assistance Act of 1983 (P.L. 98-92, as amended; 7 U.S.C. 7501 *et seq.*).

Tender — As a verb, tender announces the intention of delivering a notice or an actual commodity (i.e., XYZ Grain Growers, Inc., tenders six cars to North Pacific at the time of shipment for application on an open sales contract). As a noun, tender normally denotes a notice of an intent to buy. The tender usually spells out in detail quantities to be purchased, desired quality, time of shipment, country of origin, and all inspection, weighing, and payment terms. Overseas buyers usually issue tenders to ensure the maximum competition for a given piece of business. Also, the action of receiving offers, determining the best one, deciding whether, and how much, to buy, and announcing the awards of contracts. Ordinarily, a buyer reserves the right to reject any or all of the offers submitted.

Teratogen — A chemical that causes nonhereditary birth defects in a developing fetus. Teratogencity is taken into account in assessing the toxicity of pesticides and other chemicals. Both level and timing of exposure to teratogens determine health effects.

Terminal dwell time — The time loaded railcars spend in a terminal awaiting continued movement.

Terminal elevator — A large elevator (warehouse) facility with the capacity to transfer grain to rail cars, barges, or ships for transport to domestic or foreign markets. Terminal elevator markets are used as base locations for posted county prices.

Terminal market — A central site, often in a metropolitan area, that serves as an assembly and trading place for agricultural commodities. Terminal markets are usually at or near major transportation hubs.

Terminator gene — A genetic use restriction technology (GURT) that prevents the germination of seeds produced by a transgenic plant. GURTs also can be used as a means of intellectual property protection. See Terminator seeds.

Terminator seeds — A descriptive term for seeds that have been genetically engineered to produce a crop whose first generation produces sterile seeds, thus preventing a second generation from being grown from seeds saved from the first. This technology was developed under a cooperative research and development agreement between the Agricultural Research Service (ARS) and a private seed company. Supporters of the technology state that it is a way to build patent protection directly into high-value, genetically engineered crop varieties and thus recoup high research investment costs. Opponents maintain that the technology unfairly disadvantages farmers in developing countries who rely on saved seed for replanting and for developing locally adapted varieties, and also has potentially harmful environmental and public health effects.

Terrace — An embankment, ridge, or leveled strip constructed across sloping soils on the contour, or at right angle to the slope. The terrace intercepts surface runoff so that it can soak into the soil or flow slowly to a prepared outlet, decreasing rates of soil erosion.

Test and hold — A new requirement that federal meat inspectors not apply the "inspected and passed" mark on the carcasses of cattle that have been sampled by the Animal and Plant Health Inspection Service (APHIS) for its BSE surveillance program until the sample is determined to be negative (69 FR 1892, January 12, 2004). Formerly, unless prohibited by a Food Safety and Inspection Service (FSIS) veterinary medical officer, carcasses of BSE-sampled animals were allowed to be stamped inspected and passed, and packing plants were allowed to process them for human food before the sample results were received.

Test weight — The average weight of a cereal as measured in pounds per bushel (1bu. = 8 gallons or 2150.42 cu. inches). Test weight is an important predictor of milling yield for rice and flour extraction rate for wheat. USDA's official weight per bushel for the highest grade for major cereals and oilseeds include: wheat and soybeans (60 lbs./bu.); corn, sorghum, and rye (56 lbs./bu.); barley (48 lbs./bu.); oats (32 lbs./bu.); and rice (45 lbs./bu.). When producers deliver grain that is significantly below the official weights the prices are discounted. Official U.S. Grain Standards (developed and used under authority of the United States Grain Standards Act, USGSA) include test weight criteria.

TFP — Thrifty Food Plan.

Third country dumping — A situation in which exports of a product from one country are being injured or threatened with injury because of exports of the same product from a second country into a third country at less than fair value. Section 1318 of the Omnibus Trade and Competitiveness Act of 1988 (P.L. 100-418) establishes procedures for U.S. industries to petition the U.S. Trade Representative to request a foreign government that is a signatory to the GATT Anti-Dumping Code to initiate an antidumping investigation on behalf of a U.S. industry that claims it is being injured by dumping in that country's market.

Third party providers (vendors) — Provisions in the 2002 farm bill (P.L. 107-171, Sec. 2701) require the Natural Resources Conservation Service (NRCS) to create a certification program to approve non-USDA experts to provide technical assistance to landowners for federal conservation programs, and a process to reimburse the vendors through USDA. Also known as technical service providers.

Threatened species — Species listed by regulation under ESA (16 U.S.C. 1531 *et seq.*), and are generally given a lower level of protection than endangered species.

Three entity rule — Federal law currently sets an annual cap on the amount of payments that a person may receive from the major farm programs. A provision in this law permits a person to receive payments up to the full cap on the first farm in which the person has a substantial beneficial interest, and up to half the full cap on each of two additional farms; hence the so-called three-entity rule.

Threshold price — Used in agricultural policy, particularly in the EU Common Agricultural Policy, to denote a minimum price for a commodity or product.

Threshold — In toxicology, the lowest non-zero dose of a chemical at which a specified measurable effect occurs. Sometimes used to refer to the income level above which an applicant for food stamp benefits would be ineligible.

Thrifty Food Plan (TFP) — The TFP is one of four USDA-designed food plans specifying foods and amounts of foods to provide adequate nutrition. It is used as the basis for designing Food Stamp Program benefits. It is the cheapest food plan and is calculated monthly using data collected for the consumer price index (CPI). However, it is not the same as the food components of the CPI. The monthly cost of the TFP used for the Food Stamp Program represents a national average of expenditures (four-person household consisting of an adult couple and two school-age children) adjusted for other household sizes through the use of a formula reflecting economies of scale. For food stamp purposes, the TFP as priced each June sets maximum benefit levels for the fiscal year beginning the following October.

Tier II sugar — See Second-tier Mexican sugar. Sometimes referred to as high- tier sugar.

Timber sale — A contract for the sale of federal timber to a private purchaser with the right to cut and remove trees for an agreed-upon stumpage price; the contract includes an estimated volume of wood and an appraised stumpage price, which is the basis for competitive bidding by purchasers.

Timberland Suitability — Refers to national forest land where timber harvesting is feasible , based on biological and economic factors, and is allowed on a regulated and sustained basis under forest plans developed by the Forest Service.

Timberland — Forestland that can grow annually a minimum amount of wood (20 cubic feet per acre per year) that can be used to produce commercial wood products; excludes lands where timber cutting is prohibited by law or by executive decision.

TMDL — Total maximum daily load.

Tobacco Price Support Program — The tobacco price support program used a combination of marketing quotas and nonrecourse loans to keep prices stable and higher than they would be otherwise. The tobacco quota limited production in order to raise prices. Nonrecourse loans allowed producers to hold tobacco stocks for long periods in order to balance supplies with market demand conditions. By law after 1982, tobacco loan program operations were required to function at no net cost to taxpayers (P.L. 97-218). A no net cost assessment was collected on all leaf tobacco sold to build a reserve fund that reimbursed the Commodity Credit Corporation (CCC) for any losses of loan principal and interest. Adoption of the tobacco quota buyout in P.L. 108-357, Title VI, ended also the price support program for the 2005 crop and subsequent years.

Tobacco quota — Poundage quotas (and in some cases acreage allotments) were a supply control feature of federal price support for tobacco. Burley tobacco was subject to marketing quotas and flue-cured tobacco was subject to marketing quotas and acreage allotments. Tobacco quota owners (owners of farmland to which quota is assigned) voted every three years on whether or not to continue with price support (through no-net-cost loans) and marketing quotas. Producers of several minor tobaccos (including Maryland (type 32), Pennsylvania cigar-filler (type 41), and Connecticut Valley cigar-binder (types

51-52)) had disapproved federal support. The national marketing quota (basic quota) was calculated according to a formula specified by law that included consideration of intended purchases by domestic manufacturers, average exports over the preceding three years, and reserve stock requirements. The effective quota was the basic quota plus and minus temporary adjustments for allowable previous year under and over marketings. The Fair and Equitable Tobacco Reform Act of 2004 (P.L. 108-357, Title VI) ended tobacco quotas for 2005 crop and subsequent years.

Tobacco Settlement Agreement — The tobacco settlement agreement, officially known as the Master Settlement Agreement (MSA), is an accord reached in 1998 between the state Attorneys General of 46 states, five U.S. territories, the District of Columbia and America's major tobacco companies concerning the advertising, marketing and promotion of tobacco products. Payments by the tobacco companies to the states amount to over $206 billion over 25 years. As a followup, but separate from the MSA, the tobacco companies agreed in 1999 to pay 14 tobacco producing states $5.15 billion over 12 years to offset the adverse economic on farmers and rural communities of the MSA. This is called Phase II of the tobacco settlement.

Tolerance fees — The Federal Food, Drug, and Cosmetic Act (FFDCA), requires EPA to collect fees to cover the costs of administering the tolerance-setting program. (Sec. 408(m); 21 U.S.C. 6a). This authority has existed since 1954 (P.L. 83-518), but the Food Quality Protection Act (FQPA) of 1996 (P.L. 104-170) directed EPA to deposit tolerance fees collected into the Reregistration and Expedited Processing Fund until they are needed to process tolerances. The FQPA also directed the Agency to review all tolerances set prior to 1996 to ensure that they are consistent with the safe standard set by the FQPA.

Tolerance — The maximum pesticide residue level allowed on or in food or animal feed. Tolerances are set by EPA under the Federal Food, Drug, and Cosmetic Act (FFDCA; 21 U.S.C. 321 *et seq.*), as amended by the Food Quality Protection Act of 1996 (FQPA; P.L. 104-170). Tolerances must be set at a level that is safe, which Congress defined to mean there will be a "reasonable certainty of no harm" from all pesticide exposures. A tolerance must be established before a pesticide can be registered (or reregistered) for use on a food crop.

Total maximum daily load (TMDL) — An identification of the total amount of a specific pollutant or the property of a pollutant from all sources (point sources, nonpoint sources, and background sources) and including a margin of safety, that may be discharged into a water body while insuring that water body attains water quality standards. Most water bodies are affected by multiple pollutants. The TMDL program, authorized in the Clean Water Act (33 U.S.C. 1251-1387), is administered by states, which may require industries, cities, and farmers, and others to use new pollution controls to meet requirements. TMDL program impacts on agricultural and other nonpoint source discharges have been controversial.

Total support estimate (TSE) — OECD indicator of the annual monetary value of all gross transfers from taxpayers and consumers arising from policy measures that support agriculture, net of the associated budgetary receipts, regardless of their objectives and impacts on farm production and income, or consumption of farm products. The TSE can be expressed in monetary terms or as a percentage of the gross domestic product. In addition to the TSE, other measures used to compare levels of support to agriculture across counties include the producer support estimate (PSE), consumer support estimate (CSE), and general services support estimate (GSSE).

Toxic Substances Control Act (TSCA) of 1976 — P.L. 94-469, as amended, authorizes the EPA to regulate toxic substances (any chemical that may present a risk of unreasonable harm to man or the environment). By definition, however, the Act excludes from EPA regulation under TSCA certain substances (15 U.S.C. 2601), including pesticides (as defined by and regulated under the Federal Insecticide, Fungicide, and Rodenticide Act (7 U.S.C. 136 *et seq.*)), tobacco or tobacco products, and any food or food additive (as defined by and regulated under the Poultry Products Inspection Act (21 U.S.C. 451 *et seq.*), the Federal Meat Inspection Act (21 U.S.C. 601 *et seq.*), the Egg Products Inspection Act (21 U.S.C. 1031 *et seq.*), or the Federal Food, Drug, and Cosmetic Act (21 U.S.C. 321 *et seq.*)).

TPRG — Trade Policy Review Group.

Traceability — The ability to identify and track a food or feed commodity from point of purchase or consumption back, through all stages of production and marketing, to its origins on the farm. Traceability arises as an issue in a number of regulatory initiatives, such as the labeling and tracking of biotechnology-derived products (particularly in the European Union); country of origin (for which more stringent labeling of meats and produce was mandated by the 2002 U.S. farm bill); and food safety (so that food-related disease outbreaks could be traced to their source).

Traceback — See Animal identification and traceback.

Trackage rights — Rights obtained (for a fee) by one railroad to operate its trains over the tracks of another railroad.

Trade Act of 1974 — P.L. 93-618 provided the President with tariff and nontariff trade barrier negotiating authority for the Tokyo Round of multilateral trade negotiations. It also gave the President broad authority to counteract injurious and unfair foreign trade practices. Section 201 of the Act requires the International Trade Commission to investigate petitions filed by domestic industries or workers claiming injury or threat of injury due to expanding imports. Investigations must be completed within 6 months. If such injury is found, restrictive measures may be implemented. Action under Section 201 is allowed under the GATT escape clause, GATT Article XIX. Section 301 was designed to eliminate unfair foreign trade practices that adversely affect U.S. trade and investment in both goods and services. Under Section 301, the President must determine whether the alleged practices are unjustifiable, unreasonable, or discriminatory and burden or restrict U.S. commerce. If the President determines that action is necessary, the law directs that all appropriate and feasible action within the President's power should be taken to secure the elimination of the practice.

Trade Adjustment Assistance for Farmers — Created in 2002 by wide-ranging trade legislation (P.L. 107-210, Sec. 141), the program authorizes the expenditure of up to $90 million per year through FY2007. Under the program, certain agricultural producers can each receive payments of up to $10,000 per year if price declines for their commodity were at least partly caused by imports. To be eligible for such assistance, such producers must be members of certified groups and meet a number of criteria specified by the law. The program is administered by the USDA.

Trade Agreements Act of 1979 — P.L. 96-39 provided the implementing legislation for the Tokyo Round of multilateral trade agreements in such areas as customs valuation, standards, subsidies, and government procurement.

Trade and Tariff Act of 1984 — P.L. 98-573 clarified the conditions under which unfair trade cases under Section 301 of the Trade Act of 1974 (P.L. 93-618) can be pursued. It also provided bilateral trade negotiating authority for the U.S.-Israel Free Trade Agreement and the U.S.-Canada Free Trade Agreement, and set out procedures to be followed for congressional approval of future bilateral trade agreements.

Trade barriers — Laws, regulations, policies that protect domestic products from foreign competition or artificially stimulate exports of domestic products. The USTR classifies trade barriers into eight categories: (1) Import policies (tariffs and other import charges, quantitative restrictions, import licensing, and customs barriers); (2) standards, testing, labeling, and certification; (3) government procurement; (4) export subsidies; (5) lack of intellectual property protection; (6) service barriers; and, (7) investment barriers. USTR's annual *National Trade Estimates Report* catalogs trade barriers applied by foreign countries to U.S. products.

Trade creation — Increases in trade and the more efficient production of goods resulting from the removal of trade barriers.

Trade diversion — A shift in production and trade from a relatively efficient country to a relatively inefficient country, caused by the latter's participation in a preferential trade agreement.

Trade liberalization — With respect to agricultural trade, trade liberalization means expanding market access, reducing and/or eliminating agricultural export subsidies, and reducing or eliminating trade-distorting domestic farm subsidies. Trade liberalization so defined is the overarching aim of the Doha Development Agenda (DDA), the current round of multilateral trade negotiations.

Trade Policy Committee — The TPC is the senior U.S. Government interagency trade committee established to provide broad guidance on trade issues. It is chaired by the U.S. Trade Representative (USTR) and is comprised of other cabinet officers, including the Secretary of Agriculture. The Trade Policy Review Group (TPRG) that reports to the TPC is chaired by the Deputy USTR and is comprised of sub-cabinet representatives, including the Under Secretary of Agriculture for Farm Services and Foreign Agriculture. The Trade Policy Staff Committee, the level at which position papers are initiated, is chaired by a Deputy Assistant USTR and has representation from other cabinet departments including USDA.

Trade promotion authority (TPA) — Statutory authority for the President to negotiate trade agreements, with expedited procedures for legislation to implement those agreements. This authority is sometimes referred to as fast-track authority. The most recent TPA applies to agreements entered into before July 1, 2005, but there is a mechanism to extend it 2 additional years. TPA was part of a wide-ranging trade bill (H.R. 3009) signed into law (P.L. 107-210) on August 6, 2002. TPA is intended to facilitate the negotiation of numerous bilateral free trade agreements, a Free Trade Agreement of the Americas (FTAA), and agreements reached through the World Trade Organization. Many agricultural and food industry groups have been among the export-oriented interests in support of TPA, anticipating that its use will lead to new trade agreements opening and expanding markets throughout the world. Some farm groups disagree, contending that past agreements have not yielded the benefits that supporters had promised.

Trade relief; trade remedy — Statutorily-prescribed procedures intended to assist U.S. industries, including agricultural producers, deemed to be injured by imports or unfair trade practices. Currently, for example, federal law provides for three primary remedies that producers can pursue to seek relief from imports: safeguards (which can be temporary duties, quotas, or import restrictions), countervailing duties, and anti-dumping duties.

Transfer of development rights (TDR) — Property rights that may not be used on the land from which they were derived. Usually they are sold from areas where uses are restricted, such as active agricultural areas, to receiving or growth areas where development is being encouraged.

Transgenic crop — A crop modified to exhibit desired traits by inserting genetic material from another organism. Recombinant DNA techniques (DNA formed by combining segments of DNA from different organisms) are usually used.

Transition Yields, or T-yields — When a producer participating in the federal crop insurance program does not have at least four years of actual production history, the insurer can offer coverage based on a countywide 10-year average of production yields, or transition yields, as determined by USDA.

Transmissible spongiform encephalopathies (TSEs) — The name of a number of degenerative brain diseases that infect humans and animals. For example, bovine spongiform encephalopathy (BSE) infects cattle; scrapie infects sheep and goats; Creutzfeldt-Jakob disease (CJD) infects humans.

Transparency — A World Trade Organization principle stipulating that a country's policies and regulations affecting foreign trade should be clearly communicated to its trading partners. For example, out of recognition that sanitary and phytosanitary measures may (sometimes deliberately) be unclear, arbitrary, or capricious, recent international trading agreements have provisions calling on countries to notify others, in advance, about any measures that could affect trade, to fully explain them, and to provide a means for commenting on them.

Transshipment — A cargo moved from its originating country through one or more intermediary countries before reaching its final destination country. U.S. agricultural import and export statistics usually can be adjusted to account for such movements. The term transshipment also can refer to a cargo that is transferred from one mode of transport to another (e.g., a cargo of produce that arrives by ship and is moved inland by truck).

Tree Assistance Program (TAP) — A disaster assistance program, administered by the Farm Service Agency, that makes payments for lost orchard trees and vines that produce annual crops. The program has been funded on an *ad hoc* basis, usually by emergency supplemental appropriations. Most recently, the program was funded for losses incurred between October 1, 1997, and September 30, 1998.

Tree measurement sales — A timber sale where purchasers pay the total bid value (the estimated timber volume times the stumpage price) regardless of the volume of timber actually removed. Compare with scaled sales.

Trickle irrigation / drip irrigation — An irrigation method in which water drips to the soil from perforated tubes or emitters. This irrigation technology is water conserving compared to flooding, furrows, and sprinklers.

Triple base plan — Also called the flexible base plan. A proposal under which farmers who raise program crops would receive program payments only on a certain percentage of their permitted acreage. A producer participating in a federal price support program actually would have three categories of base acres for program purposes: 1) permitted acres on which deficiency payments would be made; 2) permitted acres on which no federal payments would be made, but could be planted to other crops, either specified or unspecified; 3) idled acres (those required to be set aside under acreage reduction rules) where no crops other than those for conservation could be planted. Triple base is another name for what came to be known as normal flex acres. Production flexibility contracts under the 1996 farm bill (P.L. 104-127) and the Direct and Counter-cyclical Program (DCP) agreements under the 2002 farm bill (P.L. 101-171, Sec. 1101-1108) eliminated the linkage between direct payments and actual plantings.

TRIPS — Agreement on Trade-Related Aspects of Intellectual Property Rights.

Triticale — A genetic cross between wheat and rye. Triticale was developed more than a century ago by crossing wheat with rye in hopes of creating a food grain with good baking qualities. Although the acceptance of triticale has been slow, it's a good source of protein and B vitamins. Triticale has enough gluten to be used as a bread flour. It has a nut-like flavor that is richer than wheat but not as strong as rye. Under the 2002 farm bill (P.L. 101-171, Sec. 1206), triticale that is grazed instead of harvested is eligible for "payments in lieu of LDPs" at the same rates as wheat, even though it is not eligible for marketing assistance loans or LDPs when harvested as grain.

TRQ — Tariff-rate quota.

TSCA — Toxic Substances Control Act (15 U.S.C. 2601 *et seq.*).

TSE — Total support estimate.

TSEs — Transmissible spongiform encephalopathies.

TSRA or TSREEA — Trade Sanctions Reform and Export Enhancement Act of 2000 (P.L. 106-386). See Embargo.

Two-tiered pricing — A farm program system under which commodities grown for domestic use are supported at one level and those grown for export markets at another, lower level. The peanut price support program, until policy changes made by the 2002 farm bill (P.L. 101-171, Sec. 1301-1310), used a two-tiered pricing system with a higher level of support for "quota peanuts" that could be sold domestically as edible peanuts and a lower level of support for "additional peanuts" that only could be exported or crushed if that stayed in the United States.

U.S. Trade Internet System — A comprehensive, interactive, on-line source of agricultural import and export data maintained by USDA's Foreign Agricultural Service. Users can find, organize, and customize this data (including that provided through BICO and FATUS) by commodity grouping, country, year, and related categories.

U.S. Trade Representative (USTR) — See United States Trade Representative. [www.ustr.gov].

U.S.-Canada Consultative Committee on Agriculture (CCA) — Binational panel of government officials, formed in April 1999, to improve cooperation and discuss differences on issues related to agricultural trade between the two countries. The CCA meets twice yearly.

U.S.C. — United States Code. [www.gpoaccess.gov/uscode/index.html].

UEP — United Egg Producers. [www.unitedegg.org].

UF milk — Ultra filtered milk.

UFFVA — United Fresh Fruit and Vegetable Association. [www.uffva.org].

UFW — United Farmworkers. [www.ufw.org].

UIC — Underground injection control.

Ultra filtered milk (UF milk) — A subclassification of milk protein concentrate that is produced by passing milk under pressure through a thin, porous membrane to separate the components of milk according to their size, permitting greater efficiency in cheese making. Specifically, ultra filtration allows the smaller lactose, water, mineral, and vitamin molecules to pass through the membrane, while the larger protein and fat molecule (key components for making cheese) are retained and concentrated. (Depending on the intended use of the UF milk product, the fat in whole milk may be removed before filtration.) The removal of water and lactose reduces the volume of milk, and thereby lowers its transportation and storage costs. Ultra-filtration makes cheese manufacturing more efficient and can benefit consumers if cost savings are passed on. However, U.S. milk producers are concerned that imported UF milk may displace domestically produced milk used to make cheese.

UMR — Usual market requirements.

UNCTAD — United Nations Commission on Trade and Development. [www.unctad.org].

Underground storage tank — For purposes of Subtitle I of the Resource Conservation and Recovery Act (P.L. 94-580, as amended) this is any tank used to store petroleum or hazardous substances regulated under the Comprehensive Environmental Response, Compensation, and Liability Act (P.L. 99-499) the volume of which is 10% or more beneath the surface of the ground. This does not include, among other things, any farm or residential tank of 1,100 gallons or less capacity used for storing motor fuel for noncommercial purposes, tanks used for storing heating oil for consumptive use on the premises, or septic tanks.

Underwriting Gain — Within the federal crop insurance program, a private insurance company can transfer a portion of the risk to the federal government as part of the standard reinsurance agreement. The portion of business that is not transferred to the government is said to be retained by the private company. A private company has an underwriting gain when its share of retained net premium exceeds its retained net losses.

Underwriting Loss — The amount by which a private crop insurance company's share of retained net premium exceeds its retained net losses.

Unfair trade practice — Usually, government support to firms that takes the form of subsidies or dumping, and that gives competitive advantages to those firms in international trade.

Unified Export Strategy (UES) — A single, consolidated application process that U.S. agricultural trade promotion groups use to apply for funding for a variety of USDA export promotion programs, including the market access program and the foreign market development cooperator program. USDA introduced UES in 1998 in large part to ensure that applicants for the programs have a long-term overall strategy and to make it easier for them to integrate various sources of federal funding into that strategy.

Uniform Grain and Rice Storage Agreement (USGRSA) — The contractual arrangement governing transactions between the Farm Service Agency and private grain storage companies. Commercial warehouses storing grain under a nonrecourse loan or owned by the Commodity Credit Corporation (CCC) must have a signed USGRSA.

Uniform methods and rules (UM&R) — Documents by the Veterinary Services office of USDA's Animal and Plant Health Inspection Service (APHIS) that specify the minimum standards for preventing, detecting, controlling, and/or eradicating a particular animal disease. APHIS in late 2004 had UM&Rs posted on bovine tuberculosis eradication, brucellosis, brucellosis in cervidae, equine infectious anemia, pseudorabies eradication, swine brucellosis control/eradication, and voluntary scrapie flock certification program standards. UM&Rs usually are developed with input from state animal health authorities, industry, and the U.S. Animal Health Association. Although not legally binding, as are laws and regulations, UM&Rs are widely recognized within the industry and profession as the "gold standard" for addressing an animal disease of national concern; they may be incorporated by states into their own animal health codes.

Unique farmland — Land, other than prime farmland, that has combined conditions to produce sustained high quality and high yields of specialty crops, such as citrus, nuts, fruits, and vegetables when properly managed.

Unit train — Generally refers to a string of freight cars that all carry the same commodity, frequently over long distances. Unit trains are widely used to haul such raw commodities as coal and grains, because they are less costly for railroads than mixed freight shipments. Unit trains have been a point of contention between agricultural shippers and the railroads, mainly because the shippers have increasingly been asked to fill longer unit trains (for example, 104 hoppers) that many local elevators have not been equipped to handle.

United Nations Conference on Trade and Development (UNCTAD) — A UN agency that focuses attention on international economic relations and on measures that might be taken by developed countries to accelerate economic development in developing countries. [www.unctad.org/Templates/StartPage.asp?intItemID=2068].

United Nations Convention on Biological Diversity — One of the international agreements that emerged from the 1992 Earth Summit in Rio de Janeiro, the Convention commits signatory parties to three major objectives: the conservation of biological diversity, the sustainable use of its components, and the fair and equitable sharing of the benefits from the use of genetic resources. The United States, along with 187 other countries, signed the agreement However, Congress never ratified U.S. participation. Thus, the United States is not among the 168 countries that currently are parties.

United States Agency for International Development (USAID) — An independent agency of the executive branch, established in 1961, that administers U.S. international development and humanitarian assistance programs. USAID administers commodity donations for humanitarian or development purposes under Titles II and III of P.L. 480, or Food for Peace, and commodity import programs. [www.usaid.gov].

United States Animal Identification Plan (USAIP) — Officials from approximately 70 animal industry organizations and government agencies have been working since early 2002 on a plan for a national system to identify that might follow food animals from birth to slaughter. The primary purpose is to trace animals back from slaughter through all premises within 48 hours of an animal disease outbreak, in order to determine the disease's origin and to contain it quickly. The plan calls for recording the movement of individual animals or groups of animals in a central database or in a seamlessly linked database infrastructure. USDA's Animal and Plant Health Inspection Service (APHIS) is to coordinate animal ID activities in cooperation with state animal health authorities and producers for disease tracking purposes. Congressional interest in animal ID intensified after a cow with bovine spongiform encephalopathy (BSE) was discovered in the United States in December 2003. USDA in 2004 accelerated work on animal ID, and is incorporating major elements of the USAIP into what it has termed the National Animal Identification System (NAIS). Among the issues in establishing a national program are privacy of producer records, implementation cost and who should pay, and whether animal ID should be mandatory or voluntary. [animalid.aphis.usda.gov/nais/index.shtml].

United States Code (U.S.C.) — The consolidation and codification of all the general and permanent laws of the United States. The U.S. Code is divided into 50 titles that represent broad subject areas. Title 7 is Agriculture. Each title is divided into chapters followed by subdivisions into parts covering specific areas. For example, 7 U.S.C. Chapter 45 Subchapter III deals with the Conservation Reserve Program. Regulations issued to administer the laws are first published in the Federal Register and then in the Code of Federal Regulations. [www.gpoaccess.gov/uscode/index.html].

United States Grain Standards Act (USGSA) of 1916 — P.L. 64-190, as amended (7 U.S.C. 71 *et seq.*), authorizes the Grain Inspection, Packers and Stockyards Administration to establish official marketing standards (not health and safety standards) for grains and oilseeds, and requires that exported grains and oilseeds be officially weighed and inspected. Domestically marketed grain and oilseeds may be, but are not required to be, officially inspected. Export inspections are carried out by federal inspectors or by federally supervised state inspection agencies, called delegated official inspection agencies. Official inspections of domestically traded grain is done by federally supervised state agencies and private companies, called designated official inspection agencies. Typically, marketing standards describe the physical characteristics (such as weight, damaged kernels, foreign material, shrunken and broken kernels, and defects) of the commodity and serve as contract language to facilitate marketing. Official weighing and inspection is paid for on a fee-for-service basis, not with federal funds. Major changes to the law were adopted in the USGSA Amendments of 1968, the USGSA of 1976 (P.L. 94-582), and the Grain Quality Improvement Act of 1986 (P.L. 99-641).

United States Trade Representative (USTR) — The Office of the U.S. Trade Representative, originally Office of the Special Trade Representative, is responsible for developing and coordinating international trade, commodity, and direct investment policy, and leading or directing negotiations with other countries. It is headed by the United States Trade Representative (also USTR), a Cabinet-level official with the rank of Ambassador. The agency provides trade policy leadership and negotiating expertise on all matters within the World Trade Organization (WTO); trade, commodity, and direct investment matters dealt with by international institutions such as the Organization for Economic Cooperation and Development (OECD) and the United Nations Conference on Trade and Development (UNCTAD); export expansion policy; industrial and services trade policy; international commodity agreements and policy; bilateral and multilateral trade and investment issues; trade-related intellectual property protection issues; and import policy. The agency has administrative responsibility for the Generalized System of Preferences (GSP); Section 301 complaints against foreign unfair trade practices; unlawful and unfair import competition under Section 337; and import relief cases under Section 201. [www.ustr.gov].

United States-Canada Free Trade Agreement Implementation Act of 1988 — P.L. 100-449 implemented the bilateral trade agreement between the United States and Canada, including agricultural trade. The agreement phased out tariffs between the two countries over 10 years and revised other trade rules.

Universal soil loss equation — A formula used to estimate erosion rates by considering climate, soils, and topographic conditions at a site, as well as any degree to which the use and management of the soil reduce erosion. It has been replaced by the revised universal soil loss equation (RUSLE).

Unreasonable risk — Under the Federal Insecticide, Fungicide, and Rodenticide Act (FIFRA; 7 U.S.C. 136 *et seq.*), "unreasonable adverse effects on the environment means (1) any unreasonable risk to man or the environment, taking into account the economic, social, and environmental costs and benefits of any pesticide, or (2) a human dietary risk from residues that result from a use of a pesticide in or on any food..." in excess of that allowed by a tolerance.

Upland cotton — The predominant variety of cotton grown in the United States; upland cotton has long been eligible for government price support and is eligible for support under the 2002 farm bill (P.L. 101-171, Title I) as a loan commodity and a covered commodity.

Uplands — Land at higher elevations than the alluvial plain or low stream terrace; all lands outside the riparian-wetland and aquatic zones. Used often in discussion of federal land management practices.

UR — Uruguay Round.

Urban and built-up areas — A classification in the natural resources inventory, now called developed areas, that includes cities, villages, other build-up areas of more than 10 acres, industrial sites, railroad yards, cemeteries, airports, golf courses, shooting ranges, institutional and public administration sites, and similar areas. The 1997 national resources inventory placed over 98 million acres in this category, an increase of 25 million acres since 1982.

Urban growth boundary — An agreed-upon line that allows development and urban-type services on one side and retains open space, agriculture, and other less intense uses on the other. These lines typically are set by local government at a place that can encompass many years of development at anticipated rates on the urban side of the boundry. Oregon is well-known for its use of these boundaries.

Urea — Urea is a nitrogen fertilizer used for direct application, in mixed fertilizers, and in liquid nitrogen. Plants can use nitrogen (N) in one of two forms: ammonium nitrogen ($NH4+$) or nitrate nitrogen ($NO3-$). Urea-N rapidly hydrolyzes to $NH4+$.

Uruguay Round — The 8th round of multilateral trade negotiations (MTN) conducted within the framework of the GATT. Launched in Punta del Este, Uruguay, in 1986 and concluded in December 1993, the final Uruguay Round agreement signed in Marrakech in April 1994, embraces 110 participating countries ("contracting parties") and came into effect in 1995. It has been implemented over the period to 2000 (2004 in the case of developing country contracting parties) under the administrative direction of the newly created World Trade Organization (WTO). The Uruguay Round Agreement on

Agriculture, administered by the WTO, brings agricultural trade more fully under the GATT. It provides for converting quantitative restrictions to tariffs and for a phased reduction of tariffs. The agreement also imposes rules and disciplines on agricultural export subsidies, domestic subsidies, and sanitary and phytosanitary (SPS) measures.

Uruguay Round Agreements (URA) Act of 1994 — P.L. 103-465 approved and implemented the trade agreements concluded in the Uruguay Round of multilateral trade negotiations conducted under the auspices of the GATT, including the Agreement on Agriculture, the Agreement on the Application of Sanitary and Phytosanitary (SPS) Measures, and the Agreement on Technical Barriers to Trade (TBT). The law allowed for the reduction of tariffs and government subsidies on agricultural products and prohibits the use of Section 22 fees and quotas with respect to products imported from WTO members. The law also extended the authorization of funding for the Export Enhancement Program (EEP) and Dairy Export Incentive Program (DEIP) through 2001 and eliminated the requirement that the EEP be targeted to respond to unfair trade practices. The law eliminated the requirement that the Market Promotion Program be used to counter the adverse effects of unfair trade practices. The law also included a Sense-of-Congress resolution that the President should consult with other nations to discuss appropriate levels of food aid commitments to developing countries.

USAA — U.S. Apple Association. [www.usapple.org].

USAHA — U.S. Animal Health Association. [www.usaha.org].

USAID — U.S. Agency for International Development. [www.info.usaid.gov].

USAIP — United States Animal Identification Plan. [usaip.info].

USB — United Soybean Board. [www.unitedsoybean.org].

USDA — U.S. Department of Agriculture. [www.usda.gov].

USDEC — U.S. Dairy Export Council. [www.usdec.org].

User fees — Any of various charges and assessments levied on a specifically delineated group that is directly subject to a particular government service, program, or activity (sometimes but not always voluntary); such fees are not levied on the general public. User fees are intended for use solely to support that service, program, or activity. For example, about two-thirds of the nearly $300 million budget (FY2002 estimate, program level) of the Agricultural Marketing Service (AMS), which provides a variety of inspection and grading, market news reporting, and other services to the agricultural community, comes from user fees; the other third is appropriated funds. Similarly, grain inspection by the Grain Inspection, Packers & Stockyards Administration (GIPSA) is paid for through user fees.

User marketing certificates — See step 2 payments.

USITC — U.S. International Trade Commission. [www.usitc.gov].

USMEF — U.S. Meat Export Federation. [www.usmef.org].

USTR — United States Trade Representative. [www.ustr.gov].

Usual market requirements (UMR) — A measure of the import requirement of a country met through commercial purchases; usually defined as a five-year average. The UMR is used to determine whether concessional sales (e.g., under Title I of P.L. 480) will adversely affect normal commercial agricultural trade.

Utilization rates — The percentage of milk in federal milk marketing orders that is used in each of the classes: Class IV (butter and nonfat dry milk), Class III (cheese), Class II (all other manufactured products), Class I (milk used for fluid consumption). Utilization rates serve as weights for determining the weighted average price, or blend price, received by dairy farmers within a region each month.

UV — Ultraviolet.

Value-Added Producer Grants — The Value-added Producer Grants program was authorized by the Agriculture Risk Protection Act of 2000 and amended by the 2002 farm bill (P.L.107-171, Sec. 6401). The farm bill authorizes $40 million per year in grants for the development and marketing of value-added agricultural products, including organic agricultural production. Eligible applicants are independent producers, farmer and rancher cooperatives, agricultural producer groups, and majority-controlled producer-based business ventures.

Value-added products — In general, products that have increased in value because of processing; such products include wheat flour and soybean oil. Livestock are considered value added products because they have increased the value of pasture and feed grains going into them. The terms value-added and high-value are often used synonymously, if imprecisely.

Value-added agriculture — Value-added refers most generally to manufacturing processes that increase the value of primary agricultural commodities. Value-added agriculture may also refer to increasing the economic value of a commodity through particular production processes, e.g., organic produce, or through regionally-branded products that increase consumer appeal and willingness to pay a premium over similar but undifferentiated products. This concept has gained currency in the small farm policy debate, in response to the concern that the farm value of the consumer food dollar continues to decrease. Value-added agriculture might be a means for farmers to capture a larger share of the consumer food dollar. Examples include direct marketing; farmer ownership of processing facilities; and producing farm products with a higher intrinsic value (such as identity-preserved grains, organic produce, hormone-free beef, free-range chickens; etc.), for which buyers are willing to pay a higher price than for more traditional bulk commodities. Value-added agriculture is regarded by some as a significant rural development strategy. Small-scale, organic food processing, non-traditional crop production, agri-tourism, and bio-fuels development are examples of various value-added projects that have created new jobs in some rural areas.

Value-added tax — A tax on the amount that a business firm adds to the price of a commodity during production and distribution of a good. While the tax is collected at each stage of the production-distribution chain, it is effectively passed on to the final consumer because each seller receives a credit for their tax payments. This tax is used throughout Europe, and by many countries in South America, Asia, and Africa.

Value-based pricing — Packers are increasingly using this method of determining how much to pay cattle and hog producers for animals. Rather than simply paying a fixed rate based on the weight of the animals, value-based pricing attempts to establish the individual merits of each animal (or lot) purchased, factoring quality characteristics such as yield, fat thickness, likely grade (such as choice, select, etc.) into a formula to arrive at the price that will be paid. Under this system, the producer assumes the financial responsibility that the animals, once slaughtered, will meet these criteria. In traditional pricing methods, it is the packer that bears the greater financial risks associated with the uncertain quality of the animals purchased.

Value-chains — A term denoting the increased vertical integration of agro-food production. Commodity supply-chains are evolving into value-chains to rigorously control integration and management of all phases of food manufacturing from the farm to the grocery shelf. Of central importance to a value-chain is the capacity to assure quality and traceability throughout the chain by identity-preserved (IP) production for specific end-users.

Variable import levy — A charge levied on imports that raises their price to a level at least as high as the domestic price. Such levies are adjusted frequently (hence variable) in response to changes in world market prices, and are imposed to defend administered prices set above world market prices. Under the Uruguay Round Agreement on Agriculture, the variable levies of the EU have been converted into fixed tariffs or tariff-rate quotas.

Variable rate technologies — A system that uses measurements of varying conditions across a field to control the rate at which inputs, such as seed, fertilizer, or crop protection chemicals, are applied. Variable rate technologies are used in precision agriculture.

VAT — Value-added tax.

Vegetative controls — Nonpoint source pollution control practices that involve planting cover crops to reduce erosion and minimize loss of pollutants.

VER — Voluntary export restraint agreement.

Vertical coordination — The process of ensuring that each successive stage in the production, processing, and marketing of a product is appropriately managed and interrelated to the next, so that decisions about what to produce, and how much, are communicated as efficiently as possible from the consumer to the producer. Agricultural economists believe that vertical coordination of markets is particularly important in the food industry because of its complexity, the large number of firms that participate in one or more stages, and the relative perishability of the products involved. Vertical integration is a type of vertical coordination, but the latter does not necessarily require that a single organization own or control all of the stages. For example, the use of contracts and marketing agreements between buyers and sellers, and the availability of timely, accurate price and other market information are methods for achieving vertical coordination.

Vertical integration — The integrating of successive stages of the production and marketing functions under the ownership or control of a single management organization. For example, much of the broiler industry is highly vertically integrated in that processing companies own or control the activities from production and hatching of eggs, through the growth and feeding of the chickens, to slaughter, processing, and wholesale marketing. Hog production during the 1990s and early 2000s also was rapidly moving toward vertical integration (or at least vertical coordination.).

Vesicular stomatitis — *Vesicular stomatitis* is a viral disease that can affect horses, swine, cattle, and other ruminants. It causes affected livestock to become lame, and to refuse food and water due to mouth sores. The disease is of concern because its symptoms are similar to those of foot-and-mouth disease. The only way to diagnose and differentiate these diseases is through laboratory tests. There have been occurrences of *vesicular stomatitis* in the southwestern United States in 1985, 1995, and 1998. People who handle infected animals also can become infected. The Animal Plant Health Inspection Service (APHIS) is working with state officials to identify all cases of the disease and prevent its spread.

Veterinary biologics — Vaccines, antigens, antitoxins and other preparations made from living organisms (or genetically engineered) and intended for use in diagnosing, treating, or immunizing animals. Unlike some pharmaceutical products, such as antibiotics, most biologics leave no residues in animals. Veterinary biologics are regulated by the Animal and Plant Health Inspection Service (APHIS), which licenses the facilities that produce them and conducts a program to ensure that animal vaccines and other veterinary biologics are safe, pure, potent, and effective.

Veterinary equivalency — The mutual recognition by two or more countries that each party's safety and sanitation standards for animal products, even where not identical, provide an equivalent level of protection to public and animal health. Aimed at facilitating trade, the practical effect of veterinary equivalency is that each country's individual products and facilities will not have to submit to the separate standards of importing countries and to cumbersome and costly inspections by foreign reviewers. Veterinary equivalency has been a contentious issue for the United States and European Union (EU). The two parties in 1997 agreed in principle to an agreement recognizing each other's standards, but did not sign the agreement until July 1999 due to a series of unresolved technical disputes. Some parts of this agreement remained in dispute in 2002.

Veterinary Feed Directive (VFD) — A veterinary feed directive is a written statement that authorizes the owner or caretaker of animals to obtain and use animal feed containing VFD drugs to treat their animals in accordance with the FDA-approved directions for use. A VFD drug is a new animal drug approved under section 512(b) of the Federal Food, Drug, and Cosmetic Act. VFD drugs are limited to use under the professional supervision of a licensed veterinarian. No extra-label uses of a VFD drugs are permitted. VFD drugs are a category created as part of the Animal Drug Availability Act of 1996 (P.L. 104-250) and final regulations are published by the FDA's Center for Veterinary Medicine in 21 CFR 510, 514, and 558.

VFD — Veterinary Feed Directive.

Virus-Serum-Toxin (VST) Act — The VST Act (P.L. 430 of 1913, as amended; 21 U.S.C. 151-158) forbids the preparation, sale, barter, exchange, or shipment of worthless, contaminated, dangerous or harmful animal biologicals in interstate commerce. The Act provides for licensing of products and establishments, and requires permits for the importation of animal biologicals. Regulations (9 CFR 101-117) require preparation of biologicals in accordance with an approved Outline of Production so as to meet prescribed test requirements for purity, safety, potency, and efficacy. Regulations also require approval of all labeling and claims to be made in advertising. Animal biologicals are subject to both the Federal Food, Drug, and Cosmetic Act (FFDCA; 21 U.S.C. 321 *et seq.*) and the VST Act. The USDA Animal and Plant Health Inspection Service (APHIS) enforces the VST Act.

VOC — Volatile organic compounds.

Volunteer Fire Assistance — A State & Private Forestry program under which the Forest Service provides financial and technical assistance to states for grants and agreements with volunteer fire departments that protect communities of fewer than 10,000 people.

Volunteer plant — Generally, a plant that is not intentionally grown or cultivated, but which grows spontaneously. This includes plants derived from seeds unintentionally included in planting, plants derived from seeds dropped by a previous year's crop, or an uncultivated self-sown plant. In cases of genetically modified (GM) plants, some biotechnology critics have voiced concern that a volunteer GM plant may show up in non-GM crops. Biotechnology proponents argue that USDA's existing regulations regarding GM crops are sufficient to minimize chances of unwanted GM material in the food supply.

Vomitoxin — Deoxynivalenol (DON), also referred to as vomitoxin, is a naturally occurring mycotoxin produced by several species of Fusarium fungi. It can infect barley, wheat, oats, and rye. Vomitoxin is not a threat to public health among the general population. However, in rare cases it can produce acute nausea and vomiting in humans and animals. The Food and Drug Administration (FDA), which regulates animal feeds as well as human foods, sets an advisory level for its presence in livestock rations, and advises food manufacturers to limit its presence in finished products to less than 1 part per million.

WAOB — World Agricultural Outlook Board. [www.usda.gov/agency/oce/waob/waob.htm].

Warehouse receipt — A document certifying possession of a commodity in a licensed warehouse. Some warehouse receipts are recognized for delivery purposes by a commodity exchange.

WASDE — The acronym for *World Supply and Demand Estimates*, the official USDA monthly report on supply, demand, prices and other data for major agricultural commodities published by the World Agricultural Outlook Board. [www.usda.gov/agency/oce/waob/index.htm].

Wash versus trim — USDA requires that any time fecal contamination is detected during meat and poultry processing, it must be removed from the carcass. At issue is how this rule has been applied and enforced by USDA in meat and poultry plants. For a number of years, poultry processors have been permitted to either rinse (wash) off or cut (trim) away such contamination, but beef processors have only been permitted to (trim) it with a knife. This, they argue, costs them money in lost product weight and imposes a requirement that poultry producers do not have to meet. The policy jargon for this debate is "wash versus trim." USDA, early in 1997, clarified its zero tolerance rule for poultry; a year earlier it gave beef plants permission to use a new high-temperature vacuuming method to remove fecal contamination in lieu of cutting it off.

Waste treatment pond — A shallow lagoon or similar storage facility, often man-made, used to treat liquid agricultural wastes, particularly liquid manure from livestock production farms, through the interaction of sunlight, wind, algae, and oxygen. Through natural biological processes, microscopic organisms consume wastes present in the water.

Water 2000 Initiative — The program administered by the Rural Utilities Service whose goal is to improve the quality of drinking water in distressed rural areas with the most serious safe drinking water problems.

Water Bank Program (WBP) — A program to set aside wetlands for a period of 10 years (renewable) for conservation purposes. Participants receive annual rental payments. As these contracts expire, participants are offered the opportunity to place the land in the Wetland Reserve Program (WRP). Land is no longer being enrolled in Water Bank.

Water banking — The practice of foregoing water deliveries during certain periods, and "banking" either the right to use the foregone water in the future, or saving it for someone else to use in exchange for a fee or delivery in kind. Usually used where there is significant storage capacity to facilitate such transfers of water. It is typically regulated and managed at the state level.

Water quality standards — State-adopted and EPA-approved ambient standards for water bodies. The standards prescribe the use of the water body and establish the water quality criteria that must be met to protect designated uses, and contain policies to protect against degradation of water quality once standards are attained and maintained.

Water Quality Incentives Program — This program was authorized in the 1990 farm bill (P.L. 101-624) and is administered by the Farm Service Agency. It was repealed and replaced by the Environmental Quality Incentives Program (EQIP) in the 1996 farm bill (P.L. 104-127). It provided cost-share assistance to implement comprehensive water quality protection plans and was funded by earmarking a portion of the Agricultural Conservation Program.

Water Quality Initiative — A multi-agency effort, initiated by USDA in 1990, to determine relationships between agricultural activities and water quality, and develop and implement strategies that protect surface and groundwater quality. This program, which builds on earlier USDA water quality protection efforts, includes research activities, projects involving landowners, and information and data development. Landowners participate in demonstration projects, hydrologic unit area projects, water quality special projects, and water quality incentive projects.

Water service contract — A type of contract, authorized by the Reclamation Project Act of 1939, whereby water is furnished for irrigation or municipal or miscellaneous purposes at rates to produce revenue sufficient to cover charges reimbursable to the federal government.

Waterfowl production areas — A small component of the National Wildlife Refuge System. There are over 2 million acres of this prime duck-producing land, mostly prairie potholes in the Dakotas, Minnesota, and Montana. The Fish and Wildlife Service owns, leases, or holds easements on the lands.

Watershed and Flood Prevention Operations Program — This program of the Natural Resources Conservation Service (NRCS) includes Watershed Operations (under the Flood Control Act of 1944, P.L. 78-534), Emergency Watershed Protection, and Small Watershed Operations (under the Watershed Protection and Flood Prevention Act of 1954 (P.L. 83-566)). These programs have built small watershed projects that reduce floods, protect watersheds, improve water quality, reduce soil erosion, improve water supply, and provide recreation. They involve strong partnerships with local interests. They are called small watershed projects to distinguish them from larger downstream projects built by the U.S. Army Corps of Engineers and by the Bureau of Reclamation.

Watershed Program — See Watershed and Flood Prevention Operations Program.

Watershed Protection and Flood Prevention Act of 1954 — P.L. 83-566 established USDA's Small Watershed Program administered by the Natural Resources Conservation Service (NRCS). Purposes include flood reduction, sediment and erosion control, and water conservation. Since its inception, over $4.2 billion has been appropriated to the Small Watershed Program, which has constructed more than 1,600 projects and more than 10,000 structures. Also known as the PL-566 program (16 U.S.C. 1001 *et seq.*). The Small Watershed Program is one of the programs that make up the Watershed and Flood Prevention Operations Program.

Watershed — The total land area, regardless of size, above a given point on a waterway that contributes runoff water to the flow at that point. It is a major subdivision of a drainage basin. The United States is generally divided into 18 major drainage areas and 160 principal river drainage basins containing about 12,700 smaller watersheds.

Wellhead protection area — A surface and subsurface land area regulated to prevent contamination of a well or well-field supplying a public water system. This program, established under the Safe Drinking Water Act (42 U.S.C. 330f-300j)is implemented through state governments.

WEQ — Wind erosion equation.

Western Grain Transportation Act (WGTA) — Canadian legislation under which a transport subsidy, informally called the Crow benefit or Crow subsidy, was provided by the Canadian federal government to assist the rail transportation of specified grains and grain products to specified destinations within Canada for export. Since the producer paid only a portion of the freight rate, the WGTA had the effect of increasing the prices received by grain producers and paid by livestock producers on the prairies. The subsidy encouraged the movement of grain east and west to export shipping terminals. The Act was abolished as part of budget balancing initiatives undertaken in 1995. Subsequently, grain began moving south for transshipment through the United States.

Wet-milling — A process in which feed material is steeped in water, with or without sulphur dioxide, to soften the seed kernel in order to help separate the kernel's various components. For example, wet-milling plants can separate a 56-pound bushel of corn into more than 31 pounds of starch (which in turn can be converted into corn sweeteners or ethanol), 15 pounds of animal feed, and nearly 2 pounds of corn oil.

Wetland mitigation banking — A system used to offset destruction of wetlands that are regulated under section 404 of the Clean Water Act (33 U.S.C. 1344) or state wetland law. A bank is established by creating or restoring a wetland. After the bank is in place, any entity who would modify or destroy a wetland and is required by the section 404 regulatory program to mitigate that action can meet their obligation by buying credits in the mitigation bank. The purchase of credits offsets the costs associated with establishing the bank, and if the bank is a private enterprise, can be used to generate a profit. Some agencies, especially state transportation agencies, have established banks for their projects, but private companies have established them as profit-making ventures as well. Many view mitigation banking as a more successful option to mitigation on a case by case or site by site basis.

Wetlands Reserve Program (WRP) — A program authorized by 1990 farm bill (P.L. 101-624) to provide long-term protection for wetlands. Producers enrolling in the program must agree to implement approved wetland restoration and protection plans. In return, participating producers receive payments based on the difference in the value of their land caused by placing an easement on a portion of it. The program reached its authorized enrollment ceiling of 1,075,000 acres before the 2002 farm bill (P.L. 107-171) was acted upon. The 2002 legislation reauthorized the program with mandatory funding from the Commodity Credit Corporation (CCC) through FY2007, and set a maximum enrollment ceiling of 2.275 million acres (and with an annual enrollment ceiling of 250,000 acres).

Wetlands — Areas of predominantly hydric soils that can support a prevalence of water-loving plants, know as hydrophitic vegetation. Transitional between terrestrial and aquatic systems are wetlands typified by a water table at or near the surface, or the land is covered by shallow water at least part of the year. Types of wetlands are distinguished by water patterns (the frequency and length of flooding) and location in relation to upland areas and water bodies. Wetlands perform many functions including wildlife and fish habitat, storage and conveyance of flood waters, sediment and pollution control, and recreation. Under the swampbuster program, landowners may produce crops in these areas, but only if the water patterns, or hydrology, in the wetland area is not altered and any woody vegetation is not removed.

WFP — United Nations World Food Program. [www.wfp.org].

WFPO — Watershed and Flood Prevention Operations.

WGA — Western Growers Association. [www.wga.com].

WGTA — Western Grain Transportation Act.

Wheat — Wheat is one of the so-called covered commodities and loan commodities eligible for marketing loan benefits, direct payments, and counter-cyclical payments under the 2002 farm bill (P.L. 101-171, Title I). The top five producing states account for 54% of national production (based on 2003 crop data): Kansas (20%), North Dakota (14%), Oklahoma (8%), Washington (6%), Montana (6%). In the United States, wheat is divided into five official classes that have relatively uniform characteristics related to milling and baking: hard red winter (HRW), hard red spring (HRS), soft white (SW), soft red winter (SRW), and durum. Marketing assistance loan rates are separately set for each of these classes of wheat. Hard white wheat is another class that is produced only in a small quantity in the United States, but is being encouraged by a new Hard White Wheat Incentive Program (P.L. 107-171, Sec. 1616). Wheat is milled into flour by grinding it between corrugated steel rollers and screening it to separate the endosperm from the bran and germ. Bulgur is a parched crushed wheat. Farina is a coarsely ground endosperm free of bran and germ. Flour is finely rolled and sifted farina. Semolina is a grind like farina, but from only durum wheat and is used for pasta products. Also known as the extraction rate, it is the amount of flour (represented as a percent of total weight) that can be made from a given unit of wheat kernels. The flour extraction rates (flour yields) vary considerably by grain quality and class, but tend to average about 70 to 75% of total weight from unmilled wheat in the United States. The remaining 25-30% is called millfeed (MF), wheat mill run (WMR), or wheat middlings (WM) and is normally used as livestock feed).

Whey — A collective term referring to the serum or watery part of milk that remains after the manufacture of cheese. Whey can be transformed into a dry product by different techniques, and the quality of the product varies with the technology applied.

WHIP — Wildlife Habitat Incentive Program.

WHO — World Health Organization. [www.who.org].

Whole herd buyout program — Another term for the dairy termination program.

Wholesale price index — A composite index of prices of commodities sold in primary U.S. markets. Wholesale refers to sale in large quantities by producers, not to prices received by wholesalers, jobbers, or distributors. In agriculture, it is the average price received by farmers for their farm commodities at the first point of sale when the commodity leaves the farm.

WIC Farmers' Market Nutrition Act of 1992 — P.L. 102-314 established a program authorizing projects that provide participants in the Special Supplemental Nutrition Program for Women, Infants, and Children (WIC) with food coupons that can be used to purchase fresh, unprocessed foods, such as fruits and vegetables at farmers' markets.

WIC vendors — Grocery and other stores authorized as eligible to accept WIC coupons or vouchers and to receive reimbursement from the state WIC program for purchases made with these food instruments.

WIC vouchers (coupons) — Food instruments commonly issued by WIC agencies to participants that are used in grocery and other authorized food stores to buy certain quantities and types of foods listed on the coupon, which are designated by the state as being authorized for purchase under the WIC program.

WIC — See Special Supplemental Nutrition Program for Women, Infants and Children.

Wilderness Preservation System — Federal land set aside to be managed to minimize human impact and maintain its pristine character. Congress has designated more than 104 million acres in 44 states as wilderness, and more than half of this land is in Alaska.

Wilderness study areas — Federal lands which the administering agencies are required to examine for potential additions by Congress to the national Wilderness Preservation System. The areas generally are to be managed to preserve their wilderness characteristics until Congress decides whether to include them in the system.

Wilderness — An area of pristine federally-owned public land where the impact of man is largely unnoticeable, and which is managed to minimize any impacts. Federal land managed as wilderness often has been designated by Congress as a unit in the Wilderness Preservation System.

Wildland-urban interface — Lands within and adjacent to (usually within ½ mile from) communities that abut or are surrounded by wildlands potentially subject to wildfires.

Wildlife corridors — A relatively narrow passage between two larger areas that provide habitat for wildlife. Corridors can range from lengthy passages for large animals who roam over broad territories, such as grizzly bears to relatively short passages, perhaps between ponds, for amphibians. As development, and especially highways have become barriers to wildlife movement, corridors that avoid or bypass these barriers have become increasingly important to maintaining animal populations.

Wildlife Habitat Incentive Program (WHIP) — A program established by the 1996 farm bill (P.L. 104-127) to promote voluntary implementation of on-farm conservation practices to develop habitat for wetland and upland wildlife, threatened and endangered species, fish and other types of wildlife using cost-share payments and technical assistance. Between its inception and the start of FY2002, the program enrollment included 10,729 long term agreements on over 1.6 million acres. The 2002 farm bill (P.L. 107-171, Sec. 1241) reauthorized the program through FY2007 with mandatory annual funding from the Commodity Credit Corporation (CCC), growing from $15 million in FY2002 to $85 million in FY2005 through FY2007. It also created a pilot program using up to 15% of the funding to provide additional payments to land owners who agree to enroll land for at least 15 years.

Wildlife Refuges — Units of the National Wildlife Refuge System. They may be designated under general authorities of the Migratory Bird Treaty Act (16 U.S.C. 703 *et seq.*), the Endangered Species Act (16 U.S.C. 1531 *et seq.*), or (rarely) by specific acts of Congress. There are 537 refuges, with over 92 million acres.

Wildlife Services (WS) Program — An Animal and Plant Health Inspection Service (APHIS) program to protect agriculture, natural resources, property, public health, and endangered species from unwanted and potentially harmful effects of wildlife species, particularly predators. WS also works to prevent wildlife/airplane collisions at civilian and military airports. The program was called the Animal Damage Control Program until August 1997. [www.aphis.usda.gov/ws].

William F. Goodling Child Nutrition Reauthorization Act of 1998 — This law (P.L. 105-336) extended expiring authorizations for child nutrition and commodity assistance programs, and the WIC program, through FY2003. Among other things, this law significantly expanded the availability of federal subsidies (through the school lunch program and the CACFP) for snacks served in after-school programs and authorized demonstration projects providing free breakfasts for elementary schoolchildren without regard to family income. Modest revisions were made to child nutrition and WIC program rules.

Wind erosion equation — An equation used to design wind erosion control systems, which considers soil erodibility, soil roughness, climate, the unsheltered distance across a field, and the vegetative cover on the ground.

Wind erosion — The detachment and transportation of soil by wind. Wind erosion is a cropland management concern in the Plains states. According to the 1997 National Resources Inventory, wind was eroding 840 million tons of soil per year, while water (excluding gully erosion) was eroding 1,060 million tons of soil per year from all cropland (including land enrolled in the Conservation Reserve Program (CRP).

Windbreak — A living barrier that usually includes several rows of trees, and perhaps shrubs, located upwind of a farm, field, feedlot or other area and intended to reduce wind velocities. Windbreaks, also called shelterbelts, can reduce wind erosion, conserve energy or moisture, control snow accumulations, and provide shelter for livestock or wildlife.

WMI — Wildlife Management Institute. [www.wildlifemanagementinstitute.org].

Wool Act of 1954 — See National Wool Act of 1954 (P.L. 83-690, Title VII).

Wool and mohair commodity programs — Income support was provided to producers of wool and mohair under authority of the National Wool Act of 1954 (P.L. 83-690, Title VII), as amended, through 1995. Phase down and termination of the programs was mandated in the omnibus budget reconciliation act enacted November 1, 1993 (P.L. 103-130). Income support was achieved through incentive payments that provided higher benefits to farmers who had more production and/or obtained high market prices. On an *ad hoc* basis wool and/or mohair support was provided in 1999 (P.L. 106-78, Sec. 801(h) and P.L. 106-224 Sec. 204(d)) and 2000 (P.L. 106-387, Seç. 814 and P.L. 107-25, Sec. 5). The 2002 farm bill (P.L. 107- 171, Sec. 1201) made both wool and mohair eligible for marketing assistance loans from 2002 through 2007.

WORC, WORC petition — The acronym stands for the Western Organization of Resource Councils, a private advocacy organization representing some western ranchers who want USDA to play a more prominent regulatory role in live cattle markets. In 1996, WORC submitted a controversial petition calling on the Department to initiate rulemaking to limit most forward contracting and cattle feeding by meat packers. The Department published the petition for public comment in January 1997 but, as of late 2002, had taken no further action. Turning to the 107th Congress, supporters in 2002 won an amendment in the Senate-passed version of omnibus farm legislation (S. 1731; H.R. 2646 as amended) that would have prohibited large packers from owning, feeding, or controlling livestock for more than 14 days prior to slaughter. However, conferees excluded the amendment from the final 2002 farm bill (P.L. 107-171).

World Agricultural Outlook Board (WAOB) — As part of the Office of the Chief Economist, the WAOB coordinates the commodity forecasting program; monitors global weather and analyzes its impact on agriculture; and coordinates USDA's weather, climate and remote sensing work. [www.usda.gov/oce/waob].

World Bank — A multilateral economic development institution established in 1945 to extend loans and technical assistance for development projects in developing countries. It is formally referred to as the International Bank for Reconstruction and Development. [http://www.worldbank.org].

World food security — See Food security.

World Food Program (WFP) — A UN agency that contributes commodities, services, and cash to developing countries to meet emergency food needs or to carry out economic and social development projects using food or local currencies generated from the sale of food aid commodities (monetizaton). The United States is the major contributor of commodities for use by the WFP in its emergency relief and development project activities. [www.wfp.org].

World price (rice) — As part of the rice marketing assistance loan program, USDA calculates the world price for each class of milled rice (long grain, medium grain, and short grain) based on the prevailing world market price for each of the classes, modified to reflect U.S. quality and the U.S. cost of exporting milled rice. USDA sets this prevailing market price after reviewing milled rice prices in major world markets, and taking into account the effects of supply-demand changes, government-assisted sales, and other relevant price indicators. The steps for calculating and announcing the world prices are prescribed in more detail in federal regulations.

World price; world market price — The price at which commodities will move in international trade under existing marketing conditions. The concept world price lacks precision unless quality, location, and other factors are specified. See Domestic price.

World Trade Organization (WTO) — The international organization established by the Uruguay Round of multilateral trade negotiations to oversee implementation of the General Agreement on Tariffs and Trade and the agreements arising from the Uruguay Round, including the Uruguay Round Agreement on Agriculture. [www.wto.org].

WPA — Waterfowl production areas. **WPS** — Worker protection standard. **WQIP** — Water Quality Incentives Program. **WRI** — World Resources Institute. [www.wri.org].

WRP — Wetlands Reserve Program.

WTO — World Trade Organization. [www.wto.org].

WWF — World Wildlife Fund. [www.wwf.org].

Yield monitoring — Collecting data on the amount of production at regular intervals combined with GPS readings. The resulting yield map is basic to decisions about fertilization, pest control, and other adjustments in a system of precision farming.

Yield — The number of bushels (or pounds or hundredweight) that a farmer harvests per acre. Under the Food Security Act of 1985 (P.L. 99-198), the farm program payment yield was the farmer's average yield for the 1981-1985 crop years, excluding the years when the yields were highest and lowest. Payment yields used to implement commodity programs remained frozen at the level fixed in the 1985 farm bill until 2002. Under the 2002 farm bill (P.L. 101-171, Sec. 1102(e)), payment yields could be updated only for counter-cyclical payments, not for direct payments.

Zero, 50/85-92 provisions — Refers to the 50/85 and 50/92 commodity program provisions for rice and cotton and the 0/85 and 0/92 commodity program provisions for wheat and feed grains that were in effect in various forms from 1986 through 1995. Under these provisions farmers could idle all or part of their permitted acreage, putting the land in a conserving use, and receive deficiency payments as if up to 92% of the permitted acreage had been planted. A minimum planting requirement of 50% of maximum payment acreage applied for rice and cotton.

Zero tolerance — In food safety policy, a zero tolerance standard generally means that if a potentially dangerous substance (whether microbiological, chemical, or other) is present in or on a product, that product will be considered adulterated and unfit for human consumption. In the meat and poultry inspection program, zero tolerance usually refers to USDA's rule that permits no visible signs of fecal contamination on meat and poultry carcasses. See Wash versus trim.

Zeroing — Zeroing refers to a controversial methodology used by the United States for calculating antidumping duties against foreign products. The foreign domestic price (FDP) of the product is compared with its U.S. import price (USIP) adjusted for transportation and handling costs. Under zeroing, the United States sets at zero the negative differences (that is whenever FDP minus USIP is less than zero). Critics of this methodology charge that, because negative amounts are excluded, zeroing results in the calculation of a margin and an antidumping duty in excess of the actual dumping practiced by the countries concerned. The European Union has called for establishment of a World Trade Organization dispute settlement panel to rule on the U.S. practice of zeroing.

Zoonotic diseases — Diseases that under natural conditions are communicable from animals to humans. Tuberculosis and rabies are examples of zoonotic diseases. Brucellosis in livestock is undulant fever in humans.

INDEX

A

acceptance, 91, 148, 220, 268
access, 28, 31, 32, 43, 46, 48, 54, 62, 79, 83, 85, 116, 118, 161, 169, 187, 210, 224, 228, 232, 252, 266, 269
accountability, 13
accounting, 62, 116, 120, 153
accreditation, 179
accumulation, 135
accuracy, 65
achievement, 256
acid, 2, 38, 69, 78, 81, 112, 168, 184, 204, 216, 217
acidity, 205
acquisitions., 137
ADC, 4, 17
additives, 42, 75, 114, 126
adjustment, 8, 37, 56, 59, 66, 119, 160, 188
adult education, 64
adults, 1, 120, 192
advertisements, 216
advertising, 23, 43, 65, 113, 123, 127, 161, 163, 178, 264, 276
advocacy, 282
affect, 2, 15, 17, 33, 40, 47, 48, 66, 89, 90, 136, 155, 168, 180, 188, 191, 199, 219, 265, 267, 273, 275
Africa, 38, 123, 135, 210, 244, 274
age, 44, 53, 78, 85, 234, 245, 246, 263
agent, 19, 23, 69, 77, 89, 174
aging, 93, 240
aging process, 93
agricultural exports, 24, 25, 36, 71, 84, 87, 114, 139, 149, 150, 172, 177, 181, 196
agricultural sector, 42, 84, 173
agriculture, vii, 1, 5, 7, 10, 11, 12, 15, 16, 17, 18, 25, 26, 29, 32, 34, 39, 42, 45, 46, 47, 48, 53, 54, 56, 59, 60, 61, 62, 63, 64, 73, 77, 79, 83, 87, 89, 91, 93, 95, 96, 100, 101, 103, 105, 107, 109, 114, 116, 118, 119, 120, 121, 122, 123, 126, 130, 134, 137, 140, 141, 144, 146, 152, 153, 154, 155, 165, 166, 170, 171, 173, 175, 181, 188, 191, 192, 194, 196, 198, 199, 206, 207, 211, 212, 213, 220, 222, 240, 245, 248, 249, 251, 253, 255, 256, 260, 264, 272, 274, 275, 280, 281, 283
air pollutants, 88, 175, 187
air quality, 13, 23, 90, 186
airborne particles, 47
alcohol, 15, 16, 92, 166
algae, 93, 277
alternative, 9, 16, 26, 29, 30, 44, 63, 67, 90, 116, 146, 148, 166, 167, 200, 221, 225, 229, 242, 244, 252, 255, 257
alternative energy, 252
alternatives, 144, 204
ambient air, 17
amendments, 11, 14, 47, 69, 89, 114, 160, 164, 166, 177, 191, 206, 222, 229
amino acids, 166, 215
ammonia, 17, 184
ammonium, 272
amylase, 247
anatomy, 16
anemia, 270
animal diseases, 190, 193, 207
animal husbandry, 195
animal welfare, 17, 19, 93, 168
animals, 6, 10, 15, 16, 18, 19, 20, 24, 25, 29, 30, 31, 41, 42, 43, 45, 49, 55, 71, 72, 75, 76, 78, 80, 81, 83, 88, 89, 94, 96, 100, 103, 110, 112, 113, 120, 121, 122, 127, 138, 140, 146, 148, 150, 155, 157, 158, 160, 165, 174, 182, 186, 190, 192, 197, 198, 205, 210, 213, 215, 217, 221, 230, 231, 242, 246, 250, 255, 260, 262, 267, 270, 274, 275, 276, 277, 281, 284
annual rate, 69

annuals, 239
anti-inflammatory drugs, 96
antitrust, 39, 64
APC, 210
appraisals, 241
aquatic systems, 279
aquifers, 134
Argentina, 37, 82, 135, 149
argument, 160, 254
ASEAN, 22, 23
Asia, 22, 244, 274
assessment, 23, 34, 57, 67, 72, 73, 78, 85, 95, 105, 113, 114, 121, 137, 154, 155, 163, 178, 179, 183, 185, 223, 225, 226, 263
assets, 37, 74, 99, 154, 183, 227, 239
assimilation, 43
association, 25, 44, 60, 203, 221
Association of Southeast Asian Nations, 22, 23
assumptions, 26
atoms, 45, 196
attacks, 118, 141
Australia, 22, 37, 65, 137, 145, 156, 195, 197, 227, 244
Austria, 92, 93
authority, 2, 8, 9, 10, 11, 16, 18, 20, 21, 25, 35, 38, 43, 50, 51, 52, 55, 60, 64, 66, 67, 76, 78, 80, 85, 86, 94, 99, 106, 108, 109, 113, 114, 115, 117, 119, 121, 125, 132, 133, 137, 141, 142, 146, 152, 155, 158, 159, 160, 161, 171, 175, 179, 180, 186, 188, 194, 196, 197, 198, 202, 204, 206, 210, 213, 218, 222, 233, 237, 240, 242, 243, 244, 252, 253, 258, 259, 262, 264, 265, 266, 282
availability, 78, 118, 143, 275, 282
averaging, 169, 191, 201
avian influenza, 24, 146
awareness, 105

B

bacteria, 17, 20, 31, 38, 49, 83, 106, 120, 237
bacterium, 31, 35, 157, 230, 248
balance sheet, 99
banking, 25, 177, 277, 279
banks, 86, 87, 98, 99, 116, 117, 128, 159, 243, 279
bargaining, 10
barriers, 7, 27, 56, 71, 92, 120, 123, 126, 128, 130, 145, 149, 161, 162, 172, 173, 181, 187, 188, 190, 222, 231, 233, 250, 258, 259, 260, 266, 281
barriers to entry, 162
barter, 25, 67, 276
basic needs, 209
behavior, 206
Belgium, 253
benchmarks, 160, 249, 258
beverages, 15, 78, 92, 139
binding, 35, 153, 164, 270
bioaccumulation, 29
biodiversity, 29
biological control, 30, 112, 148, 256
biological processes, 29, 39, 127
biomass, 29, 30, 31, 121, 138, 221
bioremediation, 121
biosafety, 31, 207
biosphere, 39
biotechnology, 5, 31, 32, 49, 65, 103, 248, 265, 276
birds, 24, 30, 73, 83, 92, 94, 112, 168, 189, 208, 218
birth, 18, 261, 270
bison, 35, 94, 226
blocks, 177, 196
blood, 17, 29, 110, 192, 251
blood stream, 29
bloodstream, 89, 251
blowing agent, 45
body, 23, 29, 30, 56, 89, 93, 129, 143, 149, 150, 151, 163, 173, 207, 219, 224, 232, 235, 242, 255, 264, 277
Bolivia, 37, 135
bonds, 98, 99, 108, 207
border control, 171
border inspections, 17, 109
borrowers, 98, 99, 100, 101, 158, 167, 223, 228, 234, 235, 236, 238
borrowing, 101, 141
brain, 124, 144, 212, 246, 251, 267
Brazil, 92, 252
breakdown, 15, 72
breakfast, 44, 61, 123, 139, 169, 192, 198, 219, 231, 245, 248
breastfeeding, 34
breathing, 38
breeding, 6, 19, 32, 35, 68, 205, 206, 210, 216, 217
brucellosis, 35, 270
budget deficit, 23, 163
budget resolution, 35, 219
buildings, 124, 140, 142, 167, 207, 214
Bulgaria, 89, 93, 149
burning, 47, 112, 199, 210, 239
business management, 147
butyl ether, 167, 172

C

caffeine, 44
calcium, 29, 166
Cambodia, 23, 197

Canada, 22, 27, 37, 38, 42, 88, 112, 134, 156, 162, 166, 168, 183, 188, 189, 190, 194, 195, 197, 217, 247, 266, 268, 272, 279
cancer, 15, 39, 45, 72, 75, 214
CAP, 7, 39, 53, 92, 168
capital account, 185
capital consumption, 100, 214
capital productivity, 214
carbohydrate, 15, 92
carbon, 29, 39, 45, 47, 57, 91, 92, 126, 133, 167, 176, 195, 196
carbon dioxide, 39, 133
carbon monoxide, 92, 126, 167
carcinogen, 45
Caribbean, 2, 40, 41, 66, 210
Caribbean countries, 40
carrier, 36, 53, 75, 146, 156, 257
casein, 169
cash flow, 100, 247
categorization, 229
catfish, 21, 112
cell, 124, 143, 174, 219
cellulose, 156
Census, 42, 65, 97, 101, 103, 106, 142, 151, 189, 203, 208, 209, 226, 229, 253
central bank, 93
Central Europe, 42
central nervous system, 5, 33, 166, 251
certificate, 8, 66, 88, 115, 163
cesium, 151
channels, 9, 41, 63, 77, 125, 136, 187, 224, 237, 255
chicken, 34, 172, 182, 204
children, 14, 15, 44, 53, 94, 116, 123, 181, 191, 192, 193, 231, 232, 233, 234, 245, 253, 254, 260, 263, 282
Chile, 22, 27, 37, 135, 161, 210
China, 22, 65, 77, 135, 172, 197, 259
chlorinc, 45
chlorophyll, 125
cholera, 140
cholesterol, 27, 78, 191
cholinesterase, 196
citizenship, 141
classes, 19, 32, 46, 113, 154, 250, 273, 280, 283
classification, 27, 46, 68, 77, 103, 122, 131, 137, 154, 169, 175, 189, 229, 239, 243, 272
Clayton Act, 46, 238
Clean Air Act, 2, 15, 23, 37, 47, 167, 171, 175, 186, 207, 220
clients, 111, 147
climate change, 39, 48
CNS, 5, 251
coal, 2, 16, 221, 270
cobalt, 151
collateral, 99, 122, 139, 141, 158, 161, 187, 238
colleges, 64, 93, 95, 122, 137, 154, 165, 171, 240, 248
collisions, 281
collusion, 238
Colombia, 37, 135
combustion, 79, 124, 185, 191, 197
commitment, 22, 32, 155, 172
commodity, 2, 3, 4, 5, 7, 8, 9, 11, 12, 13, 14, 15, 16, 22, 25, 27, 30, 35, 37, 40, 41, 43, 44, 45, 50, 51, 52, 54, 56, 59, 62, 63, 65, 66, 67, 68, 70, 74, 75, 76, 79, 81, 84, 85, 86, 88, 89, 94, 95, 98, 100, 101, 107, 110, 114, 115, 118, 120, 122, 125, 127, 129, 131, 134, 136, 145, 148, 150, 152, 158, 159, 160, 161, 162, 163, 164, 170, 180, 181, 182, 185, 186, 187, 188, 192, 193, 194, 195, 198, 199, 200, 201, 202, 203, 208, 210, 211, 213, 214, 215, 218, 219, 220, 221, 222, 225, 230, 238, 239, 244, 246, 247, 248, 250, 252, 253, 254, 255, 258, 259, 261, 263, 265, 268, 270, 271, 272, 274, 277, 280, 282, 283
commodity markets, 247
common external tariff, 53, 92
Common Market, 17, 166
Commonwealth of Independent States, 46, 197
communication, 53
community, 17, 49, 53, 54, 64, 77, 79, 84, 89, 97, 105, 114, 117, 128, 129, 142, 144, 167, 168, 198, 204, 227, 228, 229, 239, 242, 243, 249, 257, 259, 260, 273
community support, 79
compensation, 21, 35, 37, 80, 85, 95, 102, 133, 146, 213, 233, 257
competition, 41, 54, 61, 95, 128, 129, 145, 153, 169, 215, 230, 233, 261, 266, 272
competitive advantage, 269
competitive markets, 131
competitiveness, 66, 156, 239, 249, 260
complement, 30, 138
complexity, 275
compliance, 19, 42, 47, 56, 57, 70, 71, 91, 113, 118, 140, 142, 168, 170, 172, 193, 223, 230, 232
components, 24, 104, 116, 124, 134, 183, 202, 203, 210, 211, 213, 263, 269, 270, 279
composition, 125, 156, 251, 252
composting, 54
compounds, 15, 17, 31, 35, 38, 79, 103, 185, 195, 217
computer technology, 223
computer use, 78
computerization, 18
computing, 147
concentrates, 169

concentration, 29, 30, 47, 55, 60, 93, 123, 133, 155, 159, 197, 198, 239
concessional terms, 197
conditioning, 197
conduct, 13, 17, 20, 49, 51, 68, 107, 114, 125, 126, 131, 141, 145, 161, 166, 177, 179, 195, 208, 230, 258
confidence, 34, 144
confinement, 18, 19, 105
conflict, 118
conformity, 66, 219
congressional budget, 26
Congressional Budget Office, 26, 30
consensus, 80, 148
consent, 205, 225, 250
conservation, vii, 2, 3, 4, 10, 11, 13, 14, 25, 27, 28, 35, 36, 39, 49, 50, 55, 56, 57, 58, 59, 61, 68, 69, 70, 71, 79, 80, 81, 83, 91, 95, 97, 98, 101, 102, 107, 111, 112, 113, 118, 129, 133, 134, 136, 138, 140, 145, 147, 151, 154, 155, 156, 159, 168, 173, 178, 179, 182, 185, 186, 191, 198, 199, 200, 212, 214, 219, 222, 223, 237, 241, 242, 243, 249, 255, 260, 262, 268, 270, 277, 278, 281
consolidation, 10, 46, 50, 60, 68, 76, 102, 146, 147, 271
construction, 87, 96, 102, 124, 146, 153, 186, 218, 221, 228, 234, 236, 237, 251
consultants, 43
consulting, 180
consumer goods, 221
consumer price index, 263
consumers, 10, 19, 39, 53, 60, 61, 65, 66, 67, 71, 93, 104, 116, 117, 124, 127, 128, 148, 162, 191, 213, 216, 239, 251, 260, 264, 269
consumption, 5, 9, 12, 20, 46, 60, 73, 94, 100, 109, 113, 114, 116, 120, 126, 130, 149, 156, 168, 169, 189, 196, 208, 244, 252, 253, 255, 264, 265, 273, 284
contaminant, 34
contamination, 3, 16, 23, 49, 83, 106, 134, 138, 172, 184, 195, 202, 230, 231, 237, 248, 277, 279, 284
context, 16, 23, 37, 118, 145, 183, 219, 224, 243
contingency, 234
control, 1, 8, 11, 17, 21, 24, 25, 30, 31, 35, 47, 49, 63, 64, 66, 70, 71, 72, 77, 90, 94, 97, 109, 112, 121, 124, 125, 128, 133, 136, 138, 146, 147, 148, 150, 151, 153, 163, 166, 167, 168, 170, 181, 185, 187, 190, 193, 197, 201, 204, 207, 208, 210, 212, 216, 218, 225, 229, 230, 233, 242, 244, 250, 253, 254, 256, 263, 269, 270, 275, 278, 279, 282, 283
Controlled Substances Act, 146
conversion, 2, 7, 65, 78, 83, 105, 137, 145, 187, 188, 217, 256, 259
cooking, 39, 194
coping, 118
corn, 6, 8, 9, 15, 23, 25, 26, 28, 30, 31, 35, 36, 41, 48, 52, 62, 65, 70, 92, 110, 115, 127, 129, 131, 139, 140, 146, 152, 158, 160, 163, 164, 165, 169, 171, 174, 183, 187, 198, 201, 214, 216, 219, 244, 245, 248, 252, 253, 254, 258, 262, 279
corporations, 65, 200, 235, 236
corruption, 118
cost effectiveness, 63
cost saving, 269
Costa Rica, 135
cost-benefit analysis, 225
costs, 3, 9, 13, 15, 21, 28, 37, 38, 40, 46, 49, 51, 53, 54, 62, 66, 67, 86, 87, 95, 101, 107, 108, 116, 118, 123, 124, 134, 151, 158, 160, 162, 167, 171, 173, 183, 196, 206, 213, 219, 221, 226, 235, 236, 248, 255, 261, 262, 264, 269, 272, 277, 279, 284
Cotonou Agreement, 2, 66, 210
cotton, 3, 4, 6, 8, 9, 10, 11, 14, 25, 26, 31, 32, 35, 36, 52, 54, 62, 66, 75, 85, 95, 96, 97, 115, 119, 127, 131, 139, 145, 152, 156, 158, 162, 163, 164, 178, 182, 183, 187, 189, 194, 198, 201, 202, 208, 214, 216, 219, 245, 246, 248, 249, 254, 256, 258, 259, 272, 283
Council of Ministers, 168
counseling, 234
coupling, 168
coverage, 4, 9, 15, 19, 37, 41, 61, 70, 107, 134, 186, 210, 223, 267
covering, 7, 48, 52, 74, 75, 83, 176, 214, 224, 242, 271
credit, 13, 24, 48, 51, 52, 56, 68, 74, 76, 79, 80, 87, 96, 98, 99, 100, 101, 104, 107, 117, 126, 129, 130, 148, 155, 156, 167, 187, 197, 205, 223, 231, 234, 235, 238, 239, 254, 274
Creutzfeldt-Jakob disease, 19, 46, 212, 267
crime, 141
criticism, 100
Croatia, 89, 93
crop production, 59, 61, 69, 91, 104, 135, 169, 173, 185, 241, 256, 274
crops, 5, 8, 11, 13, 14, 15, 20, 21, 22, 23, 25, 26, 30, 31, 32, 35, 38, 43, 49, 54, 58, 59, 61, 62, 68, 69, 70, 71, 74, 80, 81, 82, 86, 96, 97, 99, 101, 103, 107, 109, 111, 112, 119, 127, 128, 131, 132, 134, 135, 139, 142, 143, 144, 146, 147, 151, 154, 159, 163, 164, 170, 173, 174, 176, 184, 185, 186, 188, 194, 195, 197, 199, 201, 203, 205, 206, 207, 209, 211, 214, 215, 216, 221, 223, 225, 226, 238, 242, 245, 250, 252, 255, 256, 260, 267, 268, 270, 275, 277, 280

CRP, 8, 25, 58, 61, 71, 90, 91, 101, 137, 138, 159, 241, 242, 282
CSCE, 71
CTA, 59
Cuba, 85, 135
cultivation, 11, 39, 59, 141, 146, 185, 195
cultural heritage, 173
culture, 143, 184, 234
currency, 123, 171, 230, 274
current prices, 134
curriculum, 248
customers, 65, 108, 213
CVD, 37, 72
cycles, 255
Cyprus, 89, 92, 93
Czech Republic, 89, 92, 93, 195

D

damage, 23, 30, 45, 66, 112, 124, 137, 154, 190, 194, 211, 247
danger, 33
database, 204, 224, 271
dating, 18
death, 28, 130, 140, 165, 184
death rate, 165
debt, 74, 98, 99, 100, 211, 219, 227, 259
debts, 74, 97
decision making, 90
decisions, 31, 40, 101, 148, 163, 167, 176, 222, 225, 232, 243, 257, 275, 283
decomposition, 54, 144, 166, 237
decoupling, 74
deduction, 74
defects, 261, 271
defense, 206
deficit, 23, 26, 247
definition, 5, 9, 12, 19, 23, 32, 55, 74, 76, 97, 103, 140, 143, 202, 206, 207, 218, 229, 239, 241, 245, 246, 253, 257
deflate, 60
degradation, 59, 121, 154, 185, 186, 222, 251, 278
delivery, 24, 39, 40, 44, 62, 64, 75, 79, 94, 122, 123, 125, 153, 159, 163, 189, 193, 215, 221, 230, 246, 247, 277
democracy, 172
Denmark, 92, 93
density, 71, 123, 226, 240, 247
Department of Defense, 44, 181
Department of Health and Human Services, 42, 116, 139, 192, 208
Department of the Interior, 36, 45, 76, 121, 201, 250
deposition, 2
depreciation, 183, 214
depression, 132, 209
destruction, 22, 197, 279
detachment, 282
detection, 16, 91
developed countries, 77, 153, 155, 197, 245, 270
development assistance, 2
deviation, 96
diarrhea, 83
dibenzo-p-dioxins, 79
diet, 78, 94, 116, 180, 245, 254
dietary intake, 78
differential treatment, 77, 135, 161, 229, 245
differentiation, 162
digestion, 183
digital divide, 78
dioxin, 79
direct investment, 271
directives, 109, 140
disability, 85
disaster, 14, 25, 43, 60, 79, 85, 86, 87, 96, 100, 101, 108, 113, 119, 133, 138, 157, 158, 186, 194, 210, 234, 267
disaster relief, 133, 234
disbursement, 141, 170
discharges, 11, 47, 84, 88, 110, 191, 207, 214, 264
discrimination, 181, 205
dislocation, 162
disposition, 234
disseminate, 121
dissolved oxygen, 29, 93
distillation, 166
distribution, 13, 44, 49, 51, 52, 64, 86, 87, 91, 109, 116, 128, 148, 160, 171, 177, 203, 228, 251, 261, 274
diversification, 10, 162, 214
diversity, 29, 31
division, 94
DNA, 81, 218, 219, 267
dogs, 5, 19, 226
Doha, 37, 81, 173, 201, 245, 257, 266
domain, 76
domestic demand, 162
domestic markets, 244
donations, 13, 28, 86, 116, 154, 180, 270
dosage, 96, 104, 124
drainage, 1, 43, 63, 74, 82, 87, 104, 152, 212, 234, 278
drinking water, 134, 184, 214, 229, 277
drought, 6, 138, 151, 177, 210, 247, 256
drug treatment, 142
drugs, 17, 18, 20, 33, 42, 48, 75, 96, 108, 146, 190, 276

dry matter, 19, 239
drying, 247
dumping, 4, 20, 37, 82, 83, 262, 267, 269, 284

E

early retirement, 170
earnings, 34, 77, 99, 103, 105
earth, 196
Eastern Europe, 7
ecology, 40
economic activity, 10
economic development, 78, 180, 197, 234, 259, 270, 283
economic indicator, 88
economic performance, 162
economic resources, 211
economic status, 90
economic theory, 54
economic welfare, 198
economics, 42, 64, 83, 220
economies of scale, 263
ecosystems, 29, 88, 92, 121, 189
Ecuador, 135
educational programs, 118
effluent, 19, 30, 55, 79, 84, 110, 237
egg, 53, 114, 127, 158
El Salvador, 135
elasticity, 85, 211
elasticity of demand, 85, 211
elderly, 41, 44, 51, 53, 74, 85, 117, 142, 181, 192, 209, 228, 235, 252
election, 118
electricity, 16, 78
elementary school, 282
ELISA, 85
elk, 17, 43, 45, 94, 226
embargo, 63, 85, 94
emergence, 31, 135, 212, 254
emerging markets, 87, 96, 216
emigration, 172
emission, 39, 79
emitters, 268
employees, 50, 68, 121, 138, 142, 182, 258
employment, 6, 97, 100, 116, 136, 168, 227
empowerment, 115
encephalopathy, 18, 27, 33, 69, 82, 110, 159, 166, 212, 221, 232, 267, 271
endocrine, 89
endocrine glands, 89
endosperm, 224, 280
England, 189
enlargement, 7, 89, 93, 101
enrollment, 36, 58, 61, 90, 103, 132, 222, 242, 279, 281
environment, 22, 38, 45, 47, 73, 79, 84, 88, 89, 90, 103, 108, 147, 161, 183, 184, 196, 203, 207, 211, 220, 225, 232, 245, 250, 265, 272
environmental contamination, 205
environmental degradation, 118
environmental effects, 90, 207
environmental factors, 135, 242
environmental impact, 177, 183
environmental policy, 177
environmental protection, vii, 79
Environmental Protection Agency, 47, 90, 91, 109, 177, 244, 248
environmental regulations, 205
environmental standards, 232
equality, 231
equipment, 7, 24, 37, 39, 42, 80, 86, 99, 100, 179, 181, 186, 214
equity, 90, 99, 161, 227
erosion, 11, 39, 48, 58, 59, 61, 63, 68, 90, 91, 111, 132, 136, 140, 173, 223, 224, 226, 238, 241, 243, 250, 251, 255, 272, 275, 278, 279, 282
estimating, 26, 95, 162, 225
Estonia, 89, 92, 93
estrogen, 78
ethanol, 15, 30, 40, 92, 126, 129, 183, 208, 221, 252
ethnicity, 90, 196
ethyl alcohol, 92, 167
EU, 2, 7, 21, 37, 43, 53, 65, 71, 92, 93, 95, 135, 138, 150, 168, 170, 171, 173, 209, 210, 215, 222, 251, 263, 275, 276
Euro, 92
Europe, 1, 34, 44, 87, 183, 189, 209, 274
European Central Bank, 93
European Commission, 92, 134
European Community, 53
European Monetary Union, 92, 93
European Regional Development Fund, 251
European Social Fund, 251
European Union, 2, 6, 7, 32, 53, 65, 66, 71, 92, 93, 134, 149, 254, 276, 284
evaporation, 63, 93, 151, 207
evidence, 209, 220, 243
excess supply, 242
exchange controls, 187
exchange rate, 92, 149
exclusion, 74
exercise, 22, 163, 249, 250
expectation, 38, 253
expenditures, 26, 45, 76, 107, 134, 208, 259, 263
expertise, 43, 150, 152, 193, 271
experts, vii, 34, 82, 110, 127, 260, 262

exploitation, 108
export promotion, 65, 95, 162, 269
export subsidies, 5, 7, 37, 50, 67, 72, 77, 95, 123, 133, 135, 170, 222, 266, 273
exposure, 4, 19, 23, 45, 159, 165, 185, 214, 225, 261
expression, 12, 204
extinction, 88
extraction, 104, 169, 262, 280

F

fabric, 112
FAD, 207
failure, 80, 162, 173
faith, 55, 129, 243
false negative, 145
false positive, 145
family, 15, 43, 44, 45, 65, 69, 81, 97, 100, 101, 103, 104, 123, 131, 155, 156, 174, 211, 216, 218, 228, 231, 234, 236, 237, 239, 258, 282
family income, 123, 231
family members, 65
famine, 116, 119
farm land, 43
farmers, 2, 3, 4, 8, 10, 11, 15, 24, 33, 35, 41, 43, 49, 50, 52, 53, 56, 61, 62, 63, 67, 68, 69, 70, 72, 73, 74, 75, 79, 81, 82, 84, 86, 87, 88, 97, 98, 100, 101, 104, 105, 107, 108, 111, 115, 123, 124, 128, 135, 142, 146, 147, 148, 152, 156, 158, 159, 162, 163, 168, 170, 171, 183, 187, 189, 194, 196, 197, 198, 199, 204, 205, 208, 211, 212, 213, 214, 218, 220, 223, 224, 225, 226, 227, 237, 239, 241, 242, 247, 251, 254, 255, 256, 257, 258, 262, 264, 268, 274, 280, 282, 284
farmland, 7, 12, 13, 58, 78, 83, 86, 99, 105, 204, 211, 230, 257, 263, 270
farms, 10, 12, 42, 47, 49, 55, 58, 63, 65, 78, 83, 90, 96, 100, 101, 102, 103, 104, 110, 112, 120, 128, 129, 133, 146, 156, 168, 170, 192, 193, 196, 200, 201, 204, 207, 213, 235, 239, 240, 251, 254, 257, 258, 263, 277
FAS, 65, 105, 120
fat, 17, 27, 30, 173, 216, 269, 274
fauna, 84
FCC, 106
FDA, 6, 17, 18, 19, 20, 31, 34, 42, 78, 93, 94, 96, 106, 108, 109, 110, 113, 114, 116, 117, 118, 127, 128, 129, 130, 138, 151, 152, 179, 181, 190, 232, 276, 277
feces, 24, 49
federal courts, 127
feedback, 47, 133
feet, 2, 27, 32, 65, 122, 136, 141, 165, 251, 263
fermentation, 92, 126, 129, 239
fertilization, 283
fertilizers, 11, 14, 17, 34, 63, 111, 155, 179, 184, 195, 209, 272
fetus, 261
fever, 35, 140, 256
fibers, 30, 112, 156
filtration, 269
finance, 25, 52, 134, 161, 168, 171, 178, 180, 228, 259
financial markets, 108
financial performance, 159
financial resources, 120, 141
financing, 23, 40, 50, 53, 65, 85, 95, 159, 247, 254, 259
Finland, 92, 93
fire suppression, 112
fires, 87, 210
firms, 55, 95, 123, 162, 197, 251, 252, 254, 269, 275
First Amendment, 127
first generation, 143, 261
fish, 2, 9, 12, 16, 28, 30, 43, 74, 80, 93, 129, 136, 144, 147, 158, 161, 174, 184, 189, 192, 204, 221, 224, 237, 279, 281
fisheries, 77, 116, 179
fishing, 193
fixed rate, 274
flavor, 153, 268
flex, 112, 188, 194, 268
flexibility, 11, 57, 62, 63, 69, 75, 106, 112, 114, 123, 140, 147, 162, 193, 194, 204, 212, 214, 220, 256, 268
flood, 87, 210, 240, 278, 279
flooding, 43, 113, 151, 257, 268, 279
flora, 84
fluctuations, 125, 195, 206
fluid, 26, 32, 46, 52, 73, 113, 130, 162, 168, 169, 173, 178, 189, 244, 273
fluorine, 45
foams, 45
focusing, 198
food, vii, 1, 3, 5, 6, 7, 9, 13, 14, 15, 16, 17, 18, 19, 21, 24, 25, 28, 29, 30, 31, 32, 33, 34, 35, 39, 40, 41, 42, 43, 44, 45, 48, 50, 51, 53, 54, 60, 61, 64, 65, 67, 74, 75, 76, 77, 78, 80, 82, 83, 84, 85, 86, 87, 89, 92, 94, 96, 98, 100, 102, 104, 107, 108, 109, 113, 114, 115, 116, 117, 118, 119, 120, 124, 125, 126, 127, 128, 130, 131, 133, 134, 137, 138, 140, 141, 142, 143, 144, 146, 147, 149, 151, 153, 155, 158, 159, 160, 165, 166, 167, 168, 171, 173, 174, 175, 177, 179, 180, 181, 183, 184, 191, 193, 195, 197, 200, 201, 203, 204, 208, 209, 210, 211, 212, 215, 216, 218, 219, 223, 225, 226, 230, 239,

243, 244, 245, 246, 248, 250, 252, 255, 260, 262, 263, 264, 265, 266, 268, 270, 272, 273, 274, 275, 277, 280, 281, 283, 284
food additives, 48, 75, 114, 116, 126
food industry, 266, 275
food production, 40, 92, 275
food products, 5, 51, 137, 146, 171, 191
forecasting, 282
foreclosure, 238
foreign aid, 52
foreign exchange, 52, 77
foreign policy, 52, 63, 85, 94, 95, 117
forest management, 64, 122, 248, 249, 256
forests, 2, 20, 47, 68, 77, 112, 121, 174, 178, 197, 230, 239, 256
fossil, 39
France, 92, 93, 134
free trade, 27, 53, 124, 135, 266
Free Trade Area of the Americas, 124
freedom, 172
freezing, 114, 136
freshwater, 39, 112, 152
frost, 135
fructose, 139, 164, 253
fuel, 16, 29, 124, 134, 138, 143, 166, 183, 252, 269
funding, 11, 20, 21, 23, 41, 42, 50, 51, 53, 58, 85, 86, 89, 91, 97, 102, 115, 119, 131, 132, 134, 137, 141, 142, 151, 152, 157, 165, 168, 170, 175, 178, 180, 189, 191, 199, 201, 202, 227, 231, 240, 248, 250, 260, 261, 269, 273, 279, 281
fungus, 153, 259

G

gait, 141
garbage, 243
gasoline, 15, 30, 92, 126, 166, 172, 186, 220
GATS, 182
GATT, 33, 95, 126, 172, 173, 181, 194, 233, 244, 245, 249, 262, 265, 272, 273
GDP, 116, 126, 134
gender, 196, 241
gene, 31, 35, 127, 153, 248, 261
gene transfer, 127
General Agreement on Tariffs and Trade, 33, 126, 172, 173, 194, 283
General Agreement on Trade in Services, 182
generation, 64, 143, 147, 228
genes, 31, 128, 206
genetic traits, 127
Germany, 92, 93, 134
germination, 247, 261
gestation, 29
gift, 2
global markets, 254
global trade, 149
glucose, 139
goals, 55, 109, 130, 177, 208, 239, 256
gold, 144, 270
goods and services, 25, 60, 67, 85, 99, 134, 188, 211, 265
government, 2, 15, 16, 21, 25, 28, 35, 37, 38, 39, 40, 44, 50, 51, 61, 63, 64, 67, 69, 75, 82, 85, 89, 90, 95, 97, 98, 99, 101, 105, 106, 108, 123, 127, 134, 145, 150, 155, 159, 160, 162, 164, 177, 179, 182, 183, 184, 187, 192, 196, 197, 198, 209, 210, 216, 217, 222, 223, 225, 233, 234, 247, 249, 250, 251, 252, 254, 257, 258, 262, 266, 268, 269, 270, 272, 273, 279
government intervention, 123
government procurement, 266
GPS, 129, 130, 283
grades, 11, 66, 110, 131, 136, 248
grading, 11, 66, 81, 118, 131, 171
grains, 3, 5, 6, 7, 14, 24, 32, 41, 43, 48, 49, 75, 78, 81, 97, 102, 110, 115, 118, 131, 139, 145, 149, 150, 155, 162, 164, 167, 170, 178, 194, 207, 214, 215, 221, 223, 255, 260, 270, 271, 274, 279, 283
grants, 14, 31, 34, 48, 52, 53, 54, 57, 60, 65, 80, 86, 91, 100, 120, 125, 138, 142, 147, 172, 177, 179, 180, 189, 191, 196, 227, 228, 229, 234, 235, 236, 248, 256, 260, 274, 276
grass, 36, 61, 131, 226
grasses, 30, 132, 139, 200, 203, 217, 237, 242
grasslands, 132, 178, 217, 241
gravity, 125
grazing, 11, 14, 16, 19, 25, 27, 48, 57, 59, 90, 105, 109, 119, 120, 131, 132, 133, 137, 138, 157, 182, 211, 212, 215, 217, 224, 225, 233, 250, 259
Great Britain, 34, 93
Great Depression, 8, 258
Greece, 92, 93
greenhouse gases, 47, 133
groundwater, 1, 21, 23, 56, 82, 132, 134, 155, 202, 237, 278
grouping, 44, 268
groups, 19, 28, 40, 52, 65, 66, 82, 90, 103, 109, 115, 117, 127, 139, 154, 161, 169, 174, 192, 227, 234, 240, 246, 249, 258, 260, 265, 266, 269, 271, 274
growth, 6, 20, 28, 33, 39, 61, 78, 89, 93, 111, 115, 135, 139, 143, 144, 159, 173, 174, 191, 204, 206, 242, 254, 256, 259, 267, 272, 275
growth hormone, 28, 33, 115
Guatemala, 37, 135
guidance, 55, 151, 175, 178, 179, 180, 231, 266

guidelines, 7, 18, 78, 84, 109, 111, 117, 120, 179, 191, 203, 208, 209, 222, 234
Guillain-Barre syndrome, 38
Guinea, 22
Gulf Coast, 260
Gulf of Mexico, 74, 144

H

habitat, 56, 58, 69, 89, 90, 112, 136, 179, 189, 209, 221, 222, 224, 230, 237, 239, 246, 279, 281
hardwoods, 22
harm, 30, 150, 195, 203, 204, 224, 264, 265
harmful effects, 4, 17, 281
harmonization, 149, 153, 170, 257
harmony, 195
harvesting, 5, 11, 61, 128, 132, 140, 144, 256, 263
Hawaii, 140, 166, 208, 209
hazardous substances, 243, 269
hazards, 16, 18, 35, 42, 90, 142, 203
health, 6, 15, 17, 18, 24, 28, 33, 39, 42, 43, 44, 45, 47, 49, 71, 78, 80, 88, 89, 91, 93, 101, 108, 112, 117, 121, 124, 130, 142, 150, 153, 165, 166, 168, 173, 176, 177, 179, 180, 185, 187, 190, 197, 203, 207, 209, 220, 225, 231, 232, 235, 239, 242, 245, 248, 250, 261, 270, 271, 276
health care, 80, 245
health effects, 45, 165, 185, 225, 261
health problems, 91, 180
heat, 135, 143, 207, 248
heating, 269
heating oil, 269
heavy metals, 29
hemolytic uremic syndrome, 83
herbicide, 23, 31, 72, 126, 139, 226
higher education, 171
highways, 281
hiring, 68, 142
hives, 102
home ownership, 234
homeowners, 10, 99, 228, 234
Hong Kong, 22, 197
horizontal integration, 254
household income, 156, 192, 203
housing, 49, 53, 60, 81, 100, 130, 142, 168, 174, 213, 228, 234, 235, 236, 237, 250
human welfare, 176
humus, 54, 242
hunting, 45
hybrid, 35, 206
hydraulic fluids, 217
hydrocarbons, 29, 45, 187
hydroelectric power, 147
hydrogen, 16, 17, 124, 143, 195, 196
hydroponics, 143

I

identification, 5, 18, 85, 223, 264, 265
identity, 120, 125, 144, 215, 241, 248, 260, 274, 275
ileum, 246
IMF, 56, 149
imitation, 248
immunity, 24
impact assessment, 85
implementation, 58, 67, 80, 90, 130, 150, 199, 229, 260, 271, 281, 283
import prices, 230, 244
import restrictions, 53, 254, 267
imports, 5, 9, 20, 23, 24, 27, 34, 37, 40, 53, 54, 67, 83, 88, 116, 118, 145, 150, 156, 161, 162, 164, 169, 171, 186, 188, 208, 210, 217, 220, 226, 230, 233, 244, 246, 252, 253, 259, 260, 265, 267, 275
in transition, 178
incentives, 27, 61, 158, 188, 199, 240, 254
incidence, 117, 171
inclusion, 17, 19
income, 1, 4, 7, 8, 9, 10, 13, 14, 21, 44, 46, 50, 52, 53, 54, 62, 69, 73, 74, 78, 81, 86, 88, 89, 94, 97, 99, 102, 105, 106, 115, 118, 120, 123, 125, 126, 134, 146, 152, 156, 158, 160, 162, 174, 181, 183, 184, 187, 192, 194, 195, 198, 200, 208, 209, 211, 213, 214, 215, 219, 225, 228, 231, 233, 234, 235, 236, 237, 239, 240, 242, 245, 247, 252, 253, 255, 258, 263, 264, 282
income support, 7, 9, 13, 50, 52, 69, 89, 106, 115, 118, 123, 146, 158, 160, 162, 211, 214, 219, 258
income tax, 78
India, 135
Indians, 96
indicators, 241, 283
indices, 135
Indonesia, 22, 23, 38, 197
industrial chemicals, 31
industrialized countries, 167, 201, 245
industry, 6, 28, 37, 43, 46, 51, 52, 55, 62, 64, 65, 67, 82, 92, 106, 123, 124, 129, 130, 131, 148, 150, 151, 153, 159, 162, 177, 179, 181, 196, 197, 211, 215, 216, 218, 230, 233, 244, 246, 262, 270, 275
inelastic, 211
infants, 44, 53, 184, 245
infectious disease, 31, 33
inflation, 60, 156, 181, 208, 209
influence, 25, 135, 171, 196, 227, 232, 240, 254
information exchange, 148
information technology, 147

infrastructure, 78, 126, 129, 140, 141, 228, 240, 271
ingestion, 200
inheritance, 201
inhibition, 45
injury, 28, 67, 81, 82, 130, 168, 195, 230, 245, 262, 265
input, 15, 16, 140, 211, 214, 247, 270
insecticide, 73
insects, 23, 35, 70, 135, 147, 173, 204
insecurity, 117, 118, 143
inspections, 138, 199, 258, 271, 276
inspectors, 5, 20, 42, 61, 131, 138, 140, 142, 150, 195, 202, 208, 230, 258, 262, 271
instability, 247
institutional change, 227
institutions, 10, 51, 64, 92, 93, 98, 99, 116, 122, 181, 208, 213, 231, 271
instruments, 43, 52, 159, 280, 281
insurance, 3, 4, 10, 11, 14, 21, 25, 35, 37, 41, 45, 61, 66, 69, 70, 76, 79, 89, 96, 101, 107, 113, 115, 129, 133, 134, 135, 146, 151, 156, 159, 160, 173, 186, 189, 210, 218, 219, 223, 225, 247, 267, 269
integration, 66, 162, 275
integrity, 47, 56, 208
intellectual property, 261, 266, 271
intensity, 192, 204
intent, 261
interaction, 92, 277
interest, vii, 4, 13, 25, 56, 57, 66, 70, 80, 86, 96, 98, 99, 100, 101, 102, 104, 124, 141, 156, 158, 159, 170, 171, 172, 183, 185, 187, 189, 192, 197, 199, 211, 215, 219, 227, 234, 235, 236, 247, 252, 263, 271
interest rates, 13, 159, 197
interface, 53, 138, 281
intermediaries, 167
Internal Revenue Service, 172, 188
international law, 243
International Monetary Fund, 56, 149
international trade, 7, 54, 123, 126, 127, 134, 193, 231, 233, 258, 269, 271, 283
interpretation, 243
interrelationships, 250
interval, 220
intervention, 21, 23, 37, 77, 150, 171, 204
intervention price, 21, 37, 150, 171
intrinsic value, 274
inventions, 64
investment, 227, 229, 262, 265, 266, 271
iodine, 168
ionizing radiation, 151
Iran, 85
Ireland, 92, 93
iron, 166, 168
irradiation, 151
isolation, 217
Israel, 173, 266
Italy, 92, 93, 134

J

Jamaica, 25
Japan, 22, 27, 65, 77, 134, 149, 168, 173, 195, 197
job training, 53
jobs, 198, 227, 274
joint stock company, 203
judgment, vii, 202
jurisdiction, 10, 36, 76, 90, 94, 151, 166, 191, 219, 259
justice, 90

K

knowledge, 10, 30, 78, 94, 219, 223

L

labeling, 3, 6, 18, 38, 65, 67, 96, 108, 110, 116, 151, 191, 260, 265, 266, 276
labor, 34, 81, 83, 92, 96, 97, 100, 105, 129, 134, 136, 162, 168, 170, 174, 183, 211, 214, 217, 226, 235, 236, 238, 239
lactose, 269
lakes, 2, 30, 47, 77, 110, 205, 207, 226
land, 2, 8, 10, 12, 14, 18, 19, 21, 25, 36, 42, 46, 48, 56, 57, 58, 59, 61, 64, 69, 70, 74, 77, 78, 81, 82, 83, 87, 88, 90, 91, 93, 94, 95, 97, 100, 101, 102, 105, 109, 119, 120, 122, 132, 133, 136, 137, 138, 139, 140, 141, 151, 152, 154, 155, 159, 160, 165, 171, 173, 174, 177, 178, 180, 182, 190, 198, 199, 200, 207, 211, 212, 214, 215, 216, 218, 224, 227, 235, 236, 237, 238, 239, 240, 241, 242, 244, 248, 250, 251, 254, 255, 256, 259, 263, 267, 272, 277, 278, 279, 281, 282, 284
land use, 10, 12, 48, 57, 77, 109, 122, 136, 155, 160
language, 5, 9, 97, 131, 212, 271
Laos, 23, 197
Latin America, 94
Latvia, 89, 92, 93
laws, vii, 1, 5, 8, 11, 12, 16, 18, 24, 28, 35, 39, 44, 49, 52, 59, 88, 89, 90, 105, 130, 145, 146, 150, 160, 172, 177, 180, 191, 196, 199, 202, 212, 220, 224, 225, 230, 231, 233, 243, 250, 258, 260, 270, 271
LDCs, 155

lead, 29, 60, 93, 95, 120, 165, 176, 241, 266
leadership, 271
leakage, 63
learning, 53, 193, 229
legal protection, 19, 28
legislation, 1, 9, 12, 13, 23, 26, 29, 37, 40, 47, 53, 59, 60, 75, 79, 93, 95, 98, 103, 107, 115, 141, 146, 155, 158, 166, 172, 175, 182, 219, 225, 238, 250, 258, 265, 266, 279, 282
legume, 216
lending, 10, 56, 59, 60, 98, 99, 100, 104, 108, 213
levees, 87
liability, 4, 28, 130, 254
liberalization, 266
licenses, 145, 187, 254, 276
lifespan, 89, 159, 203
lifestyle, 103, 240
lifetime, 185, 214
likelihood, 55, 107
limitation, 2, 4, 8, 46, 50, 84, 85, 93, 119, 124, 200, 203, 218, 245
linkage, 268
liquids, 243
Listeria monocytogenes, 120, 157
literacy, 10
Lithuania, 89, 92, 93
liver, 124
livestock, 4, 5, 9, 12, 13, 15, 16, 17, 18, 19, 20, 25, 27, 32, 33, 34, 35, 40, 42, 43, 55, 62, 71, 80, 82, 84, 85, 86, 91, 94, 99, 101, 109, 110, 116, 120, 122, 131, 132, 133, 134, 137, 138, 143, 146, 152, 153, 157, 158, 162, 163, 176, 183, 184, 189, 192, 194, 195, 198, 199, 207, 208, 210, 211, 212, 213, 215, 217, 221, 233, 238, 245, 246, 250, 251, 256, 275, 277, 279, 280, 282, 284
living conditions, 142
loan principal, 263
loans, 3, 4, 8, 10, 14, 27, 36, 38, 43, 50, 52, 53, 60, 66, 81, 85, 86, 89, 95, 96, 99, 100, 101, 102, 104, 106, 108, 110, 112, 113, 114, 122, 131, 139, 141, 142, 145, 156, 158, 159, 163, 164, 167, 170, 174, 181, 187, 193, 194, 196, 198, 200, 202, 211, 213, 219, 223, 227, 228, 229, 234, 235, 236, 237, 238, 239, 252, 255, 263, 268, 282, 283
local government, 47, 53, 57, 97, 155, 179, 259, 272
location, 4, 63, 75, 83, 111, 129, 158, 207, 228, 279, 283
long distance, 270
Louisiana, 181, 224
low risk, 190
lower prices, 71, 122, 197
lubricants, 29, 146, 217

M

Maastricht Treaty, 93
machinery, 99, 116, 153, 183, 214
macroeconomic policies, 134
macromolecules, 91
mad cow disease, 19, 33, 69, 144, 221
Malaysia, 22, 23, 38, 197
management, 11, 18, 19, 28, 30, 32, 34, 35, 36, 38, 40, 48, 49, 55, 57, 59, 63, 65, 77, 84, 101, 109, 111, 112, 119, 122, 123, 129, 131, 144, 148, 151, 153, 154, 160, 162, 163, 165, 174, 177, 178, 183, 191, 196, 203, 204, 209, 215, 221, 222, 223, 225, 226, 239, 255, 256, 272, 275, 282
mandates, 7, 11, 67, 88, 189
manipulation, 55, 62, 169
manufacturing, 32, 47, 61, 73, 103, 116, 129, 130, 146, 151, 160, 166, 195, 216, 269, 274, 275
manufacturing companies, 116
manure, 11, 18, 19, 29, 55, 153, 184, 191, 195, 204, 277
mapping, 243
marijuana, 146
market, 3, 5, 7, 8, 9, 10, 11, 15, 16, 20, 23, 27, 29, 39, 41, 46, 49, 53, 55, 57, 60, 62, 65, 67, 68, 70, 71, 73, 74, 75, 87, 91, 96, 101, 102, 104, 105, 106, 107, 110, 112, 113, 114, 122, 123, 127, 133, 134, 136, 137, 139, 145, 146, 149, 150, 152, 153, 158, 159, 161, 162, 163, 164, 168, 169, 171, 172, 184, 186, 187, 194, 195, 196, 204, 208, 211, 213, 215, 216, 220, 221, 222, 230, 232, 233, 239, 246, 247, 251, 252, 253, 254, 255, 258, 261, 262, 263, 266, 269, 273, 275, 282, 283
market economy, 172
market position, 159
market share, 246
marketability, 218
marketing, 4, 5, 7, 8, 9, 11, 12, 14, 16, 18, 22, 24, 26, 32, 34, 36, 37, 38, 40, 43, 46, 49, 50, 52, 64, 66, 67, 70, 71, 74, 76, 79, 80, 84, 87, 92, 95, 96, 98, 102, 104, 106, 110, 112, 113, 114, 115, 118, 125, 126, 130, 131, 136, 139, 141, 145, 148, 158, 159, 160, 161, 163, 164, 169, 170, 173, 177, 179, 181, 182, 184, 185, 187, 190, 193, 194, 195, 196, 198, 199, 200, 202, 208, 212, 213, 215, 216, 218, 220, 221, 238, 239, 240, 244, 249, 252, 254, 255, 263, 264, 265, 268, 271, 273, 274, 275, 280, 282, 283
markets, 3, 13, 20, 28, 34, 38, 39, 46, 53, 62, 65, 66, 67, 79, 82, 87, 95, 96, 104, 105, 111, 113, 121, 122, 130, 139, 146, 150, 160, 161, 162, 172, 178, 197, 198, 203, 208, 249, 250, 253, 260, 261, 267, 268, 275, 280, 282, 283
Marshall Islands, 147

mass, 125
maturation, 206
meals, 110, 116, 123, 142, 181, 191, 192, 194, 231, 232, 234, 252, 253
meanings, vii
measurement, 7, 69, 151, 159, 198, 213, 267
measures, 5, 7, 13, 32, 47, 48, 49, 60, 73, 74, 91, 92, 99, 106, 126, 133, 134, 137, 148, 149, 153, 162, 168, 170, 171, 174, 186, 188, 211, 213, 222, 229, 231, 240, 241, 242, 243, 253, 260, 264, 265, 267, 270, 273
meat, 5, 11, 16, 17, 20, 24, 27, 33, 34, 41, 43, 45, 55, 61, 75, 80, 82, 94, 104, 108, 109, 110, 114, 116, 117, 138, 143, 151, 153, 157, 165, 166, 191, 195, 197, 198, 202, 208, 218, 230, 231, 238, 248, 258, 262, 277, 282, 284
median, 155, 156, 234, 237
mediation, 14
Mediterranean, 69, 82, 166
Mediterranean countries, 82
melting, 48, 136, 255
membership, 64, 77, 89, 189, 244
memory, 130
men, 180
mergers, 137, 255
metabolism, 89
metabolites, 72
metals, 205
methanol, 15, 16
methodology, 18, 284
Mexico, 22, 34, 49, 96, 133, 135, 166, 168, 188, 189, 195, 197, 232, 252, 253
Miami, 118, 242
microclimate, 71
microorganism, 136, 200
microscope, 144
military, 18, 180, 281
milk, 7, 8, 11, 18, 26, 32, 33, 37, 38, 41, 44, 46, 52, 54, 62, 72, 73, 81, 107, 110, 113, 114, 115, 116, 119, 123, 130, 139, 143, 150, 160, 162, 163, 164, 165, 168, 169, 173, 177, 178, 179, 180, 183, 184, 186, 189, 209, 214, 230, 244, 245, 249, 251, 255, 269, 273, 280
milligrams, 184
minimum price, 26, 46, 73, 113, 149, 169, 189, 244, 252, 255, 263
minimum wage, 6, 96
mining, 219, 237, 243
minority, 156
missions, 167
Mississippi River, 74
mixing, 49, 54
mode, 267
models, 5, 47
modernization, 147
moisture, 6, 32, 77, 82, 97, 135, 139, 147, 151, 165, 170, 173, 211, 219, 224, 239, 242, 251, 253, 282
moisture content, 6, 139, 170, 224, 239
mold, 6
molecules, 219, 269
monetary policy, 93
money, 25, 54, 57, 67, 74, 97, 106, 124, 130, 154, 161, 179, 183, 196, 219, 247, 277
monitoring, 17, 30, 136, 138, 179, 199, 215, 216, 283
monopoly, 41, 71, 238, 249
monopoly power, 41, 249
moratorium, 115
Morocco, 165
morphology, 159
mortality, 121, 148, 165
mothers, 14, 34, 44, 245
motivation, 117
movement, 92, 109, 110, 134, 156, 161, 184, 202, 206, 223, 261, 271, 279, 281
multilateral aid, 28, 56
muscles, 38
Myanmar, 23

N

NAD, 5, 174, 176
NAFTA, 123, 161, 174, 188, 232, 233, 252
National Ambient Air Quality Standards, 15, 23, 47, 175, 186, 199
National Institutes of Health, 232
national product, 11, 38, 112, 244, 280
National Research Council, 175, 180, 190
National Security Council, 140
natural disasters, 15, 59, 86, 87, 118, 157, 194, 199
natural enemies, 22, 148
natural gas, 16, 17, 143, 172
natural hazards, 69
natural resource management, 227
natural resources, 17, 28, 57, 59, 60, 64, 77, 90, 109, 186, 222, 242, 260, 272, 281
nausea, 277
NCA, 182
needs, 9, 29, 54, 76, 77, 86, 89, 118, 119, 151, 167, 199, 209, 212, 216, 219, 220, 221, 227, 234, 240, 241, 250, 252, 283
neglect, 28, 130
negotiating, 135, 194, 210, 225, 265, 266, 271
negotiation, 266
nerve, 45
nervous system, 94

Netherlands, 92, 93
network, 68, 111, 120, 129, 176, 179, 184, 199, 205
neurological disease, 45, 232
New Zealand, 22, 38, 65, 114, 149, 184, 195, 197, 249
newspapers, 162
niche market, 212
nitrates, 93, 184
nitrogen, 2, 17, 29, 47, 111, 155, 176, 184, 191, 197, 204, 237, 255, 272
nitrogen compounds, 2, 184
nitrogen dioxide, 176
nitrogen fixation, 255
nitrogen oxides, 47, 197
noise, 207, 224
nontariff barriers, 31, 126, 187, 188, 217, 233, 259, 260
North America, 33, 82, 88, 123, 145, 174, 175, 188, 189, 194, 205, 209, 210, 232, 233, 239, 252
North American Free Trade Agreement, 88, 123, 145, 174, 188, 194, 209, 210, 232, 233, 252
North Korea, 85, 197
nucleus, 226
nursing, 53
nutraceutical, 124
nutrients, 11, 36, 48, 54, 55, 74, 93, 155, 166, 191, 204, 206, 219, 242, 245, 254
nutrition, 9, 13, 19, 21, 34, 41, 44, 48, 51, 54, 72, 73, 76, 78, 93, 98, 101, 104, 114, 116, 119, 120, 123, 124, 127, 139, 143, 160, 165, 179, 181, 191, 192, 215, 219, 232, 234, 252, 254, 260, 263, 282

O

obligation, 38, 89, 158, 172, 194, 195, 196, 279
oceans, 39
octane, 172
OECD, 60, 126, 162, 192, 195, 213, 220, 252, 264, 271
oil, 4, 29, 36, 38, 67, 69, 71, 112, 146, 156, 194, 217, 242, 244, 249, 252, 253, 274, 279
oils, 29, 39, 112, 116, 139, 148, 156, 193
Oklahoma, 280
omega-3, 112
operator, 41, 55, 68, 70, 71, 74, 97, 101, 112, 192, 213, 214, 252
organ, 192, 202
organic compounds, 45, 47, 166, 197, 276
organic matter, 29, 30, 111, 173
organism, 89, 128, 129, 159, 248, 267
organization, 23, 28, 41, 51, 53, 57, 63, 86, 92, 106, 116, 118, 122, 130, 149, 154, 179, 182, 193, 195, 203, 212, 216, 221, 235, 254, 275, 282, 283
Organization for Economic Cooperation and Development, 220, 271
organizations, vii, 22, 24, 53, 60, 83, 86, 98, 100, 118, 122, 128, 130, 161, 171, 175, 179, 180, 195, 196, 216, 231, 235, 236, 237, 245, 261, 270
output, 16, 50, 74, 100, 134, 145, 163, 174, 213, 214, 256
overtime, 96
ownership, 2, 3, 27, 42, 60, 65, 88, 93, 101, 114, 156, 163, 193, 198, 213, 251, 254, 274, 275
oxides, 2, 176
oxygen, 3, 29, 30, 32, 48, 74, 80, 81, 124, 143, 172, 196, 220, 277
ozone, 45, 47, 92, 167, 171, 172, 176, 196, 220, 242

P

Pacific, 2, 20, 22, 66, 192, 197, 210, 260, 261
packaging, 3, 108, 114, 144, 260
pain, 19, 142
Pakistan, 135
pancreas, 89
parents, 191, 260
particles, 88, 199, 239
partnership, 47, 66, 116, 190, 199, 203, 249
pasteurization, 33
pastures, 191
pathogens, 36, 49, 106, 120, 138, 177, 195, 204, 230
pathways, 95
PCA, 99, 201, 213
PCP, 208
peat, 143
peer review, 232
penalties, 162, 164, 184
pensions, 192
per capita income, 77
percolation, 147, 202
performance, 70, 84, 116, 130, 161, 179, 193, 215, 230
performance indicator, 130
permit, 25, 47, 51, 55, 92, 94, 97, 105, 132, 133, 146, 151, 152, 187, 202, 207, 220, 221, 222, 223, 224, 226, 234, 242, 248
perspective, 32
Peru, 22, 135
pest populations, 70, 204
pesticide, 3, 11, 15, 23, 31, 35, 38, 40, 44, 45, 48, 56, 63, 66, 75, 76, 108, 116, 124, 125, 135, 139, 146, 147, 148, 151, 160, 167, 185, 197, 204, 206, 220, 221, 222, 225, 226, 232, 255, 264, 272
pH, 2, 204
pharmaceuticals, 31, 103, 146, 221
Philippines, 22, 23, 38, 135, 197

phosphates, 93
phosphorus, 191, 196, 204, 205
photographs, 196, 220
physical activity, 78, 117
physical environment, 84
phytoplankton, 144
pigs, 105, 120
planning, 10, 37, 40, 55, 59, 77, 109, 130, 137, 138, 212, 217, 222, 229
plants, 10, 12, 13, 15, 18, 24, 29, 30, 31, 32, 34, 39, 45, 47, 61, 72, 75, 82, 88, 92, 93, 112, 117, 121, 127, 128, 135, 136, 142, 143, 150, 151, 155, 159, 161, 170, 181, 182, 190, 195, 197, 199, 204, 205, 206, 212, 216, 217, 220, 230, 231, 237, 242, 246, 248, 258, 260, 262, 276, 277, 279
poison, 76
Poland, 89, 92, 93, 195
police, 53
policy instruments, 259
policy makers, 5, 258
pollen, 15, 143
pollination, 31
pollutants, 1, 36, 56, 79, 84, 92, 111, 134, 185, 187, 197, 203, 208, 250, 255, 264, 275
pollution, 11, 14, 15, 18, 28, 44, 47, 48, 55, 57, 79, 82, 84, 90, 91, 105, 106, 111, 153, 160, 177, 184, 185, 187, 191, 204, 207, 208, 226, 237, 246, 255, 264, 275, 279
pools, 38, 163
poor, 28, 40, 77, 78, 86, 208, 209
population, 23, 33, 44, 49, 69, 78, 88, 90, 136, 137, 148, 165, 179, 203, 209, 226, 228, 229, 232, 235, 277
population density, 226, 229
ports, 12, 37, 109, 150, 153, 166, 232, 250
Portugal, 92, 93
potatoes, 7, 36, 52, 77, 128, 208, 248, 249
poverty, 120, 123, 156, 203, 208, 209, 219, 231
power, 10, 20, 117, 120, 124, 186, 228, 255, 265
precipitation, 2, 22
predictability, 62, 122
preference, 39, 40, 133
prejudice, 196, 241
preparation, 26, 61, 177, 185, 276
preparedness, 140, 141, 177
President Clinton, 253
pressure, 70, 204, 269
prevention, 16, 18, 78, 80, 150, 153, 240
price band, 210
price changes, 139, 211
price deflator, 134
price elasticity, 211
price floor, 81
price index, 27, 211, 280
price instability, 246
prices, 3, 5, 7, 8, 9, 14, 15, 20, 21, 25, 26, 33, 37, 38, 39, 40, 41, 43, 50, 53, 60, 66, 70, 71, 72, 73, 74, 95, 100, 106, 114, 116, 119, 122, 123, 137, 141, 146, 148, 149, 150, 156, 157, 158, 159, 160, 161, 162, 163, 169, 171, 177, 187, 189, 196, 198, 208, 209, 211, 213, 214, 215, 221, 223, 233, 238, 240, 244, 247, 252, 255, 258, 261, 262, 263, 275, 277, 279, 280, 282, 283
primacy, 112
principle, 65, 148, 172, 180, 181, 209, 220, 267, 276
prions, 144, 246
privacy, 271
private sector investment, 88
Producer Price Index, 209
producers, 3, 8, 9, 10, 14, 18, 19, 20, 21, 25, 30, 32, 33, 38, 39, 41, 43, 49, 52, 55, 57, 58, 59, 61, 62, 67, 68, 70, 72, 79, 81, 85, 86, 94, 99, 103, 104, 106, 110, 111, 113, 115, 119, 121, 122, 125, 127, 131, 133, 136, 138, 139, 140, 145, 146, 147, 151, 152, 153, 155, 157, 158, 161, 162, 163, 164, 169, 170, 173, 183, 187, 188, 194, 195, 198, 199, 200, 201, 202, 203, 210, 213, 215, 216, 218, 219, 223, 225, 227, 238, 240, 247, 249, 254, 255, 256, 258, 262, 263, 265, 267, 269, 271, 274, 277, 279, 280, 282
production, 2, 3, 5, 6, 8, 9, 10, 11, 12, 13, 14, 15, 16, 17, 18, 19, 20, 21, 24, 26, 29, 30, 31, 32, 33, 34, 36, 40, 42, 43, 44, 50, 54, 55, 57, 58, 59, 62, 63, 65, 66, 69, 70, 72, 73, 74, 75, 79, 81, 82, 83, 86, 87, 90, 91, 92, 93, 96, 97, 100, 102, 103, 104, 106, 113, 114, 116, 120, 123, 125, 126, 128, 129, 130, 133, 135, 137, 138, 139, 140, 144, 146, 147, 148, 152, 154, 156, 158, 162, 163, 164, 168, 169, 171, 172, 181, 182, 183, 184, 186, 188, 189, 193, 194, 195, 196, 198, 199, 200, 201, 202, 203, 204, 206, 208, 210, 211, 213, 214, 216, 219, 220, 222, 224, 225, 231, 238, 239, 241, 242, 251, 253, 254, 255, 256, 263, 264, 265, 266, 267, 274, 275, 277, 278, 282, 283
production costs, 100, 156, 162
production quota, 162
productive capacity, 132
productivity, 28, 38, 39, 53, 57, 82, 91, 152, 174, 242, 256
professional qualifications, 232
profit margin, 110
profitability, 41, 99, 169
profits, 246, 249
program, 1, 2, 3, 4, 8, 9, 10, 12, 13, 14, 16, 17, 18, 20, 21, 22, 23, 25, 26, 27, 28, 30, 32, 33, 34, 35, 36, 37, 38, 41, 43, 44, 45, 47, 49, 51, 52, 53, 54,

55, 56, 57, 58, 63, 64, 65, 66, 67, 69, 71, 72, 73, 74, 75, 76, 80, 81, 82, 86, 87, 89, 90, 94, 95, 96, 97, 100, 101, 102, 103, 104, 105, 107, 108, 109, 110, 112, 113, 114, 115, 116, 117, 118, 120, 121, 122, 123, 125, 126, 128, 129, 132, 133, 135, 136, 138, 139, 140, 141, 144, 146, 147, 148, 151, 152, 153, 156, 157, 158, 159, 160, 161, 162, 164, 165, 167, 168, 170, 174, 177, 178, 179, 180, 181, 182, 183, 185, 186, 187, 188, 189, 190, 191, 193, 194, 196, 197, 198, 199, 200, 201, 202, 203, 204, 205, 207, 208, 210, 212, 213, 214, 215, 216, 218, 219, 220, 222, 223, 224, 225, 226, 227, 229, 231, 232, 234, 235, 236, 237, 238, 239, 240, 241, 242, 243, 245, 247, 248, 250, 252, 253, 254, 255, 256, 258, 260, 261, 262, 263, 264, 265, 267, 268, 269, 270, 271, 273, 274, 276, 277, 278, 279, 280, 281, 282, 283, 284
programming, 29
promoter, 78
propagation, 21, 147
propane, 16
property taxes, 78, 100
prosperity, 170
proteins, 19, 34, 91, 110, 129, 144, 212
protocol, 1, 31
pruning, 148
PSD, 224
public administration, 77, 272
public awareness, 141
public health, 15, 34, 38, 116, 127, 155, 160, 191, 210, 245, 262, 277, 281
public interest, 56
public sector, 204
Puerto Rico, 13, 40, 119, 120, 140, 147, 191, 215
pulse, 102
purchasing power, 198

Q

qualifications, 98
quality assurance, 18, 153
quotas, 2, 7, 8, 9, 14, 16, 27, 36, 37, 72, 94, 96, 113, 145, 158, 164, 169, 187, 188, 201, 217, 233, 254, 259, 263, 267, 273, 275

R

race, 90, 196, 205
radiation, 45, 133, 197
rain, 2, 131, 187, 237, 271
rainfall, 47, 48, 82, 84, 151, 206
range, 14, 22, 57, 62, 117, 121, 132, 133, 173, 174, 183, 188, 196, 221, 225, 242, 274, 281
rangeland, 77, 97, 120, 131, 132, 138, 199, 211, 215, 217, 250, 259
raw materials, 39
RCD, 218
REA, 228
real estate, 10, 99, 100, 104, 108, 211, 236, 238
recognition, 267, 276
recombinant DNA, 31, 127, 206, 219
reconciliation, 26, 219, 282
recovery, 5, 17, 98, 121, 136, 140, 141, 225
recreation, 40, 165, 174, 182, 221, 222, 236, 278, 280
recycling, 219
redevelopment, 34
reduction, 2, 7, 11, 16, 22, 23, 26, 27, 29, 32, 57, 69, 76, 78, 81, 84, 106, 111, 133, 138, 145, 154, 161, 163, 168, 170, 188, 194, 198, 203, 230, 231, 245, 256, 268, 273, 278
refining, 172, 252
reforms, 7, 13, 108, 115, 168, 170
regeneration, 143
regulations, 5, 7, 9, 12, 28, 42, 48, 55, 82, 84, 90, 94, 97, 109, 110, 115, 130, 138, 143, 153, 162, 166, 167, 176, 177, 182, 186, 187, 195, 200, 222, 230, 231, 232, 234, 246, 248, 251, 252, 266, 267, 270, 276, 277, 283
regulators, 31, 206
regulatory bodies, 75
regulatory framework, 252
regulatory requirements, 18, 47, 126
rehabilitation, 236
rehabilitation program, 236
reinsurance, 247, 269
relationship, 100, 110, 134, 168, 169, 198, 211
relationships, 198, 223, 247, 255, 278
relevance, vii
remote sensing, 283
rent, 70, 100, 101, 187, 213, 228, 236, 237
repair, 228, 234, 235, 259
replacement, 9, 124, 137, 143, 156
reproduction, 44, 257
reserves, 21, 104, 138, 178, 254, 261
resistance, 29, 35, 126, 128, 148, 203, 206, 221, 226
resolution, 26, 35, 126, 172, 205, 219, 273
resource management, 55, 59, 77, 92, 121, 151, 222
resources, vii, 17, 29, 32, 36, 40, 48, 54, 57, 58, 59, 76, 77, 80, 97, 100, 107, 124, 130, 149, 151, 154, 155, 156, 174, 176, 177, 178, 180, 186, 187, 199, 212, 218, 221, 222, 235, 242, 251, 255, 256, 260, 270, 272

responsibility, 17, 21, 30, 42, 47, 93, 103, 108, 148, 177, 215, 271, 274
retail, 33, 34, 60, 67, 102, 108, 116, 157, 161, 164, 208, 213, 226, 239
retaliation, 10, 233
retention, 39
retirement, 3, 12, 58, 103, 240
returns, 13, 148, 151, 183
revenue, 4, 20, 27, 70, 134, 135, 146, 154, 157, 178, 201, 223, 225, 258, 278
rice, 3, 4, 8, 11, 14, 25, 26, 28, 48, 52, 62, 75, 97, 108, 119, 131, 143, 152, 158, 163, 164, 169, 170, 187, 198, 201, 202, 208, 214, 216, 223, 234, 245, 248, 253, 258, 262, 283
rights, 10, 57, 64, 77, 98, 105, 138, 147, 154, 212, 216, 224, 244, 265, 267
risk, 11, 14, 15, 23, 24, 27, 42, 54, 56, 62, 70, 72, 74, 76, 82, 88, 93, 94, 97, 105, 116, 124, 139, 151, 154, 155, 184, 186, 204, 210, 213, 214, 217, 225, 239, 245, 246, 247, 251, 265, 269, 272
risk assessment, 204
risk management, 54, 62, 97, 105, 225
Romania, 89, 93, 149
rotations, 147, 156
roughness, 282
rubber, 146
runoff, 1, 11, 47, 61, 111, 140, 144, 147, 187, 191, 204, 205, 224, 238, 255, 262, 278
rural areas, 78, 80, 134, 184, 226, 227, 229, 237, 277
Russia, 22, 134, 168, 197

S

safety, 1, 13, 15, 16, 18, 24, 31, 39, 42, 48, 75, 78, 92, 98, 99, 103, 113, 114, 116, 117, 118, 124, 128, 144, 155, 166, 168, 175, 176, 179, 181, 183, 187, 192, 204, 210, 235, 247, 264, 265, 271, 276, 284
sales, 13, 22, 43, 55, 62, 85, 96, 102, 103, 105, 108, 109, 115, 123, 127, 134, 149, 155, 156, 157, 159, 160, 162, 171, 189, 197, 207, 230, 231, 239, 240, 249, 261, 267, 273, 283
salinity, 49
salmonella, 230
salts, 49
Samoa, 119, 120, 147, 191
sample, 144, 179, 262
sampling, 20, 116, 179, 204, 216
sanctions, 80, 85, 162
SAP, 232
satisfaction, 230
saturated fat, 39, 191
savings, 35, 87, 97, 147, 170, 226
sawdust, 29
school, 9, 41, 44, 51, 52, 53, 54, 64, 65, 68, 79, 80, 89, 116, 119, 128, 154, 159, 165, 169, 171, 180, 181, 191, 192, 203, 209, 231, 232, 233, 234, 240, 245, 248, 260, 263, 282
sea level, 48
seafood, 28, 42, 120, 138, 157, 158, 166, 179, 191, 226
second generation, 143, 262
secondary schools, 181
securities, 99, 102, 104
security, 54, 60, 63, 74, 85, 94, 95, 102, 117, 118, 131, 133, 170, 173, 180, 186, 219, 283
sediment, 36, 82, 133, 224, 278, 279
sediments, 39, 48, 111, 237
seed, 5, 6, 31, 43, 69, 84, 100, 101, 110, 126, 127, 143, 144, 146, 170, 185, 193, 196, 206, 219, 226, 237, 239, 262, 275, 279
seeding, 128
segregation, 131
self, 38, 41, 42, 49, 51, 54, 72, 73, 100, 127, 174, 233, 236, 237, 276
Senate, 1, 35, 44, 76, 128, 203, 219, 225, 250, 254, 282
Senate approval, 1
sensing, 220
sensors, 209
separation, 49
septic tank, 185, 191, 269
series, 25, 34, 54, 73, 74, 115, 183, 231, 242, 276
serum, 280
service provider, 15, 262
services, 11, 12, 25, 34, 36, 43, 51, 53, 60, 63, 78, 80, 87, 99, 101, 102, 111, 116, 118, 126, 131, 136, 154, 176, 177, 178, 180, 189, 211, 212, 213, 227, 249, 251, 265, 271, 272, 273, 283
severity, 198
sewage, 47, 49, 54, 84, 106, 184, 204, 205, 237, 243
shares, 54, 65, 162, 163, 201, 213
sharing, 10, 13, 27, 49, 70, 86, 90, 102, 132, 145, 147, 162, 182, 189, 199, 201, 222, 238, 240, 246, 250, 270
sheep, 19, 35, 55, 78, 109, 142, 157, 158, 181, 226, 232, 238, 267
shellfish, 12, 74, 161, 181
shelter, 56, 74, 92, 282
Sherman Act, 20, 39, 46, 238
shortage, 40, 77
shrimp, 144
shrubs, 200, 203, 217, 238, 282
SIC, 175, 189, 239
sign, 58, 61, 62, 276
similarity, 232
Singapore, 22, 23, 197

single currency, 92, 93
sites, 49, 71, 77, 87, 88, 92, 97, 154, 155, 177, 236, 254, 272
situation, 54, 257, 262
skills, 94, 179
skin, 30, 39, 76
Slovakia, 89, 92, 93
sludge, 54, 106, 237, 243
small intestine, 246
smoke, 187, 199
social development, 283
social security, 85
social welfare, 171
soil erosion, 58, 61, 91, 133, 223, 262, 278
soils, 2, 46, 48, 57, 140, 143, 154, 160, 176, 190, 202, 212, 223, 224, 242, 243, 247, 262, 272, 279
solid waste, 243
solvency, 74, 99
solvents, 45, 197
South Korea, 22, 65, 195, 197
Southeast Asia, 23, 152
Soviet Union, 1, 82, 87, 124, 184
soybeans, 5, 23, 25, 30, 36, 41, 52, 62, 81, 97, 102, 115, 127, 131, 139, 155, 158, 163, 164, 165, 170, 186, 187, 193, 194, 196, 202, 214, 223, 226, 244, 249, 258, 262
Spain, 92, 93
specialization, 10, 30
species, 5, 6, 15, 16, 17, 21, 23, 29, 33, 43, 45, 69, 71, 82, 88, 94, 112, 121, 135, 136, 139, 150, 155, 159, 168, 172, 176, 179, 186, 190, 206, 207, 230, 244, 246, 257, 262, 277, 281
spectrum, 113, 226
speech, 127
spelling, 62
spinal cord, 5, 246
spontaneous abortion, 35
spot market, 156
stability, 10, 15
stabilization, 117, 183, 184, 247
stages, 88, 92, 135, 148, 265, 275
standard of living, 53
standards, 5, 7, 11, 19, 47, 48, 57, 66, 73, 84, 86, 88, 90, 96, 108, 109, 110, 114, 116, 117, 118, 120, 122, 129, 130, 131, 137, 139, 150, 153, 159, 163, 166, 168, 175, 179, 192, 193, 195, 199, 205, 208, 221, 222, 225, 229, 230, 231, 232, 248, 260, 266, 270, 271, 276, 277
starch, 139, 279
State Implementation Plans, 47
statistics, 137, 175, 186, 267
statutes, 16, 33, 44, 50, 64, 97, 117, 172, 175, 188, 230
steel, 280
sterile, 246, 261
stock, 10, 17, 65, 148, 187, 196, 206, 234, 250, 264
stomach, 45, 226
stomatitis, 275
storage, 5, 13, 30, 36, 38, 55, 85, 87, 96, 102, 104, 114, 134, 144, 153, 167, 170, 171, 215, 221, 222, 224, 250, 252, 255, 261, 269, 270, 277, 279
storms, 48, 84
strain, 83
strategies, 118, 256, 278
streams, 30, 110, 152, 205, 224, 226, 239
strength, 96
stress, 74, 135
structural adjustment, 133, 251
structural characteristics, 129
students, 10, 171
subsidization, 71, 95, 173
subsidy, 9, 15, 32, 39, 45, 50, 61, 67, 74, 95, 107, 193, 196, 200, 210, 213, 245, 249, 251, 279
substance use, 204
substitutes, 156
sucrose, 92
Sudan, 85
sugar, 2, 5, 23, 40, 52, 54, 71, 78, 102, 104, 107, 122, 139, 145, 149, 150, 158, 163, 164, 169, 185, 186, 187, 196, 214, 232, 252, 255, 259, 263
sulfur, 2, 176
sulphur, 279
summer, 44, 81, 82, 97, 144, 181, 253, 260
superiority, 54
supervision, 222, 225, 276
supervisor, 68
suppliers, 15, 18, 145, 146, 155, 221
supply, 7, 9, 15, 21, 26, 32, 34, 38, 39, 40, 42, 49, 63, 66, 72, 77, 81, 92, 94, 111, 118, 123, 130, 136, 138, 151, 153, 158, 160, 164, 187, 195, 198, 201, 208, 211, 217, 224, 228, 233, 235, 248, 250, 254, 263, 275, 277, 278, 283
support staff, 37
suppression, 22, 112, 172
Supreme Court, 127, 152, 209, 257
surfactant, 146
surplus, 3, 8, 9, 14, 18, 25, 26, 32, 37, 40, 43, 51, 89, 95, 116, 117, 128, 165, 173, 180, 184, 233, 253, 255, 261
surveillance, 42, 80, 120, 144, 176, 177, 179, 215, 262
survival, 112
susceptibility, 230
sustainability, 59, 103, 121, 255
Sweden, 92, 93
Switzerland, 126, 149

symptoms, 45, 275
synthesis, 13
systems, 16, 21, 27, 34, 39, 42, 46, 47, 49, 57, 59, 68, 71, 85, 91, 110, 111, 124, 128, 129, 133, 140, 141, 144, 159, 175, 177, 228, 229, 235, 255, 282

T

Taiwan, 22, 197
targets, 35, 163, 170, 190, 191, 215, 254
tariff, 4, 7, 20, 27, 33, 40, 53, 71, 72, 80, 83, 137, 145, 155, 161, 169, 170, 172, 173, 182, 187, 188, 209, 210, 217, 232, 233, 241, 244, 245, 246, 247, 252, 256, 258, 259, 265, 275
taxation, 78, 181
teachers, 191
teaching, 115, 148, 175, 193
technical assistance, 31, 37, 47, 49, 58, 59, 64, 65, 87, 116, 120, 127, 132, 133, 152, 153, 156, 161, 165, 180, 199, 227, 232, 236, 242, 248, 260, 262, 276, 281, 283
technical efficiency, 214
technology, 22, 24, 27, 31, 34, 61, 62, 84, 147, 151, 153, 198, 204, 205, 209, 220, 225, 261, 262, 268, 280
technology transfer, 22
telecommunications, 37, 53, 78, 80, 229
temperature, 6, 47, 133, 135, 206, 277
temporary protection, 233
tenants, 142, 236
terrorism, 140
test data, 73
textiles, 40, 146
TGA, 259
Thailand, 22, 23, 38, 135, 197
theft, 19
threat, 32, 34, 35, 82, 140, 245, 257, 265, 277
threshold, 19, 41, 55, 90, 103, 135, 136, 148, 170, 186, 208, 209, 227, 229, 259
threshold level, 90
thyroid, 89
timber, 16, 20, 22, 28, 154, 165, 174, 178, 193, 218, 221, 230, 231, 239, 249, 251, 263, 267
time, 4, 8, 12, 13, 16, 18, 21, 24, 25, 27, 29, 38, 39, 41, 50, 54, 56, 59, 60, 61, 62, 67, 69, 72, 75, 76, 77, 78, 80, 81, 83, 94, 96, 97, 98, 99, 100, 101, 113, 122, 124, 130, 132, 135, 136, 141, 142, 144, 145, 153, 154, 156, 158, 170, 182, 185, 186, 187, 191, 194, 195, 198, 201, 202, 210, 211, 215, 216, 220, 221, 223, 234, 235, 238, 239, 243, 245, 246, 247, 250, 258, 259, 261, 277
time frame, 210
time periods, 27, 245
time series, 25, 60
timing, 71, 135, 220, 261
tissue, 5, 166, 251
tobacco, 2, 8, 9, 10, 11, 23, 26, 35, 36, 37, 62, 72, 113, 146, 164, 185, 189, 205, 208, 259, 263, 264, 265
tonsils, 246
tourism, 274
toxic substances, 265
toxicity, 4, 16, 45, 76, 148, 155, 174, 195, 222, 225, 261
toxicology, 263
toxin, 35, 174, 200, 204
TPA, 266
tracking, 24, 42, 265, 271
trade, 2, 6, 7, 9, 13, 16, 17, 20, 22, 24, 25, 26, 27, 28, 31, 32, 37, 40, 45, 46, 48, 50, 51, 53, 54, 56, 65, 66, 67, 71, 72, 73, 77, 80, 81, 85, 92, 98, 101, 106, 114, 120, 123, 124, 126, 127, 128, 129, 133, 135, 137, 145, 148, 149, 150, 155, 161, 162, 163, 165, 167, 170, 171, 172, 173, 179, 182, 186, 188, 191, 194, 195, 198, 207, 209, 210, 216, 217, 219, 220, 221, 222, 226, 230, 231, 233, 238, 241, 243, 245, 249, 251, 253, 257, 258, 259, 260, 265, 266, 267, 269, 271, 272, 273, 276, 283
trade liberalization, 126, 135, 266
trade policy, 6, 80, 120, 148, 230, 259, 271
trading, 51, 65, 66, 71, 91, 114, 122, 127, 131, 184, 187, 202, 210, 249, 258, 260, 261, 267
trading partners, 65, 127, 258, 267
tradition, 131
training, 14, 34, 44, 65, 74, 139, 180, 191, 227, 232, 250, 260
traits, 126, 127, 143, 205, 251, 267
transactions, 92, 122, 157, 196, 270
transesterification, 29
transformations, 127
transition, 92, 113, 162, 267
transmission, 24, 34, 110, 113, 228, 232
transparency, 71
transpiration, 93
transport, 37, 100, 114, 137, 141, 148, 261, 267, 279
transportation, 11, 12, 15, 19, 22, 40, 46, 47, 53, 60, 66, 68, 75, 101, 116, 123, 124, 143, 153, 158, 166, 168, 171, 173, 213, 255, 261, 269, 279, 282, 284
Treaty of Rome, 92
trees, 15, 22, 30, 32, 48, 68, 75, 121, 122, 128, 137, 210, 230, 238, 251, 263, 267, 282
trend, 60, 254
trial, 5
tribes, 53, 98, 235, 236
triggers, 241
trust, 28, 36, 154, 178, 202, 216, 228

tuberculosis, 270
tundra, 217
Turkey, 89, 93, 190, 195

U

U.S. Geological Survey, 74
U.S. Treasury, 37, 50, 98, 124, 167
UN, 50, 116, 225, 270, 283
uncertainty, 127
uniform, 38, 169, 238, 280
unit cost, 66
United Kingdom, 92, 93, 134
United Nations, 60, 63, 97, 116, 118, 133, 149, 153, 250, 269, 270, 271, 280
United States, 1, 6, 9, 12, 13, 15, 18, 21, 22, 25, 27, 30, 31, 32, 33, 34, 36, 38, 45, 46, 48, 49, 51, 54, 56, 64, 65, 68, 69, 73, 77, 78, 81, 83, 85, 93, 102, 103, 104, 105, 106, 108, 109, 111, 112, 117, 118, 120, 127, 131, 134, 135, 136, 137, 141, 142, 143, 145, 146, 147, 149, 150, 151, 153, 160, 163, 166, 169, 173, 176, 178, 186, 188, 189, 190, 192, 193, 195, 197, 201, 206, 209, 212, 215, 217, 220, 221, 222, 223, 225, 228, 232, 233, 240, 242, 243, 244, 245, 250, 252, 257, 258, 259, 262, 268, 269, 270, 271, 272, 273, 275, 276, 278, 279, 280, 283, 284
universities, vii, 10, 64, 151, 154, 244
updating, 3
urban areas, 78, 88
urbanization, 229
Uruguay, 7, 16, 32, 37, 45, 72, 73, 77, 91, 95, 126, 133, 145, 149, 156, 161, 165, 169, 187, 194, 201, 217, 222, 230, 231, 233, 243, 244, 245, 252, 257, 259, 272, 273, 275, 283
Uruguay Round, 7, 16, 32, 37, 45, 72, 73, 77, 91, 95, 126, 133, 145, 149, 156, 161, 165, 169, 187, 194, 201, 217, 222, 230, 231, 233, 243, 244, 245, 252, 257, 259, 272, 273, 275, 283
UV, 274

V

vaccines, 18, 30, 276
vacuum, 34
validity, 129, 148, 243
values, 25, 57, 74, 174, 185, 207, 242, 251, 256, 259
variable, 95, 96, 107, 155, 210, 275
variation, 27, 29, 64, 70, 104, 135, 161
VAT, 275
vegetables, 11, 42, 53, 61, 63, 66, 67, 72, 78, 102, 105, 106, 112, 116, 139, 142, 145, 152, 157, 159, 163, 164, 166, 167, 170, 188, 194, 202, 204, 206, 214, 226, 245, 246, 270, 280
vegetation, 36, 40, 56, 77, 111, 120, 132, 143, 160, 176, 199, 217, 224, 238, 239, 279
vehicles, 24, 31, 53, 143, 172, 186, 197, 214
velocity, 36
Venezuela, 135
venture capital, 16
vertebrates, 33
vertical integration, 63, 71, 275
vessels, 39, 192, 250
victims, 141
Vietnam, 22, 23, 197
viruses, 24, 31, 120
viscera, 192
vision, 45
vitamin A, 168
vitamins, 268
vocabulary, vii
volatilization, 11
vomiting, 277
voters, 214
voting, 130, 164
vulnerability, 202

W

wage rate, 6, 136, 211
wages, 100, 136, 168, 192
war, 63, 94, 95
waste disposal, 228, 229, 235
waste water, 23, 84
wastewater, 24, 55, 80, 84, 140, 151
water, 1, 2, 11, 13, 14, 18, 21, 22, 23, 25, 28, 29, 32, 36, 38, 43, 46, 47, 49, 55, 56, 57, 58, 59, 61, 75, 76, 77, 79, 80, 82, 84, 87, 90, 91, 92, 93, 94, 106, 111, 124, 125, 132, 134, 136, 138, 143, 146, 147, 151, 152, 153, 154, 155, 159, 160, 166, 170, 173, 176, 180, 182, 184, 187, 190, 191, 202, 203, 205, 207, 208, 212, 214, 217, 218, 219, 221, 222, 223, 224, 228, 229, 232, 235, 237, 238, 240, 241, 242, 246, 250, 251, 255, 256, 264, 268, 269, 275, 277, 278, 279, 282
water policy, 2
water quality, 11, 28, 29, 47, 49, 55, 57, 58, 59, 77, 90, 222, 224, 264, 277, 278
water quality standards, 77, 264
water resources, 25, 57, 61, 79, 190, 240, 242
watershed, 49, 74, 82, 138, 165, 174, 240, 278
waterways, 36, 56, 61
wealth, 99
weapons, 140
weapons of mass destruction, 140

wearing apparel, 40
weight loss, 29
welfare, 1, 15, 19, 155, 168, 203, 236, 243
well-being, 198
Western Europe, 195
wetland restoration, 279
wetlands, 47, 58, 59, 63, 68, 87, 92, 103, 104, 125, 152, 182, 185, 189, 199, 207, 209, 212, 234, 256, 257, 277, 279
wheat, 3, 5, 8, 9, 11, 14, 24, 25, 26, 28, 36, 38, 43, 48, 52, 54, 62, 63, 75, 81, 82, 88, 95, 97, 104, 115, 118, 119, 129, 131, 132, 137, 148, 149, 153, 158, 163, 164, 169, 178, 187, 194, 198, 201, 203, 214, 215, 223, 245, 247, 248, 258, 260, 262, 268, 274, 277, 280, 283
White House, 65
wholesale, 37, 99, 209, 275
wilderness, 109, 138, 174, 281
wildfire, 121
wildland, 53, 138
wildlife, 12, 16, 17, 40, 43, 45, 56, 57, 58, 90, 92, 136, 151, 165, 173, 174, 189, 203, 221, 222, 224, 237, 239, 279, 281, 282
wind, 36, 71, 91, 206, 221, 237, 238, 244, 250, 251, 277, 282
windstorms, 87
winter, 9, 145, 166, 200, 215, 259, 280
withdrawal, 96, 227
withdrawals, 109, 152, 183
women, 44, 53, 142
wood, 9, 22, 30, 65, 112, 137, 165, 166, 212, 263
words, 65, 103, 128, 193
work, 1, 34, 40, 64, 74, 87, 92, 101, 118, 120, 131, 134, 136, 152, 175, 182, 183, 192, 193, 203, 210, 217, 256, 271, 283
workers, 6, 34, 60, 96, 100, 116, 136, 168, 191, 214, 220, 225, 227, 235, 236, 265
working conditions, 136
World Bank, 60, 283
World Health Organization, 48, 280
World Trade Center, 152
World Trade Organization, 7, 33, 37, 67, 77, 80, 82, 126, 149, 165, 173, 193, 210, 251, 266, 267, 271, 272, 283, 284
World War I, 44, 180, 249
WTO, 7, 20, 37, 48, 66, 77, 80, 81, 91, 126, 128, 135, 170, 172, 173, 181, 201, 210, 233, 253, 271, 272, 273, 283

Y

yeast, 129
yield, 3, 22, 41, 46, 61, 63, 66, 70, 75, 84, 107, 121, 125, 135, 169, 173, 174, 178, 187, 188, 195, 198, 200, 201, 214, 221, 223, 224, 239, 256, 262, 274, 283